Innovations and Trends in Environmental and Agricultural Informatics

Petraq Papajorgji
University of New York Tirana, Albania

Francois Pinet
Irstea Centre de Clermont-Ferrand, France

A volume in the Advances in Environmental Engineering and Green Technologies (AEEGT) Book Series

IGI Global
DISSEMINATOR OF KNOWLEDGE

Published in the United States of America by
IGI Global
Engineering Science Reference (an imprint of IGI Global)
701 E. Chocolate Avenue
Hershey PA, USA 17033
Tel: 717-533-8845
Fax: 717-533-8661
E-mail: cust@igi-global.com
Web site: http://www.igi-global.com

Library of Congress Cataloging-in-Publication Data

Names: Papajorgji, Petraq J., editor.
Title: Innovations and trends in environmental and agricultural informatics / Petraq Papajorgji and Francois Pinet, editors.
Description: Hershey, PA : Engineering Science Reference, [2018] | Includes bibliographical references.
Identifiers: LCCN 2017059205| ISBN 9781522559788 (hardcover) | ISBN 9781522559795 (ebook)
Subjects: LCSH: Agricultural informatics. | Environmental sciences--Data processing.
Classification: LCC S494.5.D3 I55 2018 | DDC 630.2085--dc23 LC record available at https://lccn.loc.gov/2017059205

This book is published in the IGI Global book series Advances in Environmental Engineering and Green Technologies (AEEGT) (ISSN: 2326-9162; eISSN: 2326-9170)

British Cataloguing in Publication Data
A Cataloguing in Publication record for this book is available from the British Library.

For electronic access to this publication, please contact: eresources@igi-global.com.

Advances in Environmental Engineering and Green Technologies (AEEGT) Book Series

ISSN:2326-9162
EISSN:2326-9170

Editor-in-Chief: Sang-Bing Tsai, University of Electronic Science and Technology of China Zhongshan Institute, China & Ming-Lang Tseng, Lunghwa University of Science and Technology, Taiwan & Yuchi Wang, University of Electronic Science and Technology of China Zhongshan Institute, China

MISSION

Growing awareness and an increased focus on environmental issues such as climate change, energy use, and loss of non-renewable resources have brought about a greater need for research that provides potential solutions to these problems. Research in environmental science and engineering continues to play a vital role in uncovering new opportunities for a "green" future.

The **Advances in Environmental Engineering and Green Technologies (AEEGT)** book series is a mouthpiece for research in all aspects of environmental science, earth science, and green initiatives. This series supports the ongoing research in this field through publishing books that discuss topics within environmental engineering or that deal with the interdisciplinary field of green technologies.

COVERAGE

- Policies Involving Green Technologies and Environmental Engineering
- Cleantech
- Air Quality
- Electric Vehicles
- Green Technology
- Biofilters and Biofiltration
- Industrial Waste Management and Minimization
- Alternative Power Sources
- Pollution Management
- Sustainable Communities

IGI Global is currently accepting manuscripts for publication within this series. To submit a proposal for a volume in this series, please contact our Acquisition Editors at Acquisitions@igi-global.com or visit: http://www.igi-global.com/publish/.

The Advances in Environmental Engineering and Green Technologies (AEEGT) Book Series (ISSN 2326-9162) is published by IGI Global, 701 E. Chocolate Avenue, Hershey, PA 17033-1240, USA, www.igi-global.com. This series is composed of titles available for purchase individually; each title is edited to be contextually exclusive from any other title within the series. For pricing and ordering information please visit http://www.igi-global.com/book-series/advances-environmental-engineering-green-technologies/73679. Postmaster: Send all address changes to above address.

Titles in this Series

For a list of additional titles in this series, please visit:
https://www.igi-global.com/book-series/advances-environmental-engineering-green-technologies/73679

Effective Solutions to Pollution Mitigation for Public Welfare
Ahmet Ozan Gezerman (Yildiz Technical University, Turkey) Burcu Didem Corbacioglu (Yildiz Technical University, Turkey) and Bhola R. Gurjar (Indian Institute of Technology Roorkee, India)
Engineering Science Reference • ©2018 • 274pp • H/C (ISBN: 9781522533795) • US $180.00

Handbook of Research on Microbial Tools for Environmental Waste Management
Vinay Mohan Pathak (Gurukul Kangri University, India) and Navneet (Gurukul Kangri University, India)
Engineering Science Reference • ©2018 • 335pp • H/C (ISBN: 9781522535409) • US $215.00

Food Systems Sustainability and Environmental Policies in Modern Economies
Abiodun Elijah Obayelu (Federal University of Agriculture – Abeokuta (FUNAAB), Nigeria)
Engineering Science Reference • ©2018 • 371pp • H/C (ISBN: 9781522536314) • US $195.00

Microbial Biotechnology in Environmental Monitoring and Cleanup
Pankaj (G. B. Pant University of Agriculture and Technology, India) and Anita Sharma (G. B. Pant University of Agriculture and Technology, India)
Engineering Science Reference • ©2018 • 427pp • H/C (ISBN: 9781522531265) • US $235.00

Handbook of Research on Renewable Energy and Electric Resources for Sustainable ...
Valeriy Kharchenko (Federal Scientific Agroengineering Center VIM, Russia) and Pandian Vasant (Universiti Teknologi PETRONAS, Malaysia)
Engineering Science Reference • ©2018 • 672pp • H/C (ISBN: 9781522538677) • US $345.00

Green Production Strategies for Sustainability
Sang-Binge Tsai (University of Electronic Science and Technology of China (Zhongshan Institute), China & Civil Aviation University of China, China) Bin Liu (Shanghai Maritime University, China) and Yongian Li (Nankai University, China)
Engineering Science Reference • ©2018 • 325pp • H/C (ISBN: 9781522535379) • US $235.00

For an entire list of titles in this series, please visit:
https://www.igi-global.com/book-series/advances-environmental-engineering-green-technologies/73679

IGI Global
DISSEMINATOR OF KNOWLEDGE

701 East Chocolate Avenue, Hershey, PA 17033, USA
Tel: 717-533-8845 x100 • Fax: 717-533-8661
E-Mail: cust@igi-global.com • www.igi-global.com

Table of Contents

Preface xiv

Chapter 1
Mental Informatics and Agricultural Issues: Global Change vs. Sustainable Agriculture 1
Attila Gere, Szent István University, Hungary
Dalma Radványi, Hungarian Academy of Sciences, Hungary
Richard Sciacca, Vox Populi Golbal, USA
Howard Moskowitz, Mind Genomics Advisors, USA

Chapter 2
Could NoSQL Replace Relational Databases in FMIS? 38
Giancarlo Rodrigues, State University of Ponta Grossa, Brazil
Alaine Margarete Guimarães, State University of Ponta Grossa, Brazil

Chapter 3
InWaterSense: An Intelligent Wireless Sensor Network for Monitoring Surface Water Quality to a River in Kosovo 58
Figene Ahmedi, University of Prishtina, Kosovo
Lule Ahmedi, University of Prishtina, Kosovo
Brendan O'Flynn, University College Cork, Ireland
Arianit Kurti, Linnaeus University, Sweden
Sylë Tahirsylaj, Hydrometeorological Institute of Kosovo, Kosovo
Eliot Bytyçi, University of Prishtina, Kosovo
Besmir Sejdiu, University of Prishtina, Kosovo
Astrit Salihu, University of Prishtina, Kosovo

Chapter 4
Semantic Web-Based Agricultural Information Integration 86
Kaladevi Ramar, Jerusalem College of Engineering, India
Geetha Gurunathan, Jerusalem College of Engineering, India

Chapter 5
Processing and Visualizing Floating Car Data for Human-Centered Traffic and Environment Applications: A Transdisciplinary Approach........................105
Patrick Voland, University of Potsdam, Germany
Hartmut Asche, University of Potsdam, Germany

Chapter 6
New Design Approach to Handle Spatial Vagueness in Spatial OLAP Datacubes: Application to Agri-Environmental Data..129
Elodie Edoh-Alove, Irstea Centre de Clermont-Ferrand, France & Université Laval, Canada
Sandro Bimonte, Irstea Centre de Clermont-Ferrand, France
François Pinet, Irstea Centre de Clermont-Ferrand, France
Yvan Bédard, Université Laval, Canada

Chapter 7
Discriminating Biomass and Nitrogen Status in Wheat Crop by Spectral Reflectance Using ANN Algorithms 156
Claudio Kapp Jr., State University of Ponta Grossa, Brazil
Eduardo Fávero Caires, State University of Ponta Grossa, Brazil
Alaine Margarete Guimarães, State University of Ponta Grossa, Brazil

Chapter 8
Discovering Regularity Patterns of Mobility Practices Through Mobile Phone Data...173
Paolo Tagliolato, Politecnico di Milano, Italy
Fabio Manfredini, Politecnico di Milano, Italy

Chapter 9
Environmental Monitoring Based on the Wireless Sensor Networking Technology: A Survey of Real-World Applications 196
Eirini Karapistoli, University of Macedonia, Greece
Ioanna Mampentzidou, University of Macedonia, Greece
Anastasios A. Economides, University of Macedonia, Greece

Chapter 10
Introducing Activity-Based Costing in Farm Management: The Design of the FarmBO System...252
Giacomo Carli, Open University, UK
Maurizio Canavari, University of Bologna, Italy
Alessandro Grandi, University of Bologna, Italy

Compilation of References 273

About the Contributors 307

Index 316

Detailed Table of Contents

Preface... xiv

Chapter 1
Mental Informatics and Agricultural Issues: Global Change vs. Sustainable Agriculture...1
Attila Gere, Szent István University, Hungary
Dalma Radványi, Hungarian Academy of Sciences, Hungary
Richard Sciacca, Vox Populi Golbal, USA
Howard Moskowitz, Mind Genomics Advisors, USA

This chapter presents an approach to understanding the importance of connected aspects of a topic, such as the relevance of issues for global change or for sustainable agriculture. The approach, Mind Genomics, identifies a specific topic, creates a battery of related questions which in concert "tell a story," requires the researcher to provides several alternative answers to those questions, and then tests the answers as combinations, as vignettes. Respondents rate the vignettes on judgmental attributes, such as the degree to which the respondent "agrees" with the story being told by the vignette, or the emotion that the respondent feels when reading the vignette. The analysis of such data shows the impact of each of the answers, the "communication elements," as a drive of "agreement with the values of the respondent," and the linkage of each element to a set of emotions. Mind Genomics provides a new tool to understand responses to agricultural issues, creating in its wake the possibility of a new "mental informatics."

Chapter 2
Could NoSQL Replace Relational Databases in FMIS?.....................................38
Giancarlo Rodrigues, State University of Ponta Grossa, Brazil
Alaine Margarete Guimarães, State University of Ponta Grossa, Brazil

FMIS (farm management information systems) is the computational tool responsible to process data to get information that improves farmers' decision support. The data manipulated in FMIS is originated from diverse sources, stored, and read whenever necessary without subsequent modifications, thus dismissing the necessity of complex data storage systems such as offered by the relational model. Due to its capability to handle with high performance, a large amount of unstructured data and to reduce the complexity of applications, the NoSQL data storage model, a convenient alternative to relational model, recently gained a lot of attention in the information systems market. This way, this chapter discusses how NoSQL models could improve the FMIS architecture and performance when used as data storage. Some works that have already benefited from NoSQL model adoption are reviewed and convenient use cases where both data storage models could be well used in FMIS's architecture are advised and discussed.

Chapter 3
InWaterSense: An Intelligent Wireless Sensor Network for Monitoring Surface Water Quality to a River in Kosovo..58
Figene Ahmedi, University of Prishtina, Kosovo
Lule Ahmedi, University of Prishtina, Kosovo
Brendan O'Flynn, University College Cork, Ireland
Arianit Kurti, Linnaeus University, Sweden
Sylë Tahirsylaj, Hydrometeorological Institute of Kosovo, Kosovo
Eliot Bytyçi, University of Prishtina, Kosovo
Besmir Sejdiu, University of Prishtina, Kosovo
Astrit Salihu, University of Prishtina, Kosovo

A shift in the water monitoring approach from traditional grab sampling to novel wireless sensors is gaining in popularity not only among researchers but also in the market. These latest technologies readily enable numerous advantageous monitoring arrangements like remote, continuous, real-time, and spatially dense and broad in coverage measurements, and identification of long-term trends of parameters of interest. Thus, a WSN system is implemented in a river in Kosovo as part of the InWaterSense project to monitor its water quality parameters. It is one of the first state-of-the-art technology demonstration systems of its kind in the domain of water monitoring in developing countries like Kosovo. Water quality datasets are transmitted at pre-programmed intervals from sensing stations deployed in the river to the server at university via the GPRS network. Data is then made available through a portal to different target groups (policymakers, water experts, and citizens). Moreover, the InWaterSense system behaves intelligently like staying in line with water quality regulatory standards.

Chapter 4
Semantic Web-Based Agricultural Information Integration 86
Kaladevi Ramar, Jerusalem College of Engineering, India
Geetha Gurunathan, Jerusalem College of Engineering, India

A huge volume of information is available in the worldwide web. However, the demand is on relevant information rather than available information, which are often heterogeneous and distributed. Agriculture is one such domain, which includes information like soil, crops, weather, etc. under one roof. This information is in different representations and structures (e.g., weather). This scenario leads to a challenge of how to integrate the available and heterogeneous agricultural information to deliver better production. As the information on the web is syntactically structured, the need is to provide semantic linkage. The semantic web supports the existing web to easily process and interpret information. In this chapter, a semantic-based agricultural information system (AIS) is proposed that addresses heterogeneity issues among weather systems and integrates information like soil, weather, crop, and fertilizers. AIS helps the farmers regarding the type of crop/soil, crop/climate, fertilizer applications, diseases, and prevention methods using effective retrieval of information from integrated systems.

Chapter 5
Processing and Visualizing Floating Car Data for Human-Centered Traffic and Environment Applications: A Transdisciplinary Approach 105
Patrick Voland, University of Potsdam, Germany
Hartmut Asche, University of Potsdam, Germany

In the era of the internet of things and big data, modern cars have become mobile electronic systems or computers on wheels. Car sensors record a multitude of car- and traffic-related data as well as environmental parameters outside the vehicle. The data recorded are spatio-temporal by nature (floating car data) and can thus be classified as geodata. Their geospatial potential is, however, not fully exploited so far. In this chapter, the authors present an approach to collect, process, and visualize floating car data for traffic- and environment-related applications. It is demonstrated that cartographic visualization, in particular, is an effective means to make the enormous stocks of machine-recorded data available to human perception, exploration, and analysis.

Chapter 6
New Design Approach to Handle Spatial Vagueness in Spatial OLAP Datacubes: Application to Agri-Environmental Data..129

Elodie Edoh-Alove, Irstea Centre de Clermont-Ferrand, France & Université Laval, Canada
Sandro Bimonte, Irstea Centre de Clermont-Ferrand, France
François Pinet, Irstea Centre de Clermont-Ferrand, France
Yvan Bédard, Université Laval, Canada

Spatial OLAP (SOLAP) technologies are dedicated to multidimensional analysis of large volumes of (spatial) data. Spatal data are subject to different types of uncertainty, in particular spatial vagueness. Although several researches propose new models to cope with spatial vagueness, their integration in SOLAP systems is still in an embryonic state. Also, analyzing multidimensional data with metadata brought by the exploitation of the new models can be too complex and demanding for decision makers. To help reduce spatial vagueness consequences on the exactness of SOLAP analysis queries, the authors present a new approach for designing SOLAP datacubes based on end-users' tolerance to the risks of misinterpretation of fact data. An experimentation of the new approach on agri-environmental data is also proposed.

Chapter 7
Discriminating Biomass and Nitrogen Status in Wheat Crop by Spectral Reflectance Using ANN Algorithms 156

Claudio Kapp Jr., State University of Ponta Grossa, Brazil
Eduardo Fávero Caires, State University of Ponta Grossa, Brazil
Alaine Margarete Guimarães, State University of Ponta Grossa, Brazil

Precision agriculture has the goal of reducing cost which is difficult when it is related to fertilizer application. Nitrogen (N) is the nutrient absorbed in greater amounts by crops and the N fertilizer application presents significant costs. The use of spectral reflectance sensors has been studied to identify the nutritional status of crops and prescribe varying N rates. This study aimed to contribute to the determination of a model to discriminating biomass and nitrogen status in wheat through two sensors, GreenSeeker and Crop Circle, using the resilient propagation and backpropagation artificial neural networks algorithms. As a result, a strong correlation to the sensor readings with the aboveground biomass production and N extraction by plants was detected. For both algorithms a satisfactory model for estimating wheat dry biomass production was established. The best backpropagation and resilient propagation models defined showed better performance for the GreenSeeker and Crop Circle sensors, respectively.

Chapter 8
Discovering Regularity Patterns of Mobility Practices Through Mobile Phone Data 173
Paolo Tagliolato, Politecnico di Milano, Italy
Fabio Manfredini, Politecnico di Milano, Italy

The chapter addresses the issue of analyzing and mapping mobility practices by using different kinds of mobile phone network data that provide geo-located information on mobile phone activity at a high spatial and temporal resolution. The authors present and discuss major findings and drawbacks based on an application carried out on the Milan urban region (Lombardy, Northern Italy) and suggest possible implications for policies.

Chapter 9
Environmental Monitoring Based on the Wireless Sensor Networking Technology: A Survey of Real-World Applications 196
Eirini Karapistoli, University of Macedonia, Greece
Ioanna Mampentzidou, University of Macedonia, Greece
Anastasios A. Economides, University of Macedonia, Greece

This chapter investigates real-life environmental monitoring applications based on wireless sensor networks (WSNs). Wireless sensor networking is an emerging technology, which has been adopted by many scientific fields in order to accurately and effectively monitor climate phenomena such as air pollution, destruction phenomena, etc. It has also been widely used in agriculture as well as in horticulture for field monitoring. In this chapter, the authors provide a critical overview of the basic components existing WSN deployments use. They also categorize these deployments, 111 in total, into five different field categories in order to provide a general view of the technologies used, the conditions under which the deployments were conducted, and much more. Then, five easy-to-use guides are provided discussing basic considerations for deploying WSNs in each of these fields. In order to showcase the usefulness of consulting the resulted guides, the authors consider representative application scenarios for each of these field deployments.

Chapter 10
Introducing Activity-Based Costing in Farm Management: The Design of the FarmBO System..252

Giacomo Carli, Open University, UK
Maurizio Canavari, University of Bologna, Italy
Alessandro Grandi, University of Bologna, Italy

Recent research indicates that farm managers do not rely on adequate informative support in their decision-making processes. The authors propose a model of a farm management information system which integrates the activity-based costing approach. In describing the design and development of the "FarmBO" system, the authors provide a detailed functional requirement definition and the description of a working system prototype. The solution is designed to show the impact of general costs on the different crops, allocating them on the basis of the production cycle complexity. It includes a report section directly linked to the database which provides crop balance sheets and simulations in terms of what-if analyses. The system allows farm managers to 1) analyze deviations between budgeted and actual costs, 2) compare crop balance sheets across different years, and 3) perform sensitivity analyses. The authors account for prototype validation in two farms and discuss results and possible developments.

Compilation of References 273

About the Contributors 307

Index 316

Preface

In recent years, the role of ICT in the development of agriculture and environment has received significant attention in different types of international forums. Not long ago the world enjoyed the advent of the Internet that completely changed the way people to people, business to business communicate. Nowadays, there are new technologies coming up and it looks like we are having a hard time to catch up with the progress: communication in 5G, Internet of Things, powerful mobiles and a new business philosophy based on mobiles.

How this progress affects Agriculture and Environment? This volume is trying to provide some of the answers to this fundamental question. We are putting together chapters describing the most advanced applications of new technologies in Agriculture and Environment. We hope participating authors to enlighten us with their hard work and interesting applications. Thus, we hope this publication will fill a vacuum and will provide large audiences with a much needed knowledge. Following is a short description of the chapters of this volume.

The chapter "Mental Informatics and Agricultural Issues: Global Change vs. Sustainable Agriculture" (A. Gere, D. Radvanyi, R. Sciacca, H. Moskowitz) aims to explore two topics of interest to agricultural economics, topics treated from the point of view of the person, not from the traditional point of view of economics. These two topics are laden with emotion, and therefore it is relevant to understand consumer's perspective. The first is reactions of people to messaging about global change, and the implication of the issue of global change for the correct policy and messaging. The second is the reaction to sustainability of agriculture, especially protein, put into concrete, real form, by focusing on other sources of protein, and not just on general issues.

The authors try to establish the applicability of a newly emerging science, Mind Genomics, to be used to understand the 'mind of people' who are confronted with agriculture-relevant information and/or decisions to be made. With Mind Genomics, they present an affordable approach for the agricultural profession to understand the 'mind of the consumer,' while at the same time understanding the objective, external situation in the so-called 'real world.'

Chapter "Could NoSQL Replace Relational Databases in FMIS?" (G. Rodrigues, A. M. Guimarães) discuss the possibility that the new approach to databases referred to as NOSQL Databases, could replace the existing and well-established Relational Model. As the ICT applications in agriculture and environment are increasingly more present, a myriad of sensors and devices were developed. These applications produce or collect an immense quantity of unstructured geo-referenced digital data about soil, plants, weed, pests, weather, equipment and machinery, management practices and others. The existing Relational Model, cannot handle unstructured data, thus the need for another approach, namely the NOSQL Databases, will fill the gap. It is time that this new approach to databases be present in university courses rather soon.

Chapter "InWaterSense: An Intelligent Wireless Sensor Network for Monitoring Surface Water Quality to a River in Kosovo" (F. Ahmedi, L. Ahmedi, B. O'Flynn, A. Kurti, S. Tahirsylaj, E. Bytyçi, B. Sejdiu, A. Salihu) discusses issues of water monitoring approach from traditional grab sampling to novel wireless sensors. This new approach is gaining in popularity not only among researchers but also in the market. These latest technologies *readily* enable numerous advantageous monitoring arrangements like remote, continuous, real-time, and spatially-dense and broad in coverage measurements, and identification of long-term trends of parameters of interest. Authors have designed a WSN system that is implemented in a river in Kosovo as part of the InWaterSense project to monitor its water quality parameters. It is one of the first state of the art technology demonstration systems of its kind in the domain of water monitoring in developing countries like Kosovo. Water quality datasets are transmitted at pre-programmed intervals from sensing stations deployed in the river to the server at university via the GPRS network. Data is then made available through a portal to different target groups (policy-makers, water experts, and citizens). Moreover, the InWaterSense system behaves intelligently like staying in line with water quality regulatory standards.

"Semantic Web-Based Agricultural Information Integration" (K. Ramar) presents the need to establish some level of integration among data that are often heterogeneous and distributed. Agriculture, occuring in time and space, is one such domain, which includes information like soil, crops, weather etc. This information is in different representations and structures e.g weather. The challenge is how to integrate the available and heterogeneous agricultural information to deliver better production. The authors present the need for a semantic link among the heterogeneous data. The semantic web supports the existing web to easily process and interpret information. They propose a semantic based Agricultural Information System (AIS) which addresses heterogeneity issues among weather systems and integrates information like soil, weather, crop and fertilizers. AIS will help farmers regarding the type of crop/soil, crop/climate, fertilizer applications, diseases and prevention methods using effective retrieval of information from integrated systems.

"Processing and Visualizing Floating Car Data for Human-Centered Traffic and Environment Applications: A Transdisciplinary Approach" (P. Voland and H. Asche) points out that in our wired world, vehicles have developed into complex electronic systems. There is a saying that we don't have cars anymore; we have computers we ride in. Cars are equipped with a large number of sensor devices essential for smooth technical operation (monitoring, e.g., vehicle speed or engine revolutions) and environmental monitoring (measuring, e.g., barometric air pressure, ambient air temperature). Such constantly acquired data are indispensable for a growing number of so-called Advanced Driver Assistance Systems (ADAS). Automotive sensor data are a prerequisite for semi-autonomous and autonomous driving systems currently in development and testing. Authors present an approach to collect, process and visualize floating car data for traffic- and environment-related applications. They demonstrate that cartographic visualization, in particular, is as effective means to make the enormous stocks of machine-recorded data available to human perception, exploration and analysis.

"New Design Approach to Handle Spatial Vagueness in Spatial OLAP Datacubes: Application to Agri-Environmental Data" (E. Edoh-Alove, S. Bimonte, F. Pinet et Y. Bédard) discusses the concept of Spatial-OLAP (SOLAP) technologies. These technologies are dedicated to multidimensional analysis of large volumes of (spatial) data. Spatial data are subject to different types of uncertainty, in particular spatial vagueness. Authors point out that although several researches propose new models to cope with spatial vagueness, their integration in SOLAP systems is still in an embryonic state. Also, analyzing multidimensional data with metadata brought by the exploitation of the new models can be too complex and demanding for decision-makers. To help reduce spatial vagueness consequences on the exactness of SOLAP analysis queries, authors present a new approach for designing SOLAP datacubes based on end-users' tolerance to the risks of misinterpretation of fact data. An experimentation of the new approach on agri-environmental data is also proposed.

"Discriminating Biomass and Nitrogen Status in Wheat Crop by Spectral Reflectance Using Artificial Neural Networks" (C. K. Junior, E. F. Caires and A. M. Guimarães) studies the use of spectral reflectance sensors in order to identify the nutritional status of crops and prescribe varying N rates as nitrogen (N) is the nutrient absorbed in greater amounts by crops and the N fertilizers application present significant costs. This study contributes to the determination of a model to discriminating biomass and nitrogen status in wheat through two sensors, GreenSeeker and Crop Circle, using the Resilient Propagation and Backpropagation Artificial Neural Networks algorithms. They point out that a strong correlation to the sensor readings with the aboveground biomass production and N extraction by plants was detected. For both algorithms have established a satisfactory model for estimating wheat dry biomass production. The best Backpropagation and Resilient Propagation models defined showed better performance for the GreenSeeker and Crop Circle sensors, respectively.

"Discovering Regularity Patterns of Mobility Practices Through Mobile Phone Data" (P. Tagliolato, F. Manfredini and P. Pucci) addresses the issue of analyzing and mapping mobility practices by using different kinds of mobile phone network data, which provide geo-located information on mobile phone activity at a high spatial and temporal resolution. Authors state that by overlaying the boundary of the institutional management of local public transport in the Milan area with the areas of mobility practices, taken from the mobile phone data, one can observe the deep structural effects of the mobility of people on urban policies and the obvious disconnection between fixed jurisdictions and mobile factors.

Authors conclude that the presented data and methodology they used, let the recognition of effective mobile populations in the urban environment. This knowledge can be exploited by decision makers for the definition of specific policies directed to temporary populations, which are more and more important in contemporary cities, otherwise ignored.

"Environmental Monitoring Based on the Wireless Sensor Networking Technology: A Survey of Real-World Applications" (E. Karapistoli, I. Mampentzidou and A. A. Economides) investigates real-life environmental monitoring applications based on Wireless Sensor Networks (WSNs). Authors state that during the last decades, this wireless networking technology has been adopted by many scientific fields in order to accurately and effectively monitor climate phenomena such as air pollution, destruction phenomena (i.e., landslides), etc., with uses in agriculture as well as in horticulture for field monitoring. Authors provide a critical overview of the basic components existing WSN deployments use. They also categorize these deployments, 111 in total, into five different field categories, namely agricultural monitoring, environmental monitoring, air-water pollution monitoring, monitoring of destruction phenomena, as well as monitoring of livestock, and wild animal, in order to provide a general view of the technologies used, the conditions under which the deployments were conducted, and much more.

They conclude that technological advances in sensors, sensor data logging and communication, and software management of sensor networks will continue to provide transformative potential for new and innovative avenues of WSN-based ecological research in ways previously not possible.

"Introducing Activity-Based Costing in Farm Management: The Design of the FarmBO System" (G. Carli, M. Canavari and A. Grandi) argues that farm managers do not rely on adequate informative support in their decision making processes. Thus, they propose a model of a Farm Management Information System which integrates the Activity-Based Costing approach. In describing the design and development of the "FarmBO" system, they provide a detailed functional requirement definition and the description of a working system prototype. The system allows farm managers to 1) analyze deviations between budgeted and actual costs; 2) compare crop balance sheets across different years; 3) perform sensitivity analyses.

They conclude that a more reliable, accurate, and timely costing method based on Activity-Based Costing, therefore, would be beneficial to farm managers who need financial information to effectively support strategic and tactical decision-making in several farm management areas, such as crop rotations, introduction of new crops, farming process improvement, investment and disinvestment decisions.

Chapter 1
Mental Informatics and Agricultural Issues:
Global Change vs. Sustainable Agriculture

Attila Gere
Szent István University, Hungary

Dalma Radványi
Hungarian Academy of Sciences, Hungary

Richard Sciacca
Vox Populi Golbal, USA

Howard Moskowitz
Mind Genomics Advisors, USA

ABSTRACT

This chapter presents an approach to understanding the importance of connected aspects of a topic, such as the relevance of issues for global change or for sustainable agriculture. The approach, Mind Genomics, identifies a specific topic, creates a battery of related questions which in concert "tell a story," requires the researcher to provides several alternative answers to those questions, and then tests the answers as combinations, as vignettes. Respondents rate the vignettes on judgmental attributes, such as the degree to which the respondent "agrees" with the story being told by the vignette, or the emotion that the respondent feels when reading the vignette. The analysis of such data shows the impact of each of the answers, the "communication elements," as a drive of "agreement with the values of the respondent," and the linkage of each element to a set of emotions. Mind Genomics provides a new tool to understand responses to agricultural issues, creating in its wake the possibility of a new "mental informatics."

DOI: 10.4018/978-1-5225-5978-8.ch001

INTRODUCTION: A LOT KNOWN ABOUT THE PHYSICAL WORLD BUT WHAT ABOUT THE MENTAL WORLD?

During the past century, the expansion of science has occurred at an ever-accelerating pace. The amount and depth of theoretical and practical knowledge manifest themselves in the published information to be found in the archival scientific literature, and at the same time in the practical consequences – more food, for more people, at lower costs. To many, the world of agriculture stands as a living disproof to the conjectures of that most demoralizing of all classical thinkers and economist, Thomas Malthus (Petersen, 1999), who opined centuries ago that we would soon starve because the growth of population is exponential, whereas the growth of resources to feed the population is linear.

We know a great deal about the science of agriculture, about weather changes, about the positives and negatives of various agricultural policies. When we talk about these so-called 'positives' and 'negatives,' we talk about points of view established using economic measures, measures without emotions, measures without taking into account to the soul of the most important factor of all, the person.

Our objective in this chapter is to explore two topics of interest to agricultural economics, topics treated from the point of view of the person, not from the traditional point of view of economics, of dollar and cents, of policy. These two topics are laden with emotion, and therefore relevant. The first is reactions of people to messaging about global change, and the implication of the issue of global change for the correct policy and messaging. The second is the reaction to sustainability of agriculture, especially protein, put into concrete, real form, by focusing on other sources of protein, and not just on general issues.

Our goal is to establish the applicability of a newly emerging science, Mind Genomics, to be used to understand the 'mind of people' who are confronted with agriculture-relevant information and/or decisions to be made. With Mind Genomics, we present an affordable approach for the agricultural profession to understand the 'mind of the consumer,' while at the same time understanding the objective, external situation in the so-called 'real world.'

BACKGROUND

Just What Is Informatics, and How Does Knowing the Mind of the Person Fit With Today's Informatics?

The rise of computing, and the study of information itself and its application, has led to this new field called 'informatics,' the field to which Mind Genomics wishes to contribute. The definition of Informatics below, from Indiana University (source: https://soic.iupui.edu/about/what-is-informatics/) follows:

The Indiana University School of Informatics and Computing defines the field as "the study and application of information technology to the arts, science and professions, and to its use in organizations and society at large." Informatics students build new computing tools and applications. They study how people interact with information technology. They study how information technology shapes our relationships, our organizations, and our world.

Informatics is a new and rapidly developing field. It uses computing to solve the big problems: privacy, security, healthcare, education, poverty, and challenges in our environment. All Informatics applications are computer-based. Those applications are enhanced with tools and techniques from fields such as communication, mathematics, multimedia, and human-computer interaction design. Informatics differs from computer science and computer engineering because of its strong focus on the human use of computing.

Students of Informatics learn skills that allow them to harness the power of computing to solve real problems that directly impact our lives and the lives of those around us. They use their technology and problem-solving skills to make a difference in the world. For students interested in a career with infinite potential, Informatics stands out as a strong, flexible and dynamic field of study.

The foregoing expansive definition of Informatics suggests that in the grand field of agriculture and environment, there is room for an informatics of the mind. It is to this opportunity that we now turn.

The World Today: Too Many Inputs, Too Many Attention Deserts, the Politicization of Messages

The information that people have today available to them increases with increasingly rapid speed. It is impossible to know everything, it is impossible to believe much of what one hears, and of course it is equally impossible to think particularly critically about what is being said. A perfectly plausible sounding article on global change may be real, may be a slanted article coming from a public relations firm paid to 'spin' regular news, or indeed may be totally false, *disinformatzia*, disinformation in the wonderful way the Russians have to explain and to christen.

With all this information, how could we ever be able to create the informatics of the mind, with respect of important issues of agriculture where the 'mind is the measure of all things?' We are not talking about the informatics of actual behaviors around the world, behaviors that can be measured by machines. These measures of emitted behavior, in the language of the behavioral psychologist, have little or no 'emotion' tied to them, even though undoubtedly emotion led to many of the behaviors in the first place.

We can measure behavior. We have a hard time measuring attitude. And when we open up the science to the vision of informatics, we may discover that we have ended up in even in a worse situation.

All is not hopeless, however. Rather than attempting to make sense of the blooming, buzzing confusion of the world which confronts a person, perhaps the general world, perhaps the limited world of agriculture and the environment, might we not be better off changing our focus, moving away from measuring what is, this cloud of data points needing structure? Might we be better off doing experiments with the knowledge, to identify what types of information about agriculture and the environment are believed by people, and how do they react to these individual pieces of information? In other words, rather than attempting to master the increasingly massive amount of data as the information pertains to people, why not understand the mind of people in a simpler way, focusing our interest on how people evaluate the information?

The foregoing approach, understanding the person' response to the information, means that we no longer worry about measuring 'what's out there,' on a case-by-case basis, classifying our findings, and emerging with some generalities. Rather, we measure the structure of the *subjective response* to what's out there, uncovering perhaps rules, perhaps generalities. And in measuring the structure of responses, we are very much like the color scientist who learns about the basic rules of color from measuring a relatively limited number of samples,, creates a colorimeter, a color-measuring machines, and measures the colors of many specific objects. The color scientist looks for rules, for generalities, the basic science, and then looks at the population of different items to see how the nature of the distribution of color. There is no need for the color scientist to measure the color of everything. It suffices to know the rule of measurement, the theory, and then have the instrument to use for specifics. The color scientist is not thought to know 'less,' just because the color scientist limits himself to theory, to a machine, and to selected opportunities to do measurements.

We model Mind Genomics on this approach of color vision. For a specific topic, Mind Genomics looks for the basics, the primaries, the key groups. Having found these key groups, the next task is to determine the distribution of these primaries in the population. The primaries for Mind Genomics are not basic colors, but rather basic ideas.

MIND GENOMICS: THE NEW 'MIND INFORMATICS?'

We introduce here the notion of Mind Genomics, a new science of the everyday human experience. Mind Genomics is the science of how people react to topics and messages of their everyday experience. Mind genomics deals with the concrete, the features of the experience, and the responses to those features, whether the responses be evaluative or descriptive. Further, Mind Genomics deals with and attempts to codify and understand the differences in patterns of reactions to topics and messages, these differences often being relegated otherwise to intractable, inter-individual variation. To Mind Genomics, such variation may suggest different mind-sets, different 'mental alleles,' when dealing with the same information.

It is the structure which emerges which defines the nature of the 'mental alleles,' which, for a given topic of experience, are equivalent to the color primaries.

Mind Genomics did not emerge fully formed in the first part of the 21st Century, presented to the scientific and information technology committees as a new method to deal with old problems. Mind Genomics emerged from a half century or more of scientists grappling, on one hand, with 'how to measure' (Anderson, 2001; Luce & Tukey, 1964), and with marketing grappling on the other hand with how to convince people to buy products, and indeed how to identify what to manufacture or what services to offer (Green & Krieger, 1991; Green & Srinivasan, 1990).

The applications of Mind Genomics, even before it entered the world of informatics, were practical and goal-oriented, with the desire to solve a specific problem, and not particularly of interest to understand people. Mind Genomics emerged as experimental design of ideas. One example of particular interest to author HRM is the request in 1980 by Mr. Court Shepard, General Manager of Colgate Canada, to identify what types of messages about Colgate Dental Cream, a toothpaste, resonated most strongly with Canadian consumers. The question was practical, the answer was given and easy to understand, based upon some experiments, and thus was born this new science, quiet and unheralded.

The Basics of Mind Genomics and Why It Is Relevant to an Applied Informatics of the Mind

Over several decades of research and attempts to describe this evolving science of Mind Genomics, author HRM has found that most people do not really comprehend the notions until explained to them in steps, with an actual case history to demonstrate the results. It's not that Mind Genomics is so difficult to understand as it is that Mind

Genomics trods a different path, more experimental than simply data- analytic. That is, the intellectual history of Mind Genomics is not really statistical at all, despite the fact that Mind Genomics survives on the use of a few statistical programs. Nor in fact, is Mind Genomics likely to improve with improvements in the nature of statistical treatment of data. Rather, it is the rigor of the experiment, and the information that the experiment uncovers and renders blindingly clear, almost obvious.

To illustrate the workings of Mind Genomics, we have chosen two topics of interest to the authors, food for a surviving planet, and climate changes for a suffering planet. Both food needs and climate change are sufficiently broad. Furthermore, we don't think of either topic as pertaining directly to the human psyche. We are or we are not going to run out of food. The climate is changing or is not changing. The essence of topics themselves are not our focus, but rather the focuses is on minds of people as they are exposed to information about these two topics.

We begin first with climate change, go through the different analytical steps, emerging with the contribution of Mind Genomics to the topic of understanding the mind of the citizen facing climate change as a topic of general awareness, and facing changes in expected weather patterns as all citizens. We then repeat the exercise with new forms of protein, a topic of agricultural economics. We finish with more applications of the data, and an evaluation of just how much work was required to create the data for these two topics in Mind Genomics.

Mind Genomics studies are experiments, comprising stimuli (combinations of phrases), and responses (judgement of overall feeling, along with selections of appropriate other types). For our two illustrative studies, the responses comprise an overall evaluation of the combination of phrases (the vignette), and the selection of the best fitting single emotion from a set of seven, to describe how the respondent feels after reading the vignette. The results are processed to show the linkage between the individual elements and the responses.

Experiments, Not Questionnaires

Before we move on to the actual experiment, it's worthwhile noting that quite often Mind Genomics studies are confused with questionnaires, albeit with questionnaires that are administered over the Internet, so that the respondent can answer the questionnaire wherever and whenever is most convenient. The Mind Genomics experiment does appear to be a questionnaire, asking the respondent to provide information, but the reality upon looking deeper, is that the Mind Genomics experiment is an actual experiment, and NOT a questionnaire. The respondent is presented with systematically varied combinations of messages, and simply responds. The information about the 'mind of the respondent' emerges after one relates the presence/absence of the test elements in the vignettes to the response.

This approach is the classical stimulus-response approach favored by experimental psychologists, namely present a stimulus, acquire a response, and determine the underlying relation, if there is one.

Underscoring the difference between experiments and surveys is very important, and bears repeating and amplification. The world at large does not necessarily understand the difference between knowledge obtained by questionnaire, by asking people to introspect, versus knowledge gained by creating different test stimuli, presenting these stimuli to the respondent, observing the different behaviors, and then correlating the behavior with the antecedent stimulus.

In other words, questionnaires instruct the respondent to think about what's happening, to introspect, to give an opinion. Questionnaires do not instruct the respondent to identify the rules relating stimuli (what is presented to the respondent, or the condition of the respondent) and behavior (what the respondent actually does).

The difference between questionnaire and experiment is important, just as both approaches to knowledge are important. Questionnaires ask the respondent to think, and describes the averages across people, and the variations from person to person, from group to group. Experiments look at patterns that can be created in a response or set of responses by manipulating test stimuli or somehow manipulating something about the respondent (e.g., momentary hunger, induced simply by not eating).

The Scientific and Educational Value of a Systematic Approach

Today's world is awash in information. One can barely do any work involving people and behavior without being confronted by the colossus known as Big Data (Günther, Mehrizi, Huysman, & Feldberg, 2017), whether that colossus be all the raw data available, or the current best-practice, which is to slice and dice the mass of Big Data until it becomes the 'right data.'

Big Data, for better or worse, comprises a hodge-podge of data, information, obtained from transactions of people, from the text of blogs, opinions articles in the daily press, and so forth. There is a belief that somehow, deep within the massive, ever-increasing structure of these uncounted trillions of bits, the answer lies about behavior. Even if such an answer were available, painfully obtained after sorting through the mass of information with powerful algorithms, the results are limited to the one answer. There is no systematic structure which, when looked at afar, shows the 'system,' and hints at how the system might work, and how variables may relate to each other.

The approach presented here, Mind Genomics, is grounded on systematics, on exploring a variety of aspects of one topic, to create a big, structured picture, of how people react to topics. As we will see in the following section of this chapter, each of the individual parts of the systematic approach, each individual aspect, is easy to explore, the data easy to create, the interpretation easy to understand and to convey. In order to introduce the technical background of Mind Genomics, two case studies will be presented. The first one deals with global change and aims to define what people think about global change and how do they perceive its relevance. The second study focuses on food production problems, e.g. sustainable food production and identifies what do the respondents think about their own role in sustainable food production.

Global Change

As the first step, it is important to define global change, the topic of the first study. The Environmental Protection Agency (EPA) clearly defines the differences between climate change ("*any distinct change in measures of climate lasting for a long period of time*"), global warming ("*an average increase in temperatures near the Earth's surface and in the lowest layer of the atmosphere*") and global change ("*a broad term that refers to changes in the global environment, including climate change, ozone depletion, and land use change*") (EPA, 2013).

The effects of global change can be experienced in almost all aspects of our everyday life. Since the beginning of the 20th century, the average global temperature has risen about 0.74 °C (EPA, 2013; IPPC, 2013) and its effects have already been introduced and proven by climate scientists.

Climate change and global warming are attributed to the excess greenhouse gas emission. There are natural greenhouse gasses, e.g. water vapor or carbon dioxide (CO_2) but due to human activity extremely high amount of methane (CH_4), nitrous oxide (N_2O), hydrofluorocarbons (HFCs), perfluorocarbons (PFCs) and sulfur hexafluoride (SF_6) is produced and emitted into the air (EPA, 2013; IPPC, 2013).

In the agricultural production, for example, the key problem is that the increased temperature shifts the growing areas of certain plants. One of the many affected areas is wine production. Wine grapes are sensitive to temperature (and climate also); hence only small changes in the climate they are cultivated could lead yield loss or decrease the quality of the produced wine (Mozell & Thach, 2014). Several possible suggestions have been made for wine producers, e.g., to move planting regions one Celsius isotherm further poleward for each degree of average temperature increase (Kenny & Shao, 1992), or to change harvesting time, or to grow other species, and so forth. (Mozell & Thach, 2014).

Coffee production is another, heavily affected area. Coffee, similarly to wine grapes, is extremely sensitive to climate conditions. Arabica coffee (*Coffea arabica*) is indigenous in Ethiopia, where according to a recent study, more than 60% of the 1,000 questioned farmers experienced declined rainfalls and increased temperature over the last 20 years (Bryan, Deressa, Gbetibouo, & Ringler, 2009). Climate modelling supports the observations of local farmers and it has reported in a highly recognized paper that "*The logical conclusion is that Arabica coffee production is, and will continue to be, strongly influenced by accelerated climate change, and that in most cases the outcome will be negative for the coffee industry.*" (Davis, Gole, Baena, & Moat, 2012).

Fighting against climate change is not easy, even if there are unquestionable evidences and publications about its effects on our everyday life. In order to be able to take the necessary steps, the first order of business is to get a clear picture about people's mind on the topic. The usual, questionnaire-based approaches all have the same conclusions: people are aware of global change, but, unfortunately, their behavior still remain the same. However, the major problem of such, questionnaire-based methods is, that people are able to guess the "right" or socially more acceptable answers and their true opinion remains unknown to the researcher. The fact that people know and understand climate change, does not mean that they are willing to take actions against it. Some authors suggest that the key is early education and children in schools should receive more environmental focused education (Boyes, Stanisstreet, & Yongling, 2008). However, as several scientists suggest, time is crucial and humanity needs faster, more efficient methods to understand how to convince people about the urgent need of actions against climate change.

Mind Genomics overcomes the foregoing problems because it is a design-based method, which cannot be "solved" or "tricked". Numerous, different concepts are presented to the respondents who are unable to access the weight of each of the presented elements separately. Fortunately, the statistical models incorporated in Mind Genomics are able to complete this task. In the following, the steps of creating and evaluating a Mind Genomics study will be explained in detail through the example of the global change study.

ON TO THE PROCESS OF MIND GENOMICS

Step 1: Raw Materials

Structure the project materials as a set of questions, each question of which has several answers. All of the answers are logically appropriate for the question. The questions appear below. The questions themselves should be put into an order which 'tells a story.' Even though the questions themselves will not appear, the sequence of answers to the questions, i.e., the test concepts, will tell the necessary story, or at least a version of the story to which one can respond, and evaluate.

Question 1: *What do you think about global warming?*
Question 2: *What causes global warming?*
Question 3: *What does global warming cause?*
Question 4: *What can reduce the effects of climate change?*
Question 5: *Who should play a major role in fighting climate change?*
Question 6: *Corruption tied with response to climate change*

The particular experimental design to be used later on when combining the answers requires exactly six questions (also known as silos or categories), each question in turn generating six answers. The questions and the answers are left up to the particular focus of the experiment. Only the external structure is dictated, here the number of questions and the number of answers for each question.

Table 1 shows the example of the six questions, and the six answers regarding climate change. The questions are presented in a logical order, the order designed to tell a story. As we will see later on, the actual vignette comprises combinations of answers, without presenting the question that the element answers. The strategy of creating questions and then answering them with a number of alternative answers makes the Mind Genomics project more intuitive, and easier to do. Furthermore, moving to a question-answer format allows the Mind Genomics effort to move from research in the world of science to education of students.

Table 1. Raw material for the climate change study, comprising questions and answers

Question 1: What do you think about global warming?
Global temperature is increasing... This is not a question
Global warming is only a 'media-hype'...I do not believe it
Some parts of global warming might be true ...But not the way presented
Adverse effects of global warming are noticeable ... Even nowadays
Climate change is the problem of the future generation…Not ours
Environmental issues are not my business ... I have no influence on it
Question 2: What causes global warming?
Global warming is caused by excess agricultural activity
Industrial activities are the major contributors to global warming
Simply human activity causes global warming
Global warming is independent of us ....it is completely natural
There is no clear evidence about global warming
Global warming is caused by deforestation

continued on following page

Table 1. Continued

Question 3: What does global warming cause?
Wildlife and their habitats harmed by the changed weather conditions
Storms have become more severe in the past few decades
More droughts or water shortages are occurring
Forests and plant life continually damaged by intense fires and hot summers
Sea levels are rising ...shorelines are eroded more and more
In the past few decades, extreme temperatures were experienced during summer and winter
Question 4: What can reduce the effects of climate change
Reduce climate change… restrictions on power plant emissions needed
Reduce climate change… international agreement to limit emissions is needed
Reduce climate change… tougher fuel efficiency standards for cars needed
Reduce climate change… corporate tax incentives needed
Reduce climate change… more people should drive hybrids
Reduce climate change… people should reduce their carbon footprints
Question 5: Who should play a major role in fighting climate change?
Climate scientists should solve the problem
The general public needs more awareness to save the planet
Energy industry leaders can do the most to reduce emissions
All industrial leaders should work together globally to slow down climate change
Elected officials are responsible for environmental issues also
Public organizations have the most power to fight against global warming
Question 6: Corruption tied with response to climate change
Profit and greed overshadow the actions against global warming
The spreading of renewable energy solutions is intentionally obstructed
Some politicians want to deny global warming because their constituencies are industrial companies
Some people want to fight against climate change unethically because they see a quick profit
Corrupt politicians and industry leaders can help to fight against climate change
People wanting to benefit from global warming, and they are trying to crush it for financial reasons

Table 1 deserves a parenthetical note, one dealing with the nature of the questions and the answers to those questions. The note is prefaced by its conclusion, namely 'Do not let the perfect be the enemy of the good.' It is the nature of people, especially academics and even non-academic researchers, to seek the 'best,' even at the start of a research project. By the word 'best' we mean simply that the choice of the questions and of course the choice of the answers to those questions, should represent the optimal choice. There is the old adage, GIGO, 'garbage in, garbage out.' The reality of any human endeavor of this type is that its nature is to b incremental, building upon the knowledge of what went before. Thus, for most of the research which breaks new ground, there is no 'optimum' selection of initial test stimuli. This would be an example of searching for the 'perfect.' No one knows what 'perfect' or 'good' really is, until the experiment is run. In that spirit, the questions and the answers to those questions, both shown in Table 1, are the 'good,' not the 'perfect.' Fortunately, Mind Genomics studies are so inexpensive, rapid, and easy, that one can iterate to the 'perfect,' through many trials, not all of which are even 'good.'

Step 2: The Introductory Page for the Study

Mind Genomics works with people. People do not necessarily know what to do in one of these experiments, especially when the experiment requires the respondent to read a set of clearly disconnected phrase, and treat these phrases as a unity. That task, treating the phrases as a unity, becomes very easy after the respondent reads and rates 1-2 vignettes, but we still need the orientation at the very beginning of the study to make respondents feel comfortable.

Figure 1 shows an example of the introductory page. The respondent is introduced to the topic, with as little information as possible. It's important for the respondent to understand the basic idea behind the topic, but other than that, the less that is said, the better. The objective is to have the contents of the vignettes, the elements or messages, i.e., the answers to the questions, do all the work to interest or to disinterest the respondent.

Figure 1. The introductory page, i.e., the orientation to the project

Thank you for participating. We are interested in your reactions to events and actions related to GLOBAL WARMING. You will read a set of 48 DIFFERENT vignettes (statements) or short stories on the screen, one vignette (statement) for each screen. Just please read it quickly, and give us your opinion on two things:

How much do you AGREE on these phrases together? 1 – do not agree at all; 9 – completely agree

How do you personally FEEL when you read the combination of these phrases together? 1= Curious 2= Amused 3= Don't Care 4= Very Angry 5= Surprised 6= Fear 7= Don't Believe

The purpose of this study is to understand people's opinions and feelings about climate change and global warming.

All of the vignettes (statements) are different. Please give your immediate impression, and move on. The whole exercise should take about 10-15 minutes. Thank you!

There are two aspects of the introductory page worth remarking:

1. The topic, global change, is introduced as a simple two-word phrase, and nothing more. One need not even introduce the topic. Respondents do not need to be spoon-fed about the task. Their innate intelligence guides them. Respondents do have to know what to do, i.e., what to rate.
2. The respondent is told about the mechanics task, specifically how long it will take. Respondents are not interested in the research. Respondents are interested in respondents, in themselves, in what is expected of them, and most of all, how much time they will have to donate to the project. Occasionally, respondents complain. Sometimes they complain that the same elements keep appearing, so they are rating the same vignette twice. This is an occasional complaint. Far more frequent is the complaint that the experiment 'took too long.' The instructions address that complaint, in an attempt to ward it off by presenting a reasonable time for the respondent to go through the experiment.

Step 3: Combine Elements Into Easy-To-Read Vignettes, According to an Experimental Design

The fundamental premise of Mind Genomics is that the respondent may be able to answer a question, or assign a rating, but may not be aware of WHY? That is, a person may respond quickly and intuitively, but not be cognizant of the criteria used. The psychologist-economist and Nobel Laureate, Daniel Kahneman of Princeton University has characterized these two systems as 'fast thinking' (immediately, intuitive response) and slow thinking (rational explanations for the response.) (Kahneman, 2011).

By combining the elements into vignettes, and soliciting the 'gut' or immediate response, it becomes possible to determine what elements are immediately processed, and their degree of importance. We will visit that topic in depth when we analyze the result of the study. In the meanwhile, Figure 2 shows a vignette constructed with answers (elements) from four of the six questions. The questions and their order dictate the nature of the 'story' to be told by the vignette. The vignette shows the answers to the questions, not the questions themselves. The bottom of the screen shot shows the first rating question, 'disagree/agree'. The underlying experimental design for a 6x6 (six questions, six answers) calls for exactly 48 vignettes, each vignette comprising either three or four answers (elements), and each answer appearing five times across the respondent's 48 vignette. Figure 3 shows the same vignette, this time featuring the second rating question at the bottom of the screen.

Figure 2. An example of a vignette comprising elements from four of the six silos or categories

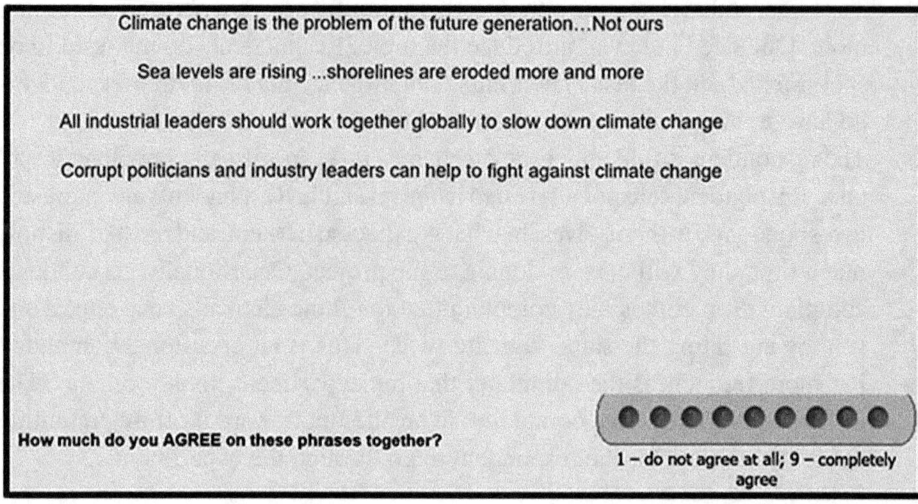

Figure 3. The same vignette, this time showing question 2, select an emotion

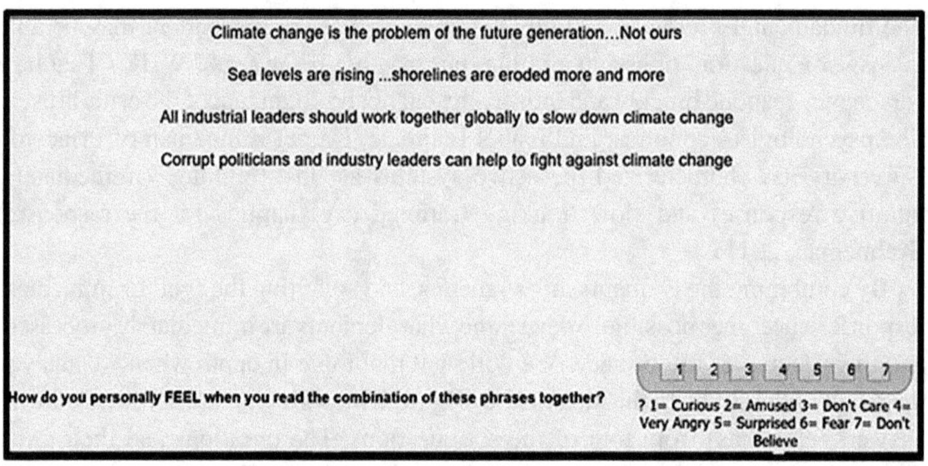

To simplify the respondent task, we present each vignette once, and ask to rating questions, first about one's values (disagree --- agree), and second about emotion (how do you feel?)

A feature worth noting in this experimental design is the feature that all respondents evaluated unique, different set of vignettes. Across all the respondents in a study, especially when there are fewer than 200, it is likely that no two vignettes would comprise the same set of elements, except by accident. And, of course, a respondent would never see duplicate the vignettes, since the vignettes are constructed according to an experimental design, valid at the level of the individual respondent.

Each respondent evaluated 48 combinations set up ahead time to ensure that the combination of elements would be appropriate for analysis by ordinary least-squares regression (Milutinovic & Salom, 2016). This strategy of creating individual sets of combinations emerges out of the belief that the replication of a study with many respondents is not simply to reduce the sampling variability of each element when analyzed by OLS regression, but also to learn more by investigating other areas of the possible stimulus space. The more typical strategy, used by most researchers, tests the SAME, limited set of vignettes many times, rather than testing new and different vignettes each time. The TRADITIONAL approach creates a single set of 48 vignettes using the experimental design, and then use only those 48 vignettes in a study, measuring the response to each of these fixed vignettes, one at a time. Of course, the researcher would randomize the ORDER of the vignettes, but in the end still uses the same vignettes.

Step 4: Analytics for Question 1 – Disagree/Agree

The ratings come in the form of an anchored 9-point scale, a so-called Likert or category scale. Although frequently used, one has a difficult time understanding the meaning of the scale, other than saying that high numbers mean a higher degree of agreement, whereas low numbers mean a lower degree of agreement. We simplify the data by converting the 9-point scale into a binary scale, with ratings of 1-6 assigned the value 0 to denote little or no agreement, and ratings of 7-9 assigned the value 100 to denote moderate to high agreement. To each transformed number is added a very small random number ($<10^{-5}$). The small random ensures that the regression model will work, and not crash should a respondent limit the ratings either to the low end (1-6) or to the high end (7-9). Both behaviors would create all 0's or all 100's, with no consequent variation in the dependent variable, causing the regression program to 'crash.'

Question 1 is not set up for OLS (ordinary least-squares) regression, also known as curve fitting (Hastie, Tibshirani, & Friedman, 2011). We do the analysis on a respondent by respondent basis, relating the presence/absence of the 36 elements (our 36 answers to the six questions) to the newly created binary rating scale (0/100). OLS regression returns with the additive constant, and 36 coefficients, one for each element.

Step 5: Extracting Mind-Sets From the Total Panel

We use the values of the 36 coefficients across each respondent to create a vector of 36 numbers, showing the pattern of responses of that individual. We will then divide the set of respondents, here 52, into smaller, complementary groups, mind-sets, with the property that the pattern of responses are more homogeneous within the mind-set. That is, the mind-set tells a story that might not be seen were we to work with the full set of 52 respondents. The division of respondents into complementary segments or mind-sets is done with the aid of a set of computer programs known collectively as clustering programs (Jain, 2010). The programs emerge with recommend groups, basing the creation of those groups on mathematical criteria, rather than interpretive criteria. That is, the clusters or segments are homogeneous in a mathematical or statistical sense. However, it is the job of the researcher to select the minimal number of clusters or segments possible (parsimony), with the proviso that the clusters or segments make intuitive sense when described (interpretability.)

A parenthetic note: We often work with studies in which hundreds of respondents participate, and from those hundreds of respondents we extract a limited number of aforementioned mind-sets. We can attempt to extract mind-sets from our 52 respondents, recognizing that with 52 respondents our mind-sets are likely to be suggestions of possible mind-sets in the population. Despite the tentativeness of the emergent mind-sets, dividing respondents into these segments is still a useful exercise to suggest the different ways people approach a topic.

Step 6: Making Sense of the Data by Looking at the Additive Constant and the Winning Elements for the Total Panel and for Each Segment

Each respondent generates a set of 36 coefficients, as well as an additive constant, these parameters based on the transformed ratings, where 1-6 were transformed to 0, and 7-9 were transformed to 100. We now begin the pleasant task of interpreting the data, to discover what nature has to tell us. Even with a small sample of 52 respondents, we will learn some important things about the mind of the person as it grapples with messages about global change. It is this ability to discover new things, to see so-called 'weak signals' emerging from small and affordable samples of respondents which makes Mind Genomics so exciting as a tool to learn, and a tool affordable to students.

The practical steps now follow. We average the corresponding additive constants and 36 coefficients across all the respondents who participated. In this particular study on global change, with a small, easy-to-work-with sample of 52 respondents, the averages appear in the column marked 'Total' in Table 2. There are three additional columns, for three emergent mind-sets, with which we deal in a moment.

The three mind-sets which emerge from this tentative analysis suggest the following three groups. The suggestion is based only on the elements which perform strongly for the mind-set segments, with strong performance represented by high positive coefficients (>8) and low negative coefficients (<-8). The high positive coefficients represent those ideas to which the respondent feels he or she has affinity, i.e., the respondent feels that the idea or element describes the person. The negative coefficients represent those ideas to which the respondent feels a lack of affinity.

Table 2. Topic = Global Change. Parameters of the function for 'Agree with the vignette.' The table shows the additive constant and the 36 coefficients for the total panel for the three emergent mind-sets.

	Agree - Global Change	**Total**	**Mind-Set 1**	**Mind-Set 2**	**Mind-Set 3**
	Base size	52	17	9	26
	Additive constant	62	56	84	60
	Mind-Set 1 – Public Knowledge and Action Is the Key				
E2	The general public needs more awareness to save the planet	2	11	-2	-2
D3	Reduce climate change… tougher fuel efficiency standards for cars needed	0	6	-6	-3
	Mind-Set 2 – Industrial Cooperation Is the Key				
E4	All industrial leaders should work together globally to slow down climate change	2	2	**10**	-1
	Mind-Set 3 – Governmental Pressure Is the Key				
E5	Elected officials are responsible for environmental issues also	0	-4	-7	4
	Less Relevant Elements				
A1	Global temperature is increasing... This is not a question	2	4	5	1
C6	In the past few decades, extreme temperatures were experienced during summer and winter	2	1	5	3
A4	Adverse effects of global warming are noticeable ... Even nowadays	2	3	4	0
E3	Energy industry leaders can do the most to reduce emissions	0	1	***-12***	3
C1	Wildlife and their habitats harmed by the changed weather conditions	0	3	-1	-3
C3	More droughts or water shortages are occurring	0	2	***-9***	0
D6	Reduce climate change… people should reduce their carbon footprints	-1	0	-1	0

continued on following page

Table 2. Continued

	Agree - Global Change	Total	Mind-Set 1	Mind-Set 2	Mind-Set 3
F6	People wanting to benefit from global warming, and they are trying to crush it for financial reasons	-1	1	***-9***	0
D1	Reduce climate change… restrictions on power plant emissions needed	-1	1	-5	-2
F2	The spreading of renewable energy solutions is intentionally obstructed	-1	3	-4	-4
B1	Global warming is caused by excess agricultural activity	-2	-1	-1	-2
C4	Forests and plant life continually damaged by intense fires and hot summers	-2	-3	***-12***	2
F4	Some people want to fight against climate change unethically because they see a quick profit	-2	-2	***-15***	1
D2	Reduce climate change… international agreement to limit emissions is needed	-2	3	***-9***	-3
E6	Public organizations have the most power to fight against global warming	-2	2	***-11***	-3
B6	Global warming is caused by deforestation	-3	-2	1	-5
C2	Storms have become more severe in the past few decades	-3	3	***-14***	-5
C5	Sea levels are rising ...shorelines are eroded more and more	-4	-4	-1	-5
B3	Simply human activity causes global warming	-4	-3	-7	-5
D4	Reduce climate change… corporate tax incentives needed	-5	0	***-14***	-6
E1	Climate scientists should solve the problem	-5	2	***-16***	-6
F5	Corrupt politicians and industry leaders can help to fight against climate change	-5	-1	***-23***	-3
D5	Reduce climate change… more people should drive hybrids	-5	-4	***-13***	-4
F1	Profit and greed overshadow the actions against global warming	-6	-1	-4	***-10***
A3	Some parts of global warming might be true ...But not the way presented	-6	-2	***-10***	-7
F3	Some politicians want to deny global warming because their constituencies are industrial companies	-6	-4	***-16***	-5
B4	Global warming is independent of us ....it is completely natural	-6	-7	***-12***	-4
B5	There is no clear evidence about global warming	***-8***	***-10***	***-12***	-6
B2	Industrial activities are the major contributors to global warming	***-8***	-7	***-12***	***-9***
A2	Global warming is only a 'media-hype'...I do not believe it	***-9***	***-8***	***-12***	***-9***
A6	Environmental issues are not my business ... I have no influence on it	***-10***	***-10***	***-12***	***-10***
A5	Climate change is the problem of the future generation…Not ours	***-12***	***-14***	***-18***	***-9***

- **Mind-set 1:** Public knowledge
- **Mind-set 2:** Industrial cooperation
- **Mind-set 3:** Governmental pressure

We interpret these average results from the Total Panel (Table 2) as follows:

1. Three mind-sets emerge from the 52 respondents. All are, of course, relatively small in size, yet clearly different from each other.
2. **Mind-set 1:** 17 of 52 respondents, responds to the one element which talks about public knowledge and action being key to dealing with global change. The highest performing element states: *The general public needs more awareness to save the planet.* Mind-set 1 disagrees with the assertion that there is no data supporting global change. Mind-set 1 is not unique in any other way
3. **Mind-set 2:** 9 of 52 respondents believes that it is the job of industry leaders to work together, presumably voluntarily, to solve the problem of global change. They react to this statement *All industrial leaders should work together globally to slow down climate change*
4. **Mind- set 3:** The remaining 26 respondents barely can be bothered to respond to the different elements. They are probably not involved. The fact that they constitute half of the population means that for global change, it is not a matter of ethics at all. They do not actively REJECT being involved in the issues of climate changes. It's just that they are really indifferent to the particulars of climate change.
5. The additive constant is the estimated likelihood of a person agreeing with a vignette in the absence of anything in the vignette. All vignettes comprised 3-4 elements (answers, messages), so the additive constant is a purely estimated parameter. Nonetheless, the additive constant gives a sense of the likelihood that a high score is due either to the basic agreement by the respondent, or in contrast, the hard work exerted by the elements.
6. The additive constant for the total panel is 62, fairly high. Respondents are likely to agree, but not overly so. The elements will have to do work as well.
7. The additive constant is moderate for Mind-Set 1 (public knowledge and action is the key), and Mind-Set 3 (Government pressure is the key). Each has one idea which they feel fits them well, a few ideas which somehow fit them (low positive coefficients). Yet they feel involved with the issues of global change. All of the elements with very low negative coefficients (does not fit me) are elements which say that global change is simply 'not our issue.'
8. The additive constant is unusually high for Mind-Set 2 (industrial cooperation is the key), with a value of 84, suggesting that they really don't find many ideas 'describing them.' There are lots of ideas which do not describe them,

all with negative coefficients, but nothing which describes them. Furthermore, the ideas which do not describe move beyond elements excusing them from concern with global change. The elements which do not describe them are elements of policy (e.g., *Energy industry leaders can do the most to reduce emissions*) and elements which describe environmental ravages due to global change (e.g., *Forests and plant life continually damaged by intense fires and hot summers.*)

9. If we were to summarize global change, we see a topic of great interest to people, as witnessed by the attention paid to it in the newspapers, and the efforts of governments and NGO's to cope with its ravages, and stop its causes. Yet, we also see few specific items. Global change is a general concept, not a specific concept with 'emotional immediacy'. People do not intimately involve themselves with the particulars of global change. Indeed, beyond the phrase, it is hard to know what global change really means. That difficulty is clearly seen by the emergence of mind-sets, with only one rallying point, one element which stands out strongly from the others. Furthermore, people know what is NOT them, such as attempts to distance oneself from the issue of global change. People know what's not them more than what's them.

Step 7: Measuring the Linkage Between the Elements and Emotions for Global Change

Question 2 instructed the respondents to select one feeling/emotion from a set of seven to describe how they felt after reading the vignette. Question 2 is not a conventional scale in the way that we are used to seeing scales. Psychologists interested in the theory of measurement call Question 2 an example of a nominal scale (Stevens, 1946). The seven different scale points correspond to seven feelings/emotions, selected to represent a range of negative to positive feelings.

Table 3 shows the linkages. The first data column presents the coefficient from Question 1 for the Total panel. The subsequent columns present the coefficients for the seven feelings/emotions, arranged in a rough descending order

By themselves the seven feelings and emotions are simply generalized responses to the vignettes, telling us very little about the 'emotional profile' of particular elements. As was the case with question 1, agreement, we have to create an easy-to-understand analysis of the linkage between the individual elements and the emotions.

When we move to the nominal scale, we must break up the one scale into seven scales, one scale for each feeling/emotion. We could also combine some of the responses into those which are positive, those which are neutral, and those which are negative. For this particular analysis we remain with the seven scales, one for each selected feeling/emotion.

Table 3. Linkage between feelings/emotions and elements for Global Change. Linkages of around +8 or over are meaningful, based upon observations from previous studies.

	Global Change	Total Panel – Agree (Constant = 62)	Very Angry	Fear	Curious	Surprised	Amused	Don't Care	Don't Believe
E2	The general public needs more awareness to save the planet	2	6	6	4	1	4	1	2
E4	All industrial leaders should work together globally to slow down climate change	2	3	2	4	6	1	3	4
A1	Global temperature is increasing... This is not a question	2	3	5	3	6	4	2	4
C6	In the past few decades, extreme temperatures were experienced during summer and winter	2	4	5	3	4	1	6	4
A4	Adverse effects of global warming are noticeable ... Even nowadays	2	5	1	3	6	5	3	4
D3	Reduce climate change… tougher fuel efficiency standards for cars needed	0	6	5	2	3	5	5	-1
E5	Elected officials are responsible for environmental issues also	0	7	4	5	4	2	4	0
E3	Energy industry leaders can do the most to reduce emissions	0	2	4	4	7	2	3	2
C1	Wildlife and their habitats harmed by the changed weather conditions	0	4	3	4	5	2	4	3
C3	More droughts or water shortages are occurring	0	4	3	2	7	2	6	1
D6	Reduce climate change… people should reduce their carbon footprints	-1	7	2	2	4	4	7	0
F6	People wanting to benefit from global warming, and they are trying to crush it for financial reasons	-1	**8**	-2	5	6	2	6	2
D1	Reduce climate change… restrictions on power plant emissions needed	-1	6	-2	4	4	0	**10**	2
F2	The spreading of renewable energy solutions is intentionally obstructed	-1	**9**	2	4	3	0	5	4
B1	Global warming is caused by excess agricultural activity	-2	7	2	3	4	5	2	3
C4	Forests and plant life continually damaged by intense fires and hot summers	-2	**9**	2	2	4	3	3	2
F4	Some people want to fight against climate change unethically because they see a quick profit	-2	**9**	4	0	6	1	6	1

continued on following page

Table 3. Continued

	Global Change	Total Panel – Agree (Constant = 62)	Very Angry	Fear	Curious	Surprised	Amused	Don't Care	Don't Believe
D2	Reduce climate change… international agreement to limit emissions is needed	-2	7	5	5	4	0	5	-1
E6	Public organizations have the most power to fight against global warming	-2	6	2	4	5	1	6	0
B6	Global warming is caused by deforestation	-3	0	5	4	7	0	5	4
C2	Storms have become more severe in the past few decades	-3	5	5	2	6	5	2	1
C5	Sea levels are rising ...shorelines are eroded more and more	-4	**10**	**8**	4	2	1	1	1
B3	Simply human activity causes global warming	-4	3	5	-1	7	5	2	6
D4	Reduce climate change… corporate tax incentives needed	-5	4	2	6	5	3	**8**	2
E1	Climate scientists should solve the problem	-5	3	6	7	3	1	1	4
F5	Corrupt politicians and industry leaders can help to fight against climate change	-5	7	0	4	5	5	5	2
D5	Reduce climate change… more people should drive hybrids	-5	3	4	3	5	1	5	4
F1	Profit and greed overshadow the actions against global warming	-6	**9**	3	4	2	4	3	3
A3	Some parts of global warming might be true ...But not the way presented	-6	5	1	5	**8**	5	4	1
F3	Some politicians want to deny global warming because their constituencies are industrial companies	-6	6	2	5	7	3	4	0
B4	Global warming is independent of us ....it is completely natural	-6	5	4	3	4	3	3	6
B5	There is no clear evidence about global warming	***-8***	5	7	1	6	0	5	3
B2	Industrial activities are the major contributors to global warming	***-8***	5	3	5	6	1	6	2
A2	Global warming is only a 'media-hype'...I do not believe it	***-9***	3	-1	1	7	6	6	6
A6	Environmental issues are not my business ... I have no influence on it	***-10***	1	0	5	**11**	0	**8**	1
A5	Climate change is the problem of the future generation…Not ours	***-12***	3	3	5	6	1	6	4

The mechanics are straightforward. We create a set of seven new variables, each variable corresponding to one of the seven choices in question 2. For any combination of an individual and a vignette, we have one and only one selection from the seven feelings/emotions. The newly-created variable corresponding to that selected answered receives the value of 100. The remaining six, newly-created variables, in turn receive the value of 0. Again, a small random number (<10-5) is added to each of these seven newly created variables, that random number being added to seven cells. For our 52 respondents, each of whom evaluated 48 vignettes, and generated 7 newly created variables from question 2, we end up with 52*48*7 or 15,288 cells, each cell with 0 or 100 augmented by the small random number.

We are now in a position to use our OLS (ordinary least-squares) regression, seven times, once for each newly-created variable. The independent variables are the 36 elements. The dependent variable is the selection of the attribute, coded as 0 (not selected) or 100 (selected), augmented by the random number. The regression model is run without an additive constant, so only the 36 elements generate results, i.e., coefficients. From previous studies, coefficients around 8-10 or higher are meaningful.

Table 3 shows the results from the OLS regression. The first column of data shows the coefficients from Question 1 (agree). The remaining columns show the seven feelings/emotions and their coefficients. The table is sorted by the value of the coefficient for Question 1 (agree).

Table 3 offers us no surprises. We saw that there is only mild emotion involvement in Global Change, despite the hoopla in the press.

1. Very angry – often the highest level of linkage to elements. The elements leading to the selection of very angry are either the crass nature of people, or the major damage to the environment.
2. Don't care – linked with a type of punitive behavior with which the respondent probably has no experience. The element with the highest linkage with 'don't care' is *Reduce climate change… restrictions on power plant emissions needed.* A second highest linkage is: *Reduce climate change… corporate tax incentives needed*
3. *Only one element is felt to 'surprise.' This element is:* Some parts of global change might be true ...But not the way presented
4. Summarizing the results from Global Change and emotion/feeling, we get a sense of very low real interest, at least interest which might lead to angry exposes of the industry called global change.

Step 8: Summing Up – Global Change

If we were to summarize Global Change, we would say, in a single word, lukewarm. There is a great deal of publicity around global change, but there is very little in the way of reasoned thought, nor are there radically different points of view (mind-set segments), nor strong emotional responses emerging from the topic.

The results from Global Change may be contrasted with those from other social issues such as crime and the failure of education to produce jobs. Unpublished results on those latter topics suggest that the elements are associated with coefficients that often have high positive values. It may well be that Global Change here, and perhaps Sustainable Agriculture in the next section reflect topics of social concern, but not deep personal involvement. Such lack of involvement reveals itself through the array of low positive coefficients, suggesting a dearth of elements with which the respondent can strongly identify!

SUMMING UP THE RESULTS OF GLOBAL CHANGE STUDY

Respondents generally agree on the statements that there are clear evidences of global change and it is not a media hype. They refuse the statements that the main contributors to global change are industrial activities and strongly believe that global change is the problem of today, and today's people.

Mind segmentation resulted three distinct mind-sets, which can be described as follows:

- **Mind-set 1:** Public knowledge (33% of the total panel)
- **Mind-set 2:** Industrial cooperation (17% of the total panel)
- **Mind-set 3:** Governmental pressure (50% of the total panel)

The most important element for members of the first mind-set states that the public awareness needs to be raised in order to save our planet. One third of the total panel belongs to this segment, which suggests that there is still insufficient information available to the people about global change.

The smallest mind-set consists of the most pessimistic or negative group. They showed the most negative coefficients, e.g. they disagree on most of the elements. Members of the second mind-set believe that the solution to global change is in the hands of industrial leaders, and believe that global international cooperation could solve the issue.

The third mind-set contains half of the respondents, who are less responsive to the presented elements, due to the high numbers of close-to-zero coefficients.

However, they trust in the politicians and think that elected officials should solve the problem. Furthermore, they disagree that profit and greed overshadow the actions against global change.

Global change is a tough topic which makes people react in different way. Results of the emotion measurement clearly showed that respondents reacted very angrily to the elements which stated that the fight against global change is obstructed by people or organizations. It is interesting to notice that they showed strong responses to elements stating that corporate restrictions or tax raises are needed.

The way of fighting against global change is extremely complex. The presented results point out to the fact, that people belong to different mind-sets, and each mind-set sees the solution differently. Using the presented approach on a broader sample may be able to exactly define the key actions which would help people to better understand global change and to show them how to fight against it on an individual level.

SUSTAINABLE FOOD PRODUCTION

Our second topic, following the same procedure as Global Change, is sustainable food production. Whether or not we accept the reality of food production as a problem, everyday life shows us that many people have taken on the cause of sustainability as something close to their heart. Perhaps the passion for sustainability comes from the fact that we eat to live. Eating is, in the end, a most profoundly intimate behavior, making the external world part of our own being.

The sad reality is that food that is healthy and of high quality simply is not available in every corner of the world. The global increase of global population left food production causing hunger and malnutrition in many parts of the globe. However, this effect is varying between countries and produced enormous inequality of food supply. Whereas many countries have to face huge food waste problems, there are other places where people have only limited access to nutritious food. This is caused partly by the problems of agricultural production.

The problem has several aspects which all need to be solved. The agricultural production need to be increased but in a way that environmental resources should be preserved and sustainable. Long-term solutions should be introduced which are able to feed the world.

Sustainable food production includes all the actions taken to preserve natural resources, such as water management, soil health and energy availability, while keeping the crop yield on a satisfying level. The foregoing is summed up by the Cambridge dictionary definition of Sustainability in environment and rural resources as:

The idea that goods and services should be produced in ways that do not use resources that cannot be replaced and that do not damage the environment.

Novel Foods and Sustainability

During the past few years several agricultural innovations have been introduced, among them those which address sustainability by producing proteins from unusual source. These innovations include algae, insect or even in-vitro meat. However, these new products are usually less accepted by the consumers. At the very start, it is important to educate, to guide consumers to understand that their food consumption habits are not sustainable at all. In 2009, the amount of food waste generated reached 32% of the world produced food (FAO, 2011). At the same time, the number of undernourished people in the world reached 1.02 billion in 2009 (FAO, IFAD, & WFP, 2015). These inequalities are mainly caused by the fact that agricultural production is unbalanced and the quality of the produced food is low. People need to change. In one parts of the world, they need to stop wasting food and throwing out huge amounts of leftovers, whereas in other parts of the world people need to learn how to include new food technologies or food products into their everyday diet.

In order to learn how and what to teach/communicate to people, we need to get further information about how and what they think about this issue globally. Whereas global change seems to be addressed by rallies, by pronouncements, and so forth, sustainable food product comes to us in the form of farmer's markets, in the form of foods available on the shelves, in the form of people's everyday eating behavior, and more important, the often-shocked looks of some people, judgmentally-prone vegans often, at what the 'other' is eating at that very minute. We hear admonitions equivalent to 'I would NEVER eat another animal,' 'I don't eat food with a face,' or some such statement.

With the foregoing in mind, we now turn to the data from the sustainable food project, again a small-scale one designed to explore some of the aspects. The reader should, by now, recognize the proclivities of the authors to explore many topics with small, affordable experiments, studies which do not definitive map out the topic and answer the questions, but rather show the different aspects and lead to deeper thinking.

Table 4 shows the six questions, and the six answers to each question. As we saw in the case of Global Change, the very breadth of the topic leads to many different questions, directing us in different ways. The very personal nature of the way we interact with food, from the society and economic to the personal, means that a study of Sustainable Food Production could be far larger, beyond the scope of these efforts. This study, a first foray into the topic, thus is necessarily sparse, covering only six issues (the questions), each with a few different aspects (the answers).

Table 4. The six questions, each with six answers for the topic of Sustainable Food Production

Question 1: What do people think about overpopulation?
Earth cannot feed so many people in the world… we need changes.
In order to provide quality protein for people all around the world… we need to find meat alternatives.
GMO plants provide the only solution to feed the world.
Developed countries must provide quality and safe food for people in third world countries .
Local farmers should be taught to produce quality and healthy food products locally.
Food supply problems should be solved locally…everyone does what they can.
Question 2: What meat substitutes would be acceptable?
I do not want to substitute meat even if it is bad for the environment.
Meat should be replaced by insect protein.
Algae would be a good meat substitute.
Vegetarian lifestyle could be a reasonable solution to reduce the carbon footprint.
Cultured meat (lab-grown meat) is a great alternative to beef.
Quality protein is provided by soy.
Question 3: How can we reduce the carbon footprint?
International food production should be replaced by local, traditional food production.
Big food companies should produce their products in a more sustainable way.
Greenhouse gas emission of food companies has to be reduced.
Food companies place production before environment protection because of profit.
Carbon footprint should be reduced individually by reducing food waste, buying local food etc.
Overconsumption of western countries raises environmental load…increasing of carbon footprint.
Question 4: What is the role of food waste?
The excess amount of food waste … wasting of resources.
Food production generates an enormous amount of waste.
Recycling of food waste as feed would give a sustainable solution.
Food waste reduction should be done individually…food waste free households.
Food shops and restaurants should sell at reduced prices food that is near its expiration date.
Wasting some food should not cause global food problems.
Question 5: What are the health effects of excess consumption that will be ameliorated?
People should eat less but quality food to protect their health.
Consumption of healthy local food helps to maintain sustainable food production.
Sustainably produced food products have higher nutritional values.
Overeating causes health-issues.

continued on following page

Table 4. Continued

Overeating is not a problem … people should eat as much as they want.
People are lazy and they want to buy processed food.
Question 6: Who should have a major role in driving the efforts for sustainable food production?
Food scientists should create new processes to feed the world.
Local farmers should be supported financially to produce more healthy, and quality local food.
Food industry leaders can do the most to reduce the carbon footprint.
Elected officials are responsible for sustainable food issues also.
Public organizations have the most power to support sustainable food production.
Celebrities should promote sustainable food production through all possible media channels.

The initial results from the total panel of 52 respondents suggest the following, as shown by Table 5:

1. The additive constant, 60, suggests a moderate basic agreement with the general topic of sustainability. If all of the vignettes were to have been rated low (e.g., 1-6), then Table 5 would show us a low additive constant, closer to 0, suggesting that the topic is not perceived as something relevant to the respondent.
2. Although the topic is relevant, for the total panel no ideas emerges which strongly 'agreed with them,' at least for the total panel. There were ideas which clearly did not agree with them. These are:

Overeating is not a problem … people should eat as much as they want. (Against common wisdom)

Algae would be a good meat substitute. (Motivated by neophobia)

Meat should be replaced by insect protein. (Motivated by neophobia)

3. The population of 52 respondents suggest a minimal number three mind-set segments, although there may be more segments to be extracted. The reason for conjecturing that there are more than three mind-sets are that the issues of food, food type, and eating calls into account many aspects of a person's life. Our study with 52 respondents again only represents an initial foray into the topic.

Table 5. Topic = Sustainable Food Product. Parameters of the function for 'Agree with the vignette.' The table shows the additive constant and the 36 coefficients for the total panel for the three emergent mind-sets.

	Sustainable Food Production (Agree)	Total	Mind-set 1	Mind-set 2	Mind-set 3
	Number of respondents	52	25	12	15
	Additive constant	60	59	65	57
	Mind-set 1 - There's a problem.. So..what do I do?				
A2	In order to provide quality protein for people all around the world… we need to find meat alternatives.	0	7	***-17***	4
	Mind-set 2 – Local but really a shallow thinker otherwise				
E2	Consumption of healthy local food helps to maintain sustainable food production.	0	-2	4	-1
	Mind-set 3 - Really into it sustainability				
C3	Greenhouse gas emission of food companies has to be reduced.	0	-3	-3	**11**
F2	Local farmers should be supported financially to produce more healthy and quality local food.	3	0	-1	**11**
A5	Local farmers should be taught to produce quality and healthy food products locally.	4	3	-1	**10**
C4	Food companies place production before environment protection because of profit.	1	0	-3	**9**
F3	Food industry leaders can do the most to reduce the carbon footprint.	0	-3	-3	**9**
E6	People are lazy and they want to buy processed food.	-2	0	***-17***	**9**
D4	Food waste reduction should be done individually…food waste free households.	-3	-4	***-16***	**8**
	Remaining elements - not strong in any mind-set				
E1	People should eat less but quality food to protect their health.	1	3	-6	5
C2	Big food companies should produce their products in a more sustainable way.	1	2	-5	6
F1	Food scientists should create new processes to feed the world.	0	5	-4	0
C5	Carbon footprint should be reduced individually by reducing food waste, buying local food etc.	0	3	***-12***	4
A1	Earth cannot feed so many people in the world… we need changes.	0	2	1	-3
F5	Public organizations have the most power to support sustainable food production.	-1	0	-7	5
B1	I do not want to substitute meat even if it is bad for the environment.	-1	-2	-1	3
C6	Overconsumption of western countries raises environmental load… increasing of carbon footprint.	-2	0	***-9***	3
E4	Overeating causes health-issues.	-2	0	***-8***	0

continued on following page

Table 5. Continued

	Sustainable Food Production (Agree)	**Total**	**Mind-set 1**	**Mind-set 2**	**Mind-set 3**
C1	International food production should be replaced by local, traditional food production.	-2	1	***-15***	6
A4	Developed countries must provide quality and safe food for people in third world countries .	-2	-1	-2	-2
E3	Sustainably produced food products have higher nutritional values.	-2	-1	***-9***	2
B6	Quality protein is provided by soy.	-3	-2	-6	0
A6	Food supply problems should be solved locally…everyone does what they can.	-3	-3	-7	0
F4	Elected officials are responsible for sustainable food issues also.	-3	0	-4	-6
D3	Recycling of food waste as feed would give a sustainable solution.	-4	-6	-2	-1
F6	Celebrities should promote sustainable food production through all possible media channels.	-4	-2	-6	-3
D1	The excess amount of food waste … wasting of resources.	-4	-2	***-14***	-1
D5	Food shops and restaurants should sell at reduced prices food that is near its expiration date.	-4	-3	-4	-6
D2	Food production generates an enormous amount of waste.	-5	-2	***-9***	***-9***
D6	Wasting some food should not cause global food problems.	-5	-1	***-12***	-7
B5	Cultured meat (lab-grown meat) is a great alternative to beef.	-7	0	***-10***	***-14***
B4	Vegetarian lifestyle could be a reasonable solution to reduce the carbon footprint.	-7	-5	***-10***	***-9***
A3	GMO plants provide the only solution to feed the world.	***-9***	-7	-6	***-13***
E5	Overeating is not a problem … people should eat as much as they want.	***-12***	***-9***	***-15***	***-13***
B3	Algae would be a good meat substitute.	***-15***	***-13***	***-22***	***-13***
B2	Meat should be replaced by insect protein.	***-17***	***-14***	***-19***	***-21***

4. The three mind-sets show similar, and modest-size additive constants, with a low of 59 to a high of 65. There is no mind-set segment with very high additive constants (80+) suggesting that there are no segments which agree, in general, with what is presented. Rather, it is the elements which do the work of driving agreement or disagreement.
5. The mind-set segments share the same responses to elements to which they do not agree. These are three elements listed above in section 2.
6. Mind-set 1, about half the respondents, recognize the problem, and suggest that there is a need to answer the question, but offers no suggestions.

7. Mind-set 2, about a quarter of the respondents, seems to be thinking 'local,' or are so-called Locavores. This respondent knows he or she does not agree, as if the person thinks only Local, and nothing else. Not only does this respondent reject the same set of elements as do the other respondents (see #2 above), but also disagrees with, a number of other elements as relevant to sustainability of food product. Mind-set 2 feels these 16 (or 36) elements simply DO NOT ADDRESS and thus CANNOT SOLVE the problem of food sustainability
    a. Algae would be a good meat substitute.
    b. Meat should be replaced by insect protein.
    c. In order to provide quality protein for people all around the world… we need to find meat alternatives.
    d. People are lazy and they want to buy processed food.
    e. Food waste reduction should be done individually…food waste free households.
    f. International food production should be replaced by local, traditional food production.
    g. Overeating is not a problem … people should eat as much as they want.
    h. The excess amount of food waste … wasting of resources.
    i. Carbon footprint should be reduced individually by reducing food waste, buying local food etc.
    j. Wasting some food should not cause global food problems.
    k. Cultured meat (lab-grown meat) is a great alternative to beef.
    l. Vegetarian lifestyle could be a reasonable solution to reduce the carbon footprint.
    m. Overconsumption of western countries raises environmental load… increasing of carbon footprint.
    n. Sustainably produced food products have higher nutritional values.
    o. Food production generates an enormous amount of waste.
    p. Overeating causes health-issues.
8. Mind-set 3, with 15 respondents, shows the strong set of elements with which the respondent agrees, doing so strongly. Mind-set 3 is really into sustainability, based upon the dramatic number of elements which have high positive values, denoting strong agreement (>8.5)

The linkage between emotions and elements from Sustainable Food Production appears in Table 6. The same rule applies as before; search for linkages of +8 or higher, based upon observations from previous studies. The emotions in Sustainable Food Production are different from that of Global Change. There is no real 'anger' or negative emotion for Sustainable Food Production, nor should there be. The emotions highlighted in Table 6 are positive, primarily 'curious.'

Table 6. Linkage between feelings/emotions and elements for Sustainable Food Production. Linkages of around +8 or over are meaningful, based upon observations from previous studies.

	Sustainable Food Production	Total (Agree)	Very Angry	Fear	Curious	Surprised	Amused	Don't Care	Don't Believe
	Additive constant	60							
A5	Local farmers should be taught to produce quality and healthy food products locally.	4	4	5	10	6	-2	4	1
F2	Local farmers should be supported financially to produce more healthy and quality local food.	3	4	3	**8**	5	2	0	4
C4	Food companies place production before environment protection because of profit.	1	2	1	**8**	4	5	5	0
E1	People should eat less but quality food to protect their health.	1	6	-2	7	**8**	0	5	1
C2	Big food companies should produce their products in a more sustainable way.	1	5	4	5	2	6	4	0
C3	Greenhouse gas emission of food companies has to be reduced.	0	6	3	**8**	5	2	1	2
F1	Food scientists should create new processes to feed the world.	0	3	5	**8**	4	3	2	1
E2	Consumption of healthy local food helps to maintain sustainable food production.	0	4	5	**8**	4	0	5	0
A2	In order to provide quality protein for people all around the world… we need to find meat alternatives.	0	5	4	**8**	1	3	4	4
C5	Carbon footprint should be reduced individually by reducing food waste, buying local food etc.	0	4	3	**11**	3	0	2	1
A1	Earth cannot feed so many people in the world… we need changes.	0	4	0	**8**	6	4	3	4
F3	Food industry leaders can do the most to reduce the carbon footprint.	0	3	4	7	6	4	1	2
F5	Public organizations have the most power to support sustainable food production.	-1	6	2	3	4	3	4	4
B1	I do not want to substitute meat even if it is bad for the environment.	-1	4	4	4	3	7	4	1
C6	Overconsumption of western countries raises environmental load…increasing of carbon footprint.	-2	3	3	4	5	3	7	0
E4	Overeating causes health-issues.	-2	6	0	**12**	7	-2	1	2
C1	International food production should be replaced by local, traditional food production.	-2	2	1	4	**8**	4	4	3

continued on following page

Table 6. Continued

	Sustainable Food Production	Total (Agree)	Very Angry	Fear	Curious	Surprised	Amused	Don't Care	Don't Believe
E6	People are lazy and they want to buy processed food.	-2	**8**	6	4	1	1	3	2
A4	Developed countries must provide quality and safe food for people in third world countries .	-2	5	2	**9**	4	1	4	1
E3	Sustainably produced food products have higher nutritional values.	-2	4	3	6	6	-1	3	3
B6	Quality protein is provided by soy.	-3	5	0	**10**	**8**	1	0	2
A6	Food supply problems should be solved locally…everyone does what they can.	-3	3	5	**9**	2	2	4	3
F4	Elected officials are responsible for sustainable food issues also.	-3	4	2	7	6	4	1	3
D4	Food waste reduction should be done individually…food waste free households.	-3	**9**	2	**10**	2	1	2	0
D3	Recycling of food waste as feed would give a sustainable solution.	-4	3	3	**11**	5	4	1	0
F6	Celebrities should promote sustainable food production through all possible media channels.	-4	0	3	**8**	1	4	7	3
D1	The excess amount of food waste … wasting of resources.	-4	4	0	5	5	2	5	5
D5	Food shops and restaurants should sell at reduced prices food that is near its expiration date.	-4	4	1	6	**9**	2	4	2
D2	Food production generates an enormous amount of waste.	-5	6	3	7	4	1	4	1
D6	Wasting some food should not cause global food problems.	-5	5	5	5	4	3	4	1
B5	Cultured meat (lab-grown meat) is a great alternative to beef.	-7	3	3	**10**	5	3	0	3
B4	Vegetarian lifestyle could be a reasonable solution to reduce the carbon footprint.	-7	4	1	3	7	3	4	4
A3	GMO plants provide the only solution to feed the world.	***-9***	4	2	2	**8**	5	4	3
E5	Overeating is not a problem … people should eat as much as they want.	***-12***	5	2	-1	3	6	6	4
B3	Algae would be a good meat substitute.	***-15***	4	2	4	5	7	2	2
B2	Meat should be replaced by insect protein.	***-17***	3	**8**	-1	4	7	2	4

Comparing the two studies, we see that when the focus changes from Global Warming to Sustainable Food Production, again several mind-sets emerge, only one mind-set of which responds strongly to the elements. The other mind-sets pick out one element from set, and respondent only modestly to that element. It may be that when it comes to social issues that only a small proportion of the population ever even stops to think about the issue. Furthermore, the emotions are either negative (Global Change) or positive (Sustainable Food Production). The emotion linkages make sense. Global Change is a negative to be solved. Sustainable Food Production is a positive to be discovered.

SUMMING UP THE RESULTS OF SUSTAINABLE FOOD PRODUCTION STUDY

Sustainable food production is a more positive topic compared to the previous topic, global change. General results of the total panel suggest that people refuse to believe that GMOs could be the only solution to feed the world and agree on that overeating is a problem of today's consumers. However, food neophobia, the fear of new or unusual food sources or products also emerges through the clear refusal of algae and/or insect protein. The identified three mind-sets all agreed on these points. In spite of these, there were clear differences found between the mind-sets.

The first mind set gives about 48% of the total panel. This mind-set sees no problem about sustainable food production and agrees on that new meat alternatives are needed. Additionally, this mind-set showed the lowest refusal of algae and insect-based meat sources. This segment is less neophobic therefore more open to new solutions, products.

The most negative mind-set proved to be the second, giving 23% of the total panel. Its members are the most neophobic, they refuse the new meat alternatives and prefer healthy local foods, but do not agree on the positive role of food waste reduction.

The remaining 29% of the respondents give the third mind-set. These people are really into sustainability and agree on all of those elements which suggest the reduction of greenhouse gas emissions, supporting local farmers financially or by education. Furthermore, they agree on that people are generally lazy and food waste reduction should be done individually.

Evaluation of the emotion results of the sustainable food production study gives different patterns compared to the global change study. The major emotions were curious and surprised. Highest values for curious were found with elements about overeating, carbon footprint and cultured meat. Cultured meat and soy showed the greatest potential among the presented meat alternatives (algae and insect). Surprise was observed with elements about the importance of quality food and local, traditional food products.

Similarly, to global change, people about sustainable food production issues should be addressed differently. There are, however, some common opinions but the identified three mind-sets have clearly different values and opinions about sustainable food production. The presented results underscore the reality that food neophobia is a huge barrier of the acceptance of new meat alternatives, which is in concordance with earlier findings (Gere, Székely, Kovács, Kókai, & Sipos, 2017; Verbeke, 2015). Furthermore, people generally are curious and surprised when reading the vignettes about the topic which implies that further education would be well accepted.

REFERENCES

Anderson, N. H. (2001). *Empirical direction in design and analysis*. London: Taylor and Francis. Retrieved from http://cds.cern.ch/record/2123907

Boyes, E., Stanisstreet, M., & Yongling, Z. (2008). Combating global warming: The ideas of high school students in the growing economy of South East China. *The International Journal of Environmental Studies*, *65*(2), 233–245. doi:10.1080/00207230701284543

Bryan, E., Deressa, T. T., Gbetibouo, G. A., & Ringler, C. (2009). Adaptation to climate change in Ethiopia and South Africa: Options and constraints. *Environmental Science & Policy*, *12*(4), 413–426. doi:10.1016/j.envsci.2008.11.002

Davis, A. P., Gole, T. W., Baena, S., & Moat, J. (2012). The Impact of Climate Change on Indigenous Arabica Coffee (Coffea arabica): Predicting Future Trends and Identifying Priorities. *PLoS One*, *7*(11), 1–13. doi:10.1371/journal.pone.0047981 PMID:23144840

EPA. (2013). *Frequently asked questions about global warming and climate change: back to basics*. Retrieved from http://www.epa.gov/climatechange/basics/

FAO. (2011). *Global food losses and food waste – Extent, causes and prevention*. Retrieved from http://www.fao.org/docrep/014/mb060e/mb060e00.pdf

FAO, IFAD, & WFP. (2015). The State of Food Insecurity in the World 2015. Meeting the 2015 international hunger targets: taking stock of uneven progress. Rome: Authors.

Gere, A., Székely, G., Kovács, S., Kókai, Z., & Sipos, L. (2017). Readiness to adopt insects in Hungary: A case study. *Food Quality and Preference*, *59*, 81–86. doi:10.1016/j.foodqual.2017.02.005

Green, P. E., & Krieger, A. M. (1991). Segmenting Markets with Conjoint Analysis. *Journal of Marketing*, *55*(4), 20–31. doi:10.2307/1251954

Green, P. E., & Srinivasan, V. (1990). Conjoint Analysis in Marketing: New Developments with Implications for Research and Practice. *Journal of Marketing*, *54*(4), 3–19. doi:10.2307/1251756

Günther, W. A., Mehrizi, M. H. R., Huysman, M., & Feldberg, F. (2017). Debating big data: A literature review on realizing value from big data. *The Journal of Strategic Information Systems*, *26*(3), 191–209. doi:10.1016/j.jsis.2017.07.003

Hastie, T., Tibshirani, R., & Friedman, J. (2011). *The Elements of Statistical Learning* (2nd ed.). New York, NY: Springer.

IPPC. (2013). *CLIMATE CHANGE 2013. The Physical Science Basis. Intergovernmental Panel on Climate Change*. Retrieved from https://www.ipcc.ch/pdf/assessment-report/ar5/wg1/WGIAR5_SPM_brochure_en.pdf

Jain, A. K. (2010). Data clustering: 50 years beyond K-means. *Pattern Recognition Letters*, *31*(8), 651–666. doi:10.1016/j.patrec.2009.09.011

Kahneman, D. (2011). *Thinking, Fast and Slow*. New York, NY: Farrar, Straus and Giroux. Retrieved from https://books.google.hu/books?id=ZuKTvERuPG8C

Kenny, G. J., & Shao, J. (1992). An assessment of a latitude-temperature index for predicting climate suitability for grapes in Europe. *Journal of Horticultural Science*, *67*(2), 239–246. doi:10.1080/00221589.1992.11516243

Luce, R. D., & Tukey, J. W. (1964). Simultaneous conjoint measurement: A new type of fundamental measurement. *Journal of Mathematical Psychology*, *1*(1), 1–27. doi:10.1016/0022-2496(64)90015-X

Milutinovic, V., & Salom, J. (2016). *Mind Genomics - A Guide to Data-Driven Marketing Strategy*. Springer International Publishing; doi:10.1007/978-3-319-39733-7

Mozell, M. R., & Thach, L. (2014). The impact of climate change on the global wine industry: Challenges & solutions. *Wine Economics and Policy*, *3*(2), 81–89. doi:10.1016/j.wep.2014.08.001

Petersen, W. (1999). *Malthus: Founder of Modern Demography*. Routledge.

Stevens, S. S. (1946). On the Theory of Scales of Measurement. *Science*, *103*(2684), 677–680. doi:10.1126cience.103.2684.677

Verbeke, W. (2015). Profiling consumers who are ready to adopt insects as a meat substitute in a Western society. *Food Quality and Preference*, *39*, 147–155. doi:10.1016/j.foodqual.2014.07.008

Chapter 2
Could NoSQL Replace Relational Databases in FMIS?

Giancarlo Rodrigues
State University of Ponta Grossa, Brazil

Alaine Margarete Guimarães
State University of Ponta Grossa, Brazil

ABSTRACT

FMIS (farm management information systems) is the computational tool responsible to process data to get information that improves farmers' decision support. The data manipulated in FMIS is originated from diverse sources, stored, and read whenever necessary without subsequent modifications, thus dismissing the necessity of complex data storage systems such as offered by the relational model. Due to its capability to handle with high performance, a large amount of unstructured data and to reduce the complexity of applications, the NoSQL data storage model, a convenient alternative to relational model, recently gained a lot of attention in the information systems market. This way, this chapter discusses how NoSQL models could improve the FMIS architecture and performance when used as data storage. Some works that have already benefited from NoSQL model adoption are reviewed and convenient use cases where both data storage models could be well used in FMIS's architecture are advised and discussed.

INTRODUCTION

The declining availability of arable land and the concern about world food production for the next years(Godfray et al., 2010) boosted the pursuit by ways to enhance the agricultural yield. Due to availability of improved electronic devices equipped with accurate Global Positioning System(GPS) at reasonable costs(Fountas et al.,

DOI: 10.4018/978-1-5225-5978-8.ch002

2015), Information Communication Technology(ICT) devices were chosen by researchers and companies as the path to achieve this goal. As a consequence, a myriad of sensors and devices were developed(Aqeel-ur-Rehman, Abbasi, Islam, & Shaikh, 2014). They produce or collect an immense quantity of unstructured geo-referenced digital data about soil, plants, weed, pests, weather, equipment and machinery, management practices and others(Peets, Mouazen, Blackburn, Kuang, & Wiebensohn, 2012; Porter et al., 2014; Steinberger, Rothmund, & Auernhammer, 2009). These data after merging and processing offer valuable information to take accurate decisions that improve income, management practices, products quality, environmental protection and comply with governmental regulations(Fountas et al., 2015; Kaloxylos et al., 2012). However, processing an enormous amount of data to get information is not a trivial task and a computational tool is used to this end.

The computational tool responsible to transport electronically, store and process data in order to get useful information that serves to support planning, execution and evaluation of farm or field operations is referred to as Farm Management Information System (FMIS)(Kaloxylos et al., 2012; Nikkilä, Seilonen, & Koskinen, 2010). Higher will be the accuracy of information produced when more data is available, for this reason all data gathered from diversified sources, including third-parties, is stored historically and handled in FMIS(Nikkilä et al., 2010; Steinberger et al., 2009). Web Services(Nash, Korduan, & Bill, 2009) are the most opportune way for third-parties to access the data, however distinct sources dispatch and consume data in diversified storage and implementation formats such as spreadsheets, images, structured XML, plain text, etc., some of them exclusive of manufacturers(Porter et al., 2014). It represents a challenge for the FMIS, which needs to be able to process all this data in order to display helpful information to the decision maker.

Establish a data format standard and suggest a software architecture capable of accepting the variety of formats were the solutions found to overcome that challenge. Two well established data standard are AgroXML (Schmitz, Martini, Kunisch, & Mösinger, 2009), a German standard for data exchange between third-party farming systems, and ISOBUS (Oksanen, Öhman, Miettinen, & Visala, 2005), an international standard for communication between machinery and agricultural implements. Although they could facilitate data exchange and technology adoption (Aubert, Schroeder, & Grimaudo, 2012), other formats still exist. For this reason some authors proposed different solutions to convert data and adjust it to a specific format(Iftikhar & Pedersen, 2011; Porter et al., 2014). This favored the adoption of Service Oriented Architecture (SOA) in FMIS in order to adjust automatically the data. Some works dealing with this issue are (Fountas et al., 2015; Kaloxylos et al., 2012; Nikkilä et al., 2010; Wolfert, Goense, & Sorensen, 2014). The FMIS with SOA has a layer responsible for adjusting data according to the desired format, easing the working with on-the-go sensors(Peets et al., 2012) and automatic machinery(Blank, Bartolein, Meyer, Ostermeier, & Rostanin, 2013) from distinct manufacturers, in addition to third-party data sharing (Wolfert et al., 2014).

The adjusting feature is not only necessary for exchanging data between distinct sources but also for tailoring it to a data storage schema. It is required when a Relational database model is used in FMIS to store data. However, using a Relational database may compromise the FMIS performance over the time due to historical data storage requirement. Since it adds extra complexity to the software and delays the development, as highlighted by Nance et al. (2013), converting data back and forth between the database and FMIS may be unnecessary, even when working with a well-defined data standard. According to Nikkilä et al. (2010), FMIS not restricted to local computers are similar to Information Systems from other industries and may benefit from research applied to them. Due to tendency of FMIS to run on cloud(Kaloxylos et al., 2012) and use Web technologies(Nikkilä et al., 2010), using already employed and well-tested technologies of Industrial Web is timely. Thus, using a data storage model already employed on Web, such as NoSQL, may bring benefits to FMIS architecture, development and performance.

The aim of this chapter is discuss how NoSQL data storage models could benefit and improve FMIS functioning and structure. We review some works and highlight a few opportunities offered by NoSQL data storage models that other authors have already explored, and suggest new opportunities that will be well accepted by the FMIS landscape. The chapter organization is as follows: Section 2 introduces some theoretical issues about Relational and NoSQL data storage, Section 3 provides a discussion about the benefits of both models, highlights works that have already used NoSQL models, and presents some use cases of storage models in the FMIS landscape. Finally, Section 4 concludes this work and suggests some future ideas to be explored.

DATA STORAGE MODELS

Databases are used to store data of an application. They are also necessary to avoid losing important data and to recover a prior state of the application. In FMIS the database is one of the main components, since it stores the majority of data used by the decision support system. The data storage models discussed here are Relational and NoSQL. We illustrate the storage models from a task planning scenario in which employees are assigned tasks to be performed.

Relational Model

A Relational database model follows the principles of the relational theory proposed by E.F. Codd(Codd, 1990), in 1969. It was designed to satisfy the majority of users and uses, thus may be considered as a one-size-fits-all model. Data is structured into attributes, tuples and relations, which remind a table structure. Figure 1 illustrates it, where attributes are table columns, tuples are table rows and the relation is represented by the table. Attributes have name and a scalar data type, both predefined, and also may have constraints that must be satisfied. Tuples without a value for a predefined attribute receive the NULL value, however is not possible to add an attribute that was not previously part of the relation. This model accepts only values for attributes defined in the design phase and almost everything is accessed by the assigned name. In fact, a modification in the table structure implies in alteration of all data stored. As a result of the mathematical operations, relations can also be merged to form a more complex set.

The whole database is manipulated by a Structured Query Language (SQL), which allows performing complex operations on data and database structure. This model also follows some transactional constraints known as ACID(Haerder & Reuter, 1983), an acronym for Atomicity, Consistency, Isolation and Durability. Atomicity stands

Figure 1. Sketch of a Relational data storage model containing three relations (Relation_Employees, Relation_Task_Planning and Relation_Tasks). Data is organized in tuples and named attributes. Empty attributes receive the NULL value, however it is not possible to add any new attributes without changing already stored data

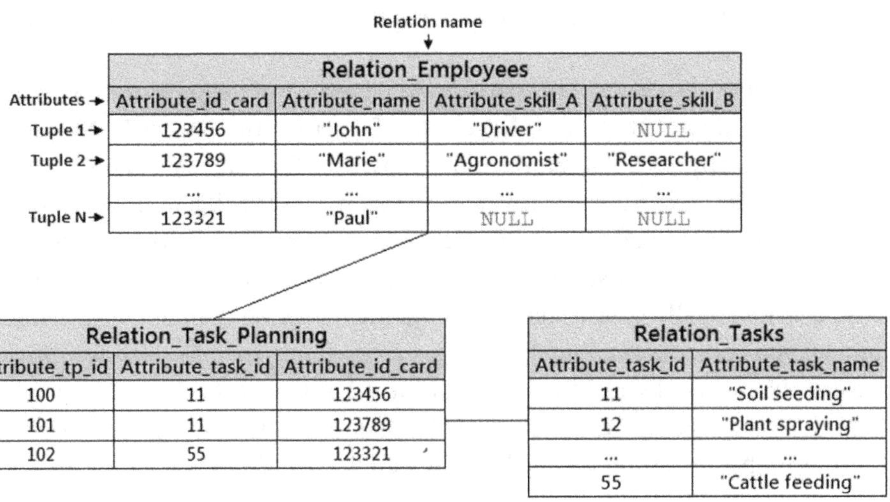

for either all or none of the transactions are executed; Consistency means that the database remains consistent after any successful transaction; Isolation means that a transaction may be executed concurrently with other transactions without change its final result; and Durability stands for after an operation all modifications will be hold even with hardware or operational system failure. These properties guarantee the security and correctness of data manipulation after each operation on the database and leave the developer free of worries about out-of-date data, locks and collisions on updates (Cattell, 2011).

When the workload volume on the database increases it is crucial to be prepared to keep the good performance of the system. This characteristic is named scalability. In general, vertical scalability is applied on Relational databases, which consists in enhancing the hardware capabilities in the server that hosts it. It results in more CPU time, RAM and disk available to the database system(Cattell, 2011). There is also the horizontal scalability that consists of distributing the data and the workload over more nodes without sharing RAM and CPU, however Cattell (2011)affirms that this kind of scalability is hard with SQL and ACID properties on Relational databases.

NoSQL Models

The NoSQL model arose as a cost-efficient alternative required by web applications support the increased read-write concurrent operations on large volumes of unstructured data. It must be noted that the Relational model was not operating in a satisfactory manner in this scenario. This new model abandons some resources present in the Relational model and implements specific solutions for a data model, which made it more efficient when handling a large volume of semi or unstructured data. As a result it is indicated to applications that need low-latency and availability when a large volume of transactions should be performed on massive datasets (Nance et al., 2013).

The Relational model did not perform well because of its ACID transactions and vertical scalability(Cattell, 2011; Nance et al., 2013). For this reason, NoSQL models replaced ACID by BASE properties, an acronym for Basically Available, Soft-State and Eventually Consistent. Basically Available means that if a node fails, only one part of data is inaccessible while the rest keeps available; Soft-State means that a pre-defined schema for data storage is not necessary, moreover it may be modified later; and Eventually Consistent means that the database will be consistent in the future. As a consequence these properties favored horizontal scalability, which is cheaper than vertical scalability since allows using low-cost commodity hardware. NoSQL also is in compliance with the CAP (Consistency, Availability and Partition Tolerance) theorem, proposed by Eric Brewer(Brewer, 2012), which states that any distributed database system satisfies simultaneously only two of the three properties (Consistency-Availability-Partition Tolerance). Second Cattell(2011), NoSQL models generally abandon Consistency.

A peculiarity of this model is the existence of distinct databases that offer data storage models in accordance with a specific data format, what maximizes its performance (Zafar, Yafi, Zuhairi, & Dao, 2016). According to Zafar et al.(2016), it implies in a developer-centered methodology, which needs to select the most appropriate database, creates the application's data model and defines the queries that will be executed. More advanced data security mechanisms also need to be established by the developer, keeping in mind that a performance reduction may occur(Srinivas & Nair, 2015). The main databases categories are key-value store, document store, wide-column store and graph store, which are summarized on list below:

- **Key-Value Store:** It is the simplest model. It consists in a unique key pre-defined by the developer that will reference the data. The only way to recover data is through this key because the data is just a BLOB that the application stores and does not know its content. Figure 2 illustrates the structure of this model highlighting the keys and their respective values. Insert, delete and lookup are operations allowed on the data, for this reason this model is suggested for applications that need fast data retrieval and have only one kind of object which is accessed by a single attribute, such as users' session, shopping cart, details of a particular product and system cache.
- **Document Store:** It is an improved key-value store which allows to lookup data by secondary indexes and also supports update operations. The document refers to an object that accepts nested documents or lists. The document structure reminds a JSON object where its attributes can be

Figure 2. Sketch of a key-value storage model. Keys are predefined by developer and are the unique way to query data

Key	Value
Employee123456	id_card:123456, name: "John", skill_A: "Driver"
Employee123789	id_card:123789, name: "Marie", skill_A: "Agronomist", skill_B: "Researcher"
Task11	task_name: "Soil seeding"
Task12	task_name: "Plant spraying"
TaskPlan100	task_name: "Soil seeding"; employees: "John","Marie"
TaskPlan102	task_name: "Cattle feeding"; employees: "Paul"

defined dynamically on runtime. Figure 3 illustrates the document structure of a document store model. Uses are similar to the Relational model, but with the freedom of an almost schema-less structure. Its use is not advised when there are many relationships or normalizations between data and complex queries need to be performed since it could worsen the database performance. Examples are blog websites, Content Management Systems and semi or no-structured data with many null values.

- **Wide-Column Store:** This model stores records which are a combination of a tuple and a document. Rather than establish the columns in a rigid database schema, each inserted row has its own columns. This way, columns may be partitioned vertically or horizontally between nodes while new attributes are added in a new column for each record. Figure 4 illustrates this storage model. Cattell(2011) suggests that the use of this model is similar to Document store, except that is more advised to high-throughput scenarios. This model offers a strong concurrency control, but adds extra complexity if compared to Document Store. It is suggested to be used for versioned data, Data Mining and analytics applications.
- **Graph Store:** It is designed to manage relationships between data and not to query it. This model is applied when need to lookup data and also to find

Figure 3. Document sketch containing keys and values, nested or not, of a document store model

```
{
  "document_id": 32326454456456464364,
  "task_planning_id":"100",
  "task_name": "Soil seeding",
  "task_employees": [
    {
      "id_card": "123456",
      "name": "John",
      "skills": [
        "Driver"
      ]
    },
    {
      "id_card": "123789",
      "name": "Marie",
      "skills": [
        "Agronomist", "Researcher"
      ]
    }
  ]
}
```

Figure 4. Sketch of a wide-column storage model. Each row has a key for identification and their columns are added in running time as data arrives

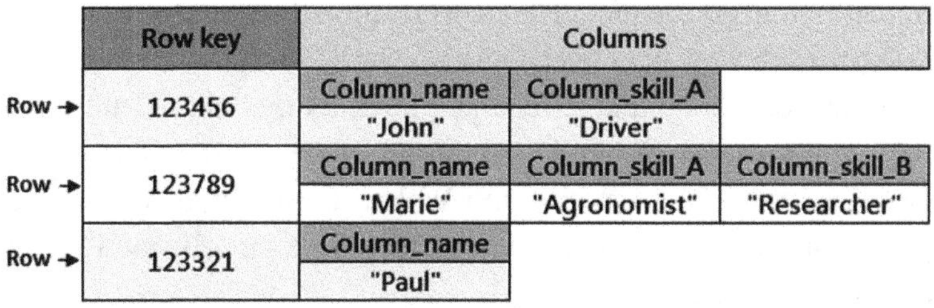

relations between them. Figure 5 illustrates the result obtained when querying all nodes in a graph database. A popular use is social networking, but it can be used in access control, bioinformatics and recommendation software.

This data storage model does not have a standardized manipulation language, such as SQL, and this is established by the tool manufacturer. A common manipulation

Figure 5. Graph example produced by a graph storage model. The nodes' colours identify the following: Green a performed task, yellow the demanded equipments and blue the employee who worked. Words between arrows name the occurring relationship between nodes

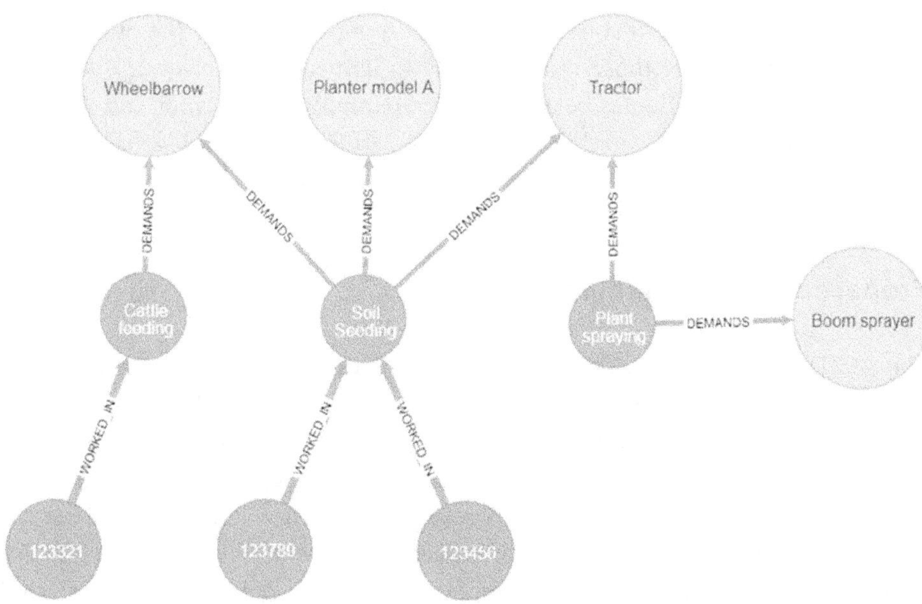

**For a more accurate representation see the electronic version.*

technique adopted in NoSQL databases to run commands is through a REST API or sending a JSON-like object, whose keys identify the command/operation and the values are the command content. The NoSQL model has a fault-tolerance design, since divides or replicates data over more nodes, supports a large number of read-write operations per second and still handle unstructured data(Cattell, 2011; Zafar et al., 2016). The majority of NoSQL solutions is open-source and big companies such as Google, Facebook, Amazon and Walmart publicly disclosed the benefits obtained with this model, which has boosted its popularity and motivated the adoption by third parties.

DISCUSSION

Comparing both models directly is not suited because each model attends a different applications set. Although the Relational model was introduced as a generic solution, NoSQL models are suggested for specific cases where the Relational model does not perform well or adds extra complexity. However, it is not obligatory to use a unique data storage model for the entire application since it is possible to benefit from what each one offers. Using more than one database model in an application is named polyglot persistence (Sadalage & Fowler, 2012) and its use is eased by SOA services separation. In fact, the architecture proposed by Nikkilä et al.(2010) already suggested dividing and storing data in several databases in order to keep the good application performance.

It is important to cite that developers are responsible to know the pros/cons of each model and choose those more suitable to its necessity, thus our discussion highlights where NoSQL may be used in order to improve functioning and structure of the FMIS. The features described in Section 2 of Relational and NoSQL models are summarized in Table 1. Then, these features are discussed related to FMIS by topics on sequence.

Properties and Scalability

The ACID properties provided by Relational model guarantee that data will be correctly and safely stored and handled. However, they may hurt the database performance when a large amount of data is stored. If we analyze how data is used in FMIS, it is stored once and read as many times as necessary. Moreover, its quantity will be increasing each time more as a result of taken decisions which affect the farming environment and produce new data to the system (Nikander et al., 2015). This way, using a technology that supports updates on data and provides other

Table 1. Comparison between Relational and NoSQL models

Item	Relational Model	NoSQL Model
Properties	ACID (Atomicity, Consistency, Isolation and Durability)	BASE (Basicaly Available, Soft-State and Eventually Consistent)
Scalability	Mainly vertical	Horizontal
Database type	Generic. Based on Relational Theory	It depends of data storage model (key-value, document, wide-column or graph)
Data schema	Rigid and static. Only accepts what was predefined (names, data type, structure)	Flexible and depends of data storage model. Accepts unstructured data and dynamic modifications on runtime
Manipulation language	SQL	Not standardized and established by application manufacturer
Indication for use	Complex operations on structured data	Simple concurrent operations on large amount of data

sophisticated resources is needless, since NoSQL model was specially elaborated to simple operations as read-write on large quantity of data as demanded by this tool. However, if an ACID property is necessary and a NoSQL model is obligatory, it is possible to implement it directly on application (Nance et al., 2013), knowing that the performance may be deteriorated.

The aspect of Basically Available and vertical scalability of NoSQL is useful to agriculture domain. Some agricultural areas are not covered with Internet signal or suffer from malfunctions of it, what could block the machinery operation or real-time data transferring. Moreover, if transmission is stopped and data are not stored we can lose important data of our running task. Storing data locally and later sync with a master server was a common practice explored by some authors(Blank et al., 2013; Iftikhar & Pedersen, 2011; Steinberger et al., 2009) to avoid losing data obtained from tasks, however it demands synchronization schemes to organize the data. Conversely, NoSQL models were designed with vertical scalability and fault-tolerance on mind and consequently automatically sync mechanisms are present. The vertical scalability feature allows data replication and partition over more nodes, so if one source stops during a task and it is possible to communicate, the data is obtained from another one. In addition, a robust structure can be mounted if each machinery or implement has an embedded computer running a NoSQL instance, such as proposed by Blank et al.(2013). They took advantage of CouchDB's native replication mechanisms, a NoSQL Document store, so that data was stored off-line and replicated to other machinery during a task and later synchronized automatically with the master server. By the reason of being a database native resource, developers do not have to worry about data synchronization, which is done automatically. Data losing and field tasks interruption can be avoided or alleviated with these strategies.

Database Type and Data Schema

Cattell (2011) affirms that the Relational schema simplifies and centralizes the data definition when working with many tables and different data types. However, when we analyze the FMIS landscape, which needs to handle sensor readings, aerial images, machinery and management data, third-party data and others, establish a schema for all these storage necessities is laborious. Furthermore, if the data format changes, more work is needed to update the rigid schema. A worst case would be if the farmer replaces its equipments by others from a distinct manufacturer, which uses another data format and is not standardized, demanding an update on the whole relational database and its data schema. These situations can be avoided or simplified if specific solutions for data models and a flexible schema are employed.

Different types of NoSQL databases favor particular data formats, since the solution developed is unique to them. It is obvious to say that a particular solution is better than a generic. In fact, a sensor collects only geo-referenced data with a format on a time-series and nothing more. Besides this, data will not be modified; it is just stored once and read whenever necessary. Thus, a simpler solution, such as key-value or document stores, is advised to sensor readings because of the simple data model obtained and also to facilitate the tool development. Another benefit is that NoSQL models can be easily executed with good performance on commodity hardware. The development of sensors with cheap embedded computer and running a database system can improve the capability of them store and process data locally, aside from minimize data loss problems through the good practice of maintaining a local copy of the data.

Storing data without knowing attribute names and supporting dynamic modifications in running time it is possible with a flexible database schema. In doing so the data is stored with its original structure, for instance AgroXML or ISOBUS, dismissing conversion layers or object mapping. Thus, the flexible schema simplifies the software development avoiding data conversions and adjustments for structures and also frees from pre-defining a specific and static structure. Despite this feature, the conversion layer must exist for communication with third-parties and to export the stored data to other existing formats, easing data sharing and compatibility between tools. Porter et al. (2014)have benefited from the flexible scheme in creating a tool to store AgMIP crop modeling exercises. They affirm that there were a wide range of specifications and quantity of records on their datasets which did not fit well on a rigid schema, so the key-value schema provided an efficient and flexible way to store inconsistent site-specific data obtained from diverse sources. These aspects enable storing only what was supplied independent of format, supports dynamic attributes inclusion and avoid NULL values such as on the Relational model, besides is optimized to a specific data format.

Manipulation Language and Popularity

One of the main benefits of the Relational model is the SQL. Likewise as in data format, a standard facilitates the tool development and this is a positive point of the Relational model. Any relational database model uses SQL and it is possible change the database tool with minor modifications on application's query commands. In its turn, some NoSQL databases use a JSON-like structure or a REST API to run commands, what is well defined and easy to understand too. Therefore, we belief that the language is not a problem, since we already are using a developer-centric methodology, which needs to know its tools in order to get the best out of each one. However, it is difficult to alternate between NoSQL database tools, primarily because of distinct data storage models and secondly because of different data manipulation language implemented by application manufacturers.

A solution for the problem of switching between data storage tools in FMIS could be surpassed employing techniques that map the application objects. This technique frees the programmer from worrying about data manipulation commands between databases because uses a specific language that is translated to other formats, however it could demand a data structure/pattern. As result, we may lose the flexible schema, therefore this technique must be well studied and investigated if is worth.

Cattell (2011)highlights that SQL is simpler than NoSQL lower-level commands in data handling and many developers are already familiar with it, so it is easier to find an experienced SQL developer. This is a consequence of Relational model popularity. Figure 6 presents a chart elaborated by DB-Engines, an initiative created and maintained by the Austrian IT consulting company Solid IT to collect and present information related to database management systems available on the market. The chart illustrates the popularity of each database category on August 2017, which was determined by number of results in search engines, IT Q&A platforms, shown interest on Google Trends, number of job offers available in job search engines, number of profiles in professional networks and the relevance in social networks. More details about the calculation method can be found in (SolidIT, 2017). Note that the chart includes other database models not mentioned in this work, such as Object Oriented DBMS and Native XML DBMS.

As view in Figure 6, at the time of writing of this work Relational model tools (Relational DBMS) had 79.8% of popularity, completing the podium with Document stores (7.2%) and Search Engines (4.1%). The popularity of the Relational model is overwhelming, so much that only 14.7% is the popularity summation of NoSQL models presented in this work. It represents 18.42% of Relational model popularity (14.7% of 79.8% is 18.42%), however we need to consider how long the tool has been available for use since its inception.

Figure 6. Chart offered by DB-Engines (SolidIT, 2017) initiative illustrating the popularity of some database models on August 2017

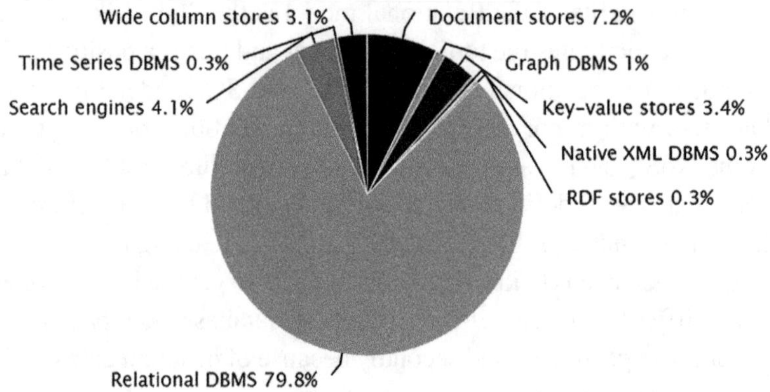

The Relational model has more than 40 years of existence and was employed in majority of developed applications while NoSQL is relatively new, with approximately 10 years of use and employed in few applications. Consequently, it was expected that, because of its time on market, the Relational model popularity was bigger than NoSQL models. This popularity also contributes to Relational model being more stable and well tested than NoSQL models, in addiction to developers more experienced with it. However, we need to consider that NoSQL models arose as an alternative to surpass Relational model problems when working mainly with Internet of Things, Big Data and unstructured data, an already prevalent reality in world agriculture (Wolfert, Ge, Verdouw, & Bogaardt, 2017). Thus, is not prudent insist in data storage technologies that had already proved insufficient on that scenario. NoSQL models are becoming popular in other industries because of its offered benefits, such as high-performance, code simplicity and easier development; thus, taking advantage of this it is a good chance to improve FMIS for developers and users.

Indications for Use

The kind of operation performed on data, either complex or simple, and its structure impact the choice of a database model. Moreover, some other criteria need to be weighted when selecting a data storage model. Cattell (2011) suggests that performance, scalability and tool/code maturity should be studied before choose one, highlighting that if the Relational model offers what you need, choose it because of its market maturity. Nance et al. (2013) say that the choice depends of the "type of application being written, the type of queries that are expected, and the regularity vs. variability of the data structure" besides evaluates cost and performance. Related to costs, NoSQL models are usually open-source which frees companies from costs with their use, but the tool support is entirely their responsibility. A positive point is that the source code is available to anybody which, if required, can adjust the database application according to its needs, fix problems and improve even more the performance. Summarizing, each solution has pros/cons and developers are in charge to evaluate each one and find the best for its application.

Figure 7 illustrates the farm organization and some operations being carried out. The letters contained in image represent specific situations in which different database models can be used. In the following topics, these situations are discussed and the most indicated database models in order to keep the good performance and simplify the FMIS development are suggested

- **Sensor Readings:** Sensors installed in greenhouses or field collect geo-referred data related to its designation. This situation is illustrated by letter A in Figure 7. The data will not be modified by FMIS, just stored and read from

Figure 7. Illustration of farm organization and some operations being performed. Letters indicate specific use cases where NoSQL models could be applied in order to improve FMIS performance and easing development

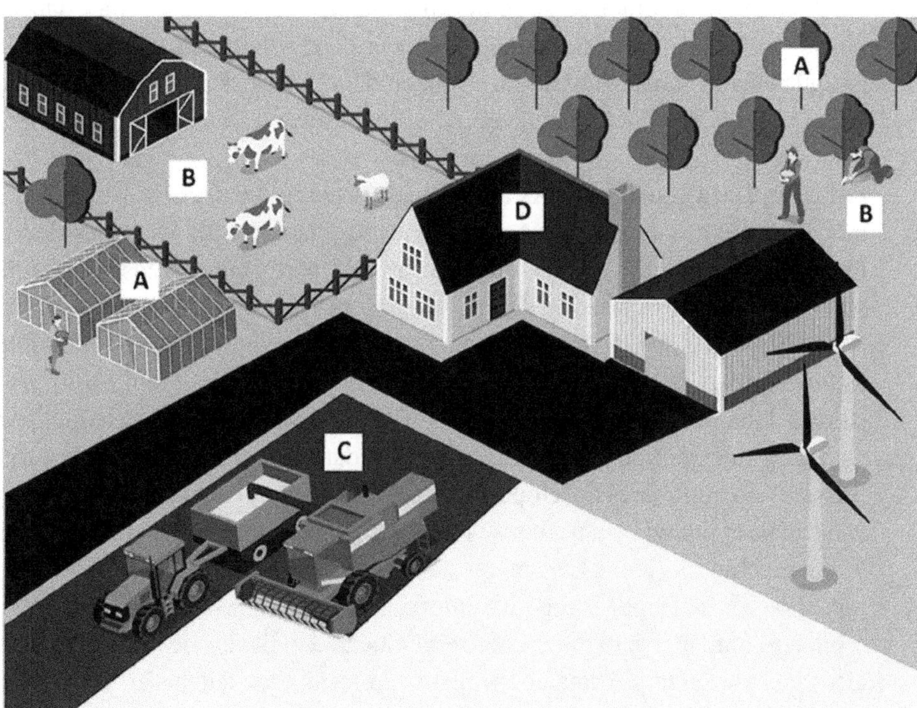

database. The data search can be done by additional attributes in addition to the reading performed, such as geographic coordinates or collect date. Thus, it is observed that complex operations will not be performed. The data's collect time interval also impacts on produced data amount and this is a recognized concern when using the relational model, because the database response time can be affected. For these reasons, simpler models such as key-value or document stores are sufficient to this situation.

- **Employees, Equipments and Animal Identification:** A farm has employees, equipments and animals that have identification, such is represented in Figure 7 letter B. Keeping them recorded in a database is a good practice to facilitate the assets control, data survey and sharing with third-parties. This kind of data may never receive a modification, besides its structure is not sophisticated. Thus, a simple key-value model is sufficient to keep record of identifications. The key may be defined as ID card for employees and identification number or name to equipments and animals, for example.
- **Task Planning and Execution:** Tasks are planned in FMIS and later transferred to machinery (Figure 7 letter C) in order to be executed or informed to workers. However, a task can vary according to its requirements, such as employee quantity, machinery and inputs demanded, etc. Predetermining all tasks' requirements in a relational model is challenging because of its rigid columns structure. If structure modifications are necessary, all data is affected. This way, a document store model is advised because it is flexible to updates and can storing all task related data. The task results also can be saved in the same document, keeping all data related to the task in a single location.
- **Farming Data Exchange:** As already discussed in Subsection 3.1, NoSQL models' native data replication and sync mechanisms allow to develop a fault-tolerant structure when machineries are performing a task, but this strategy also can be used in Wireless Sensors Networks(Aqeel-ur-Rehman et al., 2014) and any device with an embedded computer. Thus, it is possible to form a local network with all connected devices sharing, replicating and storing locally data. This minimizes problems such as data loosing and gathering interruption caused when a single data source is available. In turn, this native feature frees developers from worrying about data synchronization and can be explored by any choose NoSQL model. Taking for example Figure 7, the machinery (letter C) can get data and store locally, then replicates to greenhouses (letter A) if a communication channel is available, which also replicates until arrive to the master server located in the house (letter D). This also can be used to get data not available on local node database.

- **Data Mining Tasks:** Storing data for mining in a wide-column store may be handy because all related data to a geographic coordinate, for example, is available in a common place and can be dynamically added to columns as they arise. Thus, if you get data later, just aggregate it to columns without change the database structure or pre-establish rigid data schema. This could be implemented in a document store too, but in a data mining scenario the high-throughput is mandatory and the wide-column model is the most indicated to it. Wide-column model is also advised to store all FMIS data because of its flexibility and high-throughput. If you already store data in a database that favors data mining, your data pre-processing task may be performed easier.
- **Employees' Performance:** Despite of ample application to social networks, an interesting use to graph-store model is to find relations between employees, equipments and performed tasks. We could use, for example, execution time, efficiency and skills of an employee and relate with its tasks. If stored in other database models, these relations cannot be easily identified, but with a graph-store model the database is in charge to show us.
- **Handling Sensitive Data:** Performing an input stock update in real-time during harvesting (Figure 7 letter C) or recording employees' payment may be considered a sensitive process. Consequently, we must guarantee that these operations were performed due to its importance. That is why the more indicated to these situations is the Relational model. Its ACID properties guarantee correct data manipulation and favor the safe operation of the system. The same could be performed on NoSQL models, but the developer should implement the security mechanisms in order to perform the correct handling. As it is a relational model's native resource, well tested and already established, this model is more advised to this situation.

CONCLUSION

Analyzing the operations performed on FMIS, we conclude that a common step is store data once and read when necessary. Thus, the main operations are reading and writing. There is needless of a sophisticated structure to it, but we still must lookup data to decision-support in a fast way. The data amount increases each time more because sensors are producing it uninterruptedly together with any task performed. The Web and other industries already faced a problem with data amount when using the Relational model and NoSQL helped to overcome it, so NoSQL may be handy to the FMIS too.

This chapter highlighted some opportunities where NoSQL model could replace Relational model in FMIS. We described some differences between models, presented some works that have already explored NoSQL models and recommend some convenient situations for it on FMIS. Some of our examples could be implemented in any database model, but the applicability suggested was best fitted to specific situations presented. As already cited, the choices depend on the needs of the tool and the company. The polyglot persistence concept allows to explore the benefits of each model and consequently to develop an optimized application to specific requirements. Thus, we believe that NoSQL will not replace the Relational model completely in FMIS, but will work together to provide a more robust tool due its potential.

We have presented some examples, but it is not limited to them. As future work we suggest implementing the discussed models and evaluate the obtained results according to performance, complexity and ease of use offered by NoSQL models to FMIS. Comparative performance tests of database operations are common studies, but we also suggest employ software engineering to model data and structures of FMIS landscape and obtain an optimized architecture. Point out other situations where NoSQL could be well used in FMIS landscape is suggested too.

ACKNOWLEDGMENT

The authors would like to thank Comissão de Aperfeiçoamento de Pessoal de Nível Superior (CAPES) for the financial support.

REFERENCES

Aqeel-ur-Rehman, Abbasi, A. Z., Islam, N., & Shaikh, Z. A. (2014). A review of wireless sensors and networks' applications in agriculture. *Computer Standards & Interfaces*, *36*(2), 263–270. doi:10.1016/j.csi.2011.03.004

Aubert, B. A., Schroeder, A., & Grimaudo, J. (2012). IT as enabler of sustainable farming: An empirical analysis of farmers' adoption decision of precision agriculture technology. *Decision Support Systems*, *54*(1), 510–520. doi:10.1016/j.dss.2012.07.002

Blank, S., Bartolein, C., Meyer, A., Ostermeier, R., & Rostanin, O. (2013). iGreen: A ubiquitous dynamic network to enable manufacturer independent data exchange in future precision farming. *Computers and Electronics in Agriculture*, *98*, 109–116. doi:10.1016/j.compag.2013.08.001

Brewer, E. (2012). CAP twelve years later: How the "rules" have changed. *Computer*, *45*(2), 23–29. doi:10.1109/MC.2012.37

Cattell, R. (2011). Scalable SQL and NoSQL Data Stores. *SIGMOD Record*, *39*(4), 12–27. doi:10.1145/1978915.1978919

Codd, E. F. (1990). *The Relational Model for Database Management: Version 2*. Boston, MA: Addison-Wesley Longman Publishing Co., Inc.

Fountas, S., Carli, G., Sørensen, C. G., Tsiropoulos, Z., Cavalaris, C., Vatsanidou, A., ... Tisserye, B. (2015). Farm management information systems: Current situation and future perspectives. *Computers and Electronics in Agriculture*, *115*, 40–50. doi:10.1016/j.compag.2015.05.011

Godfray, H. C. J., Beddington, J. R., Crute, I. R., Haddad, L., Lawrence, D., Muir, J. F., ... Toulmin, C. (2010). Food Security: The Challenge of Feeding 9 Billion People. *Science*, *327*(5967), 812–818. doi:10.1126cience.1185383 PMID:20110467

Haerder, T., & Reuter, A. (1983). Principles of transaction-oriented database recovery. *ACM Computing Surveys*, *15*(4), 287–317. doi:10.1145/289.291

Iftikhar, N., & Pedersen, T. B. (2011). Flexible exchange of farming device data. *Computers and Electronics in Agriculture*, *75*(1), 52–63. doi:10.1016/j.compag.2010.09.010

Kaloxylos, A., Eigenmann, R., Teye, F., Politopoulou, Z., Wolfert, S., Shrank, C., ... Kormentzas, G. (2012). Farm management systems and the Future Internet era. *Computers and Electronics in Agriculture*, *89*, 130–144. doi:10.1016/j.compag.2012.09.002

Nance, C., Losser, T., Iype, R., & Harmon, G. (2013). Nosql vs rdbms-why there is room for both. *Proceedings of the Southern Association for Information Systems Conference*, 111–116.

Nash, E., Korduan, P., & Bill, R. (2009). Applications of open geospatial web services in precision agriculture: A review. *Precision Agriculture*, *10*(6), 546–560. doi:10.100711119-009-9134-0

Nikander, J., Koistinen, M., Laajalahti, M., Pesonen, L., Ronkainen, A., & Suomi, P. (2015). Farm Information Management Infrastructures in the Future. In *2015 26th International Workshop on Database and Expert Systems Applications (DEXA)* (pp. 104–107). IEEE. 10.1109/DEXA.2015.38

Nikkilä, R., Seilonen, I., & Koskinen, K. (2010). Software architecture for farm management information systems in precision agriculture. *Computers and Electronics in Agriculture*, *70*(2), 328–336. doi:10.1016/j.compag.2009.08.013

Oksanen, T., Öhman, M., Miettinen, M., & Visala, A. (2005). ISO 11783 – STANDARD AND ITS IMPLEMENTATION. In *IFAC Proceedings Volumes* (Vol. 38, pp. 69–74). Academic Press. 10.3182/20050703-6-CZ-1902.02102

Peets, S., Mouazen, A. M., Blackburn, K., Kuang, B., & Wiebensohn, J. (2012). Methods and procedures for automatic collection and management of data acquired from on-the-go sensors with application to on-the-go soil sensors. *Computers and Electronics in Agriculture*, *81*, 104–112. doi:10.1016/j.compag.2011.11.011

Porter, C. H., Villalobos, C., Holzworth, D., Nelson, R., White, J. W., Athanasiadis, I. N., ... Jones, J. W. (2014). Harmonization and translation of crop modeling data to ensure interoperability. *Environmental Modelling & Software*, *62*, 495–508. doi:10.1016/j.envsoft.2014.09.004

Sadalage, P., & Fowler, M. (2012). *NoSQL Distilled: A Brief Guide to the Emerging World of Polyglot Persistence* (1st ed.). Addison-Wesley Professional.

Schmitz, M., Martini, D., Kunisch, M., & Mösinger, H.-J. (2009). agroXML Enabling Standardized, Platform-Independent Internet Data Exchange in Farm Management Information Systems. In Metadata and Semantics (pp. 463–468). Boston, MA: Springer US. doi:10.1007/978-0-387-77745-0_45

Solid, I. T. (2017). *DB-Engines Ranking per database model category*. Retrieved August 29, 2017, from https://db-engines.com/en/ranking_categories

Srinivas, S., & Nair, A. (2015). Security maturity in NoSQL databases - are they secure enough to haul the modern IT applications? In *2015 International Conference on Advances in Computing, Communications and Informatics (ICACCI)* (pp. 739–744). IEEE. 10.1109/ICACCI.2015.7275699

Steinberger, G., Rothmund, M., & Auernhammer, H. (2009). Mobile farm equipment as a data source in an agricultural service architecture. *Computers and Electronics in Agriculture*, *65*(2), 238–246. doi:10.1016/j.compag.2008.10.005

Wolfert, S., Ge, L., Verdouw, C., & Bogaardt, M.-J. (2017). Big Data in Smart Farming – A review. *Agricultural Systems*, *153*, 69–80. doi:10.1016/j.agsy.2017.01.023

Wolfert, S., Goense, D., & Sorensen, C. A. G. (2014). A Future Internet Collaboration Platform for Safe and Healthy Food from Farm to Fork. In *2014 Annual {SRII} Global Conference*. IEEE. 10.1109/SRII.2014.47

Zafar, R., Yafi, E., Zuhairi, M. F., & Dao, H. (2016). Big Data: The NoSQL and RDBMS review. In *2016 International Conference on Information and Communication Technology (ICICTM)* (pp. 120–126). Academic Press. 10.1109/ICICTM.2016.7890788

Chapter 3

InWaterSense: An Intelligent Wireless Sensor Network for Monitoring Surface Water Quality to a River in Kosovo

Figene Ahmedi
University of Prishtina, Kosovo

Lule Ahmedi
University of Prishtina, Kosovo

Brendan O'Flynn
University College Cork, Ireland

Arianit Kurti
Linnaeus University, Sweden

Sylë Tahirsylaj
Hydrometeorological Institute of Kosovo, Kosovo

Eliot Bytyçi
University of Prishtina, Kosovo

Besmir Sejdiu
University of Prishtina, Kosovo

Astrit Salihu
University of Prishtina, Kosovo

ABSTRACT

A shift in the water monitoring approach from traditional grab sampling to novel wireless sensors is gaining in popularity not only among researchers but also in the market. These latest technologies readily enable numerous advantageous monitoring arrangements like remote, continuous, real-time, and spatially dense and broad in coverage measurements, and identification of long-term trends of parameters of interest. Thus, a WSN system is implemented in a river in Kosovo as part of the InWaterSense project to monitor its water quality parameters. It is one of the first state-of-the-art technology demonstration systems of its kind in the domain of water monitoring in developing countries like Kosovo. Water quality datasets are transmitted at pre-programmed intervals from sensing stations deployed in the river to the server at university via the GPRS network. Data is then made available through a portal to different target groups (policymakers, water experts, and citizens). Moreover, the InWaterSense system behaves intelligently like staying in line with water quality regulatory standards.

DOI: 10.4018/978-1-5225-5978-8.ch003

INTRODUCTION

A significant driver for Wireless Sensor Network development and implementation is legislation that is being introduced worldwide which is enabled by the real-time monitoring of and sensorization of our buildings, waterways, energy consumption and environmental conditions. Table 1 provides a snap shot of some of this legislation.

As shown in Table 1 legislation exists and continues to be drawn up to address the global challenges of climate change, the health and welfare of citizens, the reduction in energy consumption in our factories and cities and the optimisation of the use of natural resources in our society. The provision of real time data sets will help legislators monitor the impact and effectiveness of any legislation introduced on our environment and society. Real time data sets from WSN technologies in a variety of application spaces will facilitate informed decision-making around policy enforcement to meet the demands of legislation introduced by policy makers in the areas identified

Guidelines being handed down from the Water Framework Directive WFD and other government bodies worldwide in the field of water management necessitate the automation of water management systems and regular acquisition of water quality parameters. Legislation such as the WFD and other legislation (as in Table 1), will

Table 1. A selection of legislative drivers For WSN developments

Environmental, Water, Air Climate Change
EU Directive 2000/60/EC, EU Water Framework Directive - EU Policy and regulations on water Quality [1]
European Communities (Water Policy) Regulations, 2003 (S.I. No. 722 of 2003) [2]
European Communities Environmental Objectives (Surface Waters) Regulations 2009 (S.I.No.272 2009) [3]
European Communities Environmental Objectives (Groundwater) Regulations, 2010 (S.I. No. 9 of 2010):
European Communities (Good Agricultural Practice - Protection of Waters) 2010 (S.I. No. 610 of 2010) [4]
European Communities (Technical Specifications for the Chemical Analysis and Monitoring of Water Status) Regulations, 2011 (S.I. No. 489 of 2011) [5]
European Union (Water Policy) Regulations 2014 (S.I. No. 350 of 2014) [6]
EU Council Directive 92/43/EEC: Birds and Habitats Directives - the conservation of natural habitats and of wild fauna and flora aims to promote the maintenance of biodiversity [7]
EU Directive 2006/7/EC 2006 concerning the management of bathing water quality and repealing ensuring the quality of Drinking water, bathing waters and urban waste water. [8]
EU Directive 2010/75/EU on industrial emissions (integrated pollution prevention and control) [9]
EU Directive 91/676/EEC The Nitrates Directive: concerning the protection of waters against pollution caused by nitrates from agricultural sources [10]

be enabled by autonomous sensor stations (sometimes in remote areas) analysing water, air and soil quality parameters, storing the data or broadcasting it reliably under different conditions to water management headquarters.

To this end, there is a requirement for next generation, highly specified sensor systems, packaged in a reliable and robust manner to provide high temporal and spatial granularity data sets associated with environmental (Oliveira et al., 2011) parameters. These systems need to incorporate the latest technologies available, and need to operate wirelessly for extended periods of time in harsh environments.

Maintaining surface waters at high quality remains a challenge even among developed countries. Water quality monitoring practices are predominantly laborious, i.e., usually manual sampling methods are applied on the sites.

In recent years, a shift in water monitoring approach, from traditional and very limited (in terms of time and space) "stream to a bottle" measurements to novel Wireless Sensor Network (WSN) systems, is gaining in popularity not just among researchers, but also in the market. These latest technologies *readily* enable numerous advantageous monitoring arrangements, like: continuous, real-time, and spatially-dense and broad in coverage measurements, remote monitoring, flexible disposition, identification of long-term trends and fluctuations of parameters of interest, and similar (Jiang et al., 2009).

Led by the abovementioned rationale, a WSN system is designed and implemented in a river in Kosovo as part of the InWaterSense project to monitor a series of its water quality parameters, and will be introduced in this paper. To our knowledge, this is one of the first state of the art technology demonstration systems of its kind in the domain of water quality monitoring in developing countries like Kosovo. Quality data are transmitted at pre-programmed intervals from sensing stations deployed in the river to the InWaterSense server at university via the GPRS network. Data are then made available through the portal at inwatersense.uni-pr.edu to different target groups, i.e., policy-makers, water experts, as well as citizens. Moreover, the InWaterSense system provides the infrastructure to behave as an intelligent system since it supports working with data defining regulatory standards provided from water experts, data describing potential pollutant sources along the water flow, as well as with data from engineers on system specification, in addition to quality data gathered by sensors.

Related Work

Water monitoring is one such application space where sensor networks can be used to measure the chemical and biological quality of water in real-time, and to take remedial action when adverse quality conditions are experienced. The use of advanced sensor network based monitoring technologies and strategies mean that proactive measures can be implemented to improve water quality (Greenwood et al., 2008). This is driven in the EU by the requirements of the Water Framework Directive (WFD), (Council Directive 2000/60/EC), (Grath et al., 2007) (Allan et al., 2006) (Chave, 2001) (CIRCA, 2003) (EPA Ireland), as well as by global geopolitical drivers such as climate change. The establishment of high quality long-term monitoring programmes is regarded as essential if the implementation of the WFD is to be effective (Fleischmann et al., 2001). The ideal monitoring system of the near future might consist of a network of sensors deployed at key locations, capable of autonomous operation in the field for a year or more giving high temporal and spatial granularity data sets (Irvine et al., 2002) (Tateson et al., 2005) (Johnson et al., 2006) (Könaölä, 2008) (Quevauviller, 2006) (Goumopoulos et. al., 2007) (Goumopoulos et. al., 2014).

While many systems have been developed in the laboratory, they may not have been fully tested in the field and fully validated over extended periods of time. The long-term deployment of multi-sensor systems in the field is faced with many challenges. Besides fouling, on-line calibration of aquatic sensors, methods to reduce sensor drift, wireless technologies implemented and data aggregation are of high importance for remote sensors (Voigt et al., 2007) and a number of international research programs are being developed to address these challenges.

A prototype network of meteorological and hydrological sensors has been deployed in Yosemite National Park (Yang et al., 2002). Beyond ensuring drinking water safety, possible applications for the aqueous sensor network include advanced industrial process control, monitoring of aquatic biological communities and monitoring of waste-stream effluents. Fleishman and colleagues have developed an early warning system for the protection of wells and groundwater recharge (Fleischmann et al., 2002). In the United States the Environmental Protection Agency (EPA) programme EMPACT (EPA's environmental monitoring for public access and community tracking) was established in 2001 to monitor the Mystic River watershed. The project includes five water quality monitoring stations logging a number of water quality parameters (depth, temperature, pH, conductivity, dissolved oxygen and turbidity) every 15 minutes. Other water monitoring projects include MEROS in the UK [11]. CSIRO and WRON in Australia [12] are developing sensor networks in conjunction with Australia's Water boards and service providers [13]. In Canada the SmartBay group [14] are deploying sensors for resource and coastal management.

There are a number of works like SmartCoast and DEPLOY (by researchers in Ireland (O'Flynn et al., 2007, O'Flynn et al., 2010, Lawlor et al., 2012.)), EMNET (by Heliosware, USA (Ruggaber et al., 2007)), LakeNet (by Notre Dame University, USA (Seders et al., 2007)), and Fleck (by CSIRO, Australia (Stemp-Morlock, 2007)) that highlight the potential of using wireless sensor systems for monitoring water quality parameters of interest.

One of the key motivators of using WSN for water quality monitoring has been motivated from the fact that it reduces time and cost for processing as well as analysis of the data. This was also confirmed by numerous research studies, such as (Wang et al., 2009, 2010; Silva et al., 2011; Adamo et al., 2015). When it comes to design and deployment of the WSN for surface water sources, a variety of factors needs to be considered as well. These factors include the sensing capabilities of the sensor nodes, communication protocols to be used as well as considerations regarding network topology. At the research level typically for establishing WSN for fresh surface water quality-monitoring, parameters' such as temperature, pH, turbidity, and dissolved oxygen (DO) are usually taken into account. Examples of this can be found in the work done by (Rasin and Abdullah 2009), (Chaamwe 2010), (Nasser et al. 2013).

Referring to (Jiang et al., 2009) (Adu-Manu et al., 2017), the system architecture of a WSN for large-scale water environments such as for rivers consists of usually three parts (Figure 1): monitoring nodes, local database stations, and remote monitoring center. Wireless network uses the ZigBee protocol for communication between the database station and the nodes in the monitoring area it covers, whereas the communication between each of the database stations and the remote center is performed via a GPRS network.

Despite the substantial body of knowledge in the existing research when it comes to WSN applicability for water quality monitoring, with the research efforts reported in this paper we try to contribute in two directions: 1) by establishing a WSN comprised of static as well as mobile component reflecting on the specificities of Sitnica river pollution, 2) aggregated view on the collected data and possibility to exercise intelligent features with reference to regulatory standards. The differences in water quality across the relatively short distance river Sitnica, required the combined approach of static monitoring points combined with mobile solution. While the need for intelligent features, primarily targets the need for exposing the water quality issues to wider "non-expert" audiences that in a fast way could get insights regarding the water quality in the Sitnica river. Namely as one of the effects on improving water quality is also closely related on raising the social awareness of the pollution factors to wider audiences. Advanced "expert" rules are also supported by the approach in provision of useful patterns not easy to discover by the experts alone due to huge amounts of observational data usually gathered by the WSNs. In addition, modeling of the surface water quality monitoring system itself ensures the generality of the approach to adopt it to distinct deployment scenarios.

In the upcoming sections, we provide a detailed account of our approach and technical solution for reaching the project objectives.

Figure 1. Traditional "stream to a bottle" [15] *(left) vs. novel "stream to computer" water quality monitoring (right) through a typical WSN architecture for water monitoring.*

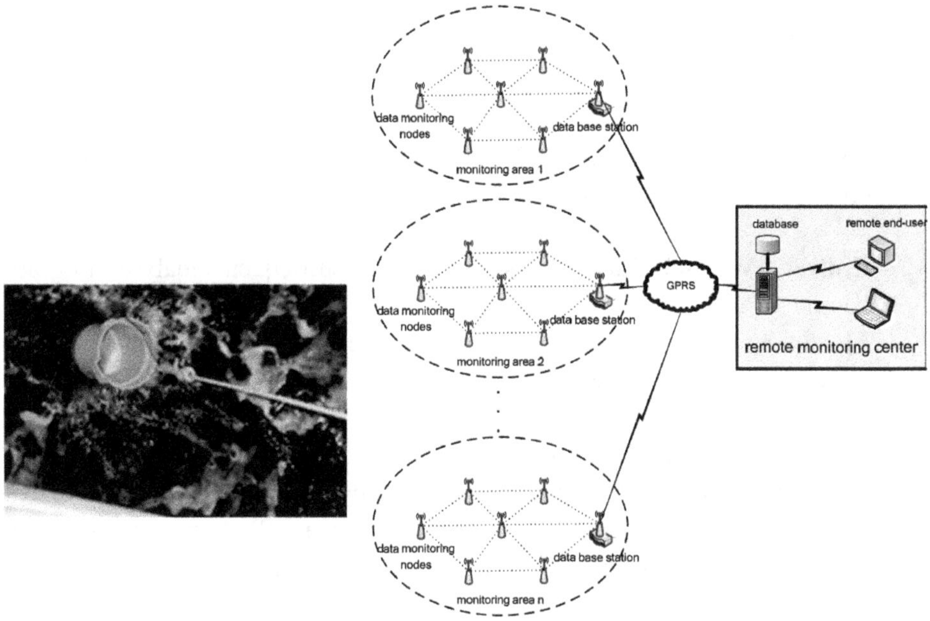

The InWaterSense Project and Its Objectives

InWaterSense is an EU funded research project aimed to apply recent advanced practices stemming from ICT in water quality monitoring for healthy environment, and strengthen Kosovo's capacity in research in national priority sectors of environment and ICT. More specifically, the project objectives were following:

Specific Objective 1: Build a WSN infrastructure in a river for monitoring water quality with the aim of providing a best practice scenario for expanding it to other surface water resources as well in the country.

Specific Objective 2: Monitor water quality in the river supported by the WSN in order to make the quality data from different locations available in real-time to the community and the decision makers.

Specific Objective 3: Transform the existing traditional WSN for water quality into an intelligent platform which is capable of automating the work usually performed by water experts on measuring water quality and identify water pollution.

Driven by these objectives drawn well in advance, the requirements specification and design of the WSN system has been defined prior to implementing the system, and will be presented next.

Kosovo Case Study

Quality monitoring for surface waters in Kosovo including rivers is a responsibility of the local water authority, Hydro-Meteorological Institute in Kosovo (HMIK), and currently involves methods which are characterized as:

- **Entirely Manual:** Based on sampling and subsequent analyses in water laboratory of HMIK.
- **Mostly Sparse in Time:** Measurements conducted usually only once per month, except in the cases of environmental accidents when the frequency of measurements might increase up to a per day basis.
- **Mostly Rare in Space:** Only few measuring points, usually some tens of them, are deployed alongside rivers in the country. Currently, there are 50 measuring points / stations in total which are installed and operate in the country alongside rivers, 5 out of them in the Sitnica river.

Why Sitnica River?

Sitnica is one of the main rivers in Kosovo with some 550000 inhabitants in its catchment. A recent study funded by the European Union (EU) emphasizes that "*Kosovo's environment has severe problems with regard to water quality*". In addition, the latest reported measurements conducted by the Hydro-Meteorological Institute of Kosovo (HMIK) show that "*Sitnica turns out to have values at a dangerous level*", and is thus estimated as the most polluted river in the country. This as some smaller rivers, which unfortunately play the role of sewerage collectors, spill into the river Sitnica. Moreover, a sub-catchment of Sitnica involves the most urbanized city in Kosovo, the capital city Prishtina, and besides, a major power plant of Kosovo lies along a sub-catchment of Sitnica. Furthermore, there is no recent water classification of the Sitnica river. Therefore, although the river Drini i Bardhë is bigger in size than Sitnica, we have chosen Sitnica as a case study, and consider the experience gained in terrain and research results obtained might be easily applied to any other river or other surface waters in Kosovo or elsewhere.

Requirements of the WSN to Monitor Water Quality in Sitnica River

In gathering the requirements of the WSN system to deploy in Sitnica to serve the objectives, following are the main issues that have thereby been addressed:

- Parameters that WSN shall support,
- Coverage sensing area, and
- Frequency in time and space of measurements.

Parameters That WSN Shall Support

The parameters to measure that have initially been considered are dissolved oxygen (DO), temperature (T), pH, BOD_5, nutrients (phosphate and nitrate), conductivity (EC), turbidity, total suspended solids (via turbidity), and optionally if possible, coliform bacteria and heavy metals. Consequently, a number of recommendations were outlined related to which parameters to measure:

- Aimed too wide coverage of parameter diversity; concentrate rather on a certain number of parameters.
- BOD5 is expensive to measure.
- Phosphate, and in general nutrients are expensive to measure:
    - Highly skilled maintenance team required to handle the probes.
    - A photometer and small-tube tests required for calibration.

Optionally, combine with manual sampling if measuring nutrients is that relevant. Measuring sulfate would also be valuable if by manual sampling.

- Coliform bacteria probe systems are yet in their infancy, and not applicable to wastewater.
- Heavy metals - no probes to measure heavy metals although it would have been important for measuring them in Sitnica due to industry alongside its catchment. No other parameters indicate the presence of heavy metals.
- Electrical conductivity is important for industrial influence.
- Dependencies between parameters:
    - Estimate total suspended solids (TSS) from turbidity measurements. Labor analyses required for calibration and validation of suspended solid values derived from turbidity: TSS analytics are required to calibrate the system.

- Electrical conductivity and chloride usually behave similar. The electrical conductivity is an indicator of the ionic (salty) compounds in the water. The chloride ion is one of them.
- Water level is important to get an idea about the loads. However, to derive flow from water level, regular measurements of water level required, and adjusting the water level to flow relation.

Based on the above rationale, the parameters required to measure through WSN sensors are dissolved oxygen (DO), temperature (T), pH, conductivity (EC), and turbidity.

Coverage Sensing Area

There are six potential sites alongside Sitnica river (Figure 2) considered for WSN deployment: the Lipjan town, Vragoli, Kozmin, Plemetin, Nedakovc, and the city of Mitrovica. The Hydro-Meteorological Institute (HMIK) conducts water quality monitoring in Sitnica at four sample stations: Lipjan, Vragoli, Plemetin, and Mitrovica.

Figure 2. Water quality monitoring stations currently in Sitnica

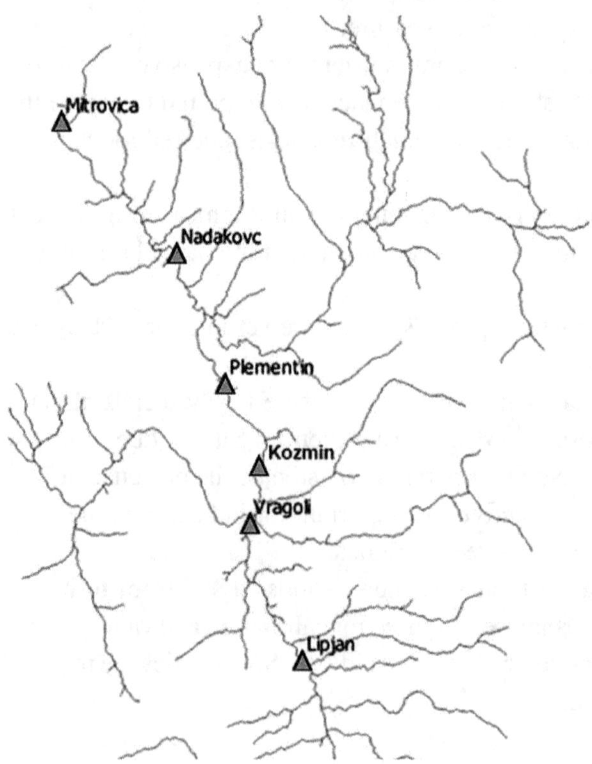

The team visiting Sitnica in all six potential deployment sites did recommend in favor of either of two sites, *Kozmin* and *Plemetin* (Figure 3, images on the left and right) for real-time continuous (non-stop) monitoring. The rationale behind that is on the "richness" and homogeneity of discharge types presumably dominating each of those two sites separately, namely domestic wastewater in case of the former, and industrial waste in case of the latter. That may ease the research in the future on identifying the pollution sources given we already presume in advance the type of polluters dominating those parts of the river, and may therefore conduct research to rather evaluate the correctness of our approach. In addition, measuring the parameters in two consecutive deployment sites enables one to also reach the more for practice important objective, i.e., to determine the trends on the health of the river in general. This aims to serve as an example scenario for later expansion of the WSN system in the future covering Sitnica in its whole breadth and depth, as well as other surface and ground waters in Kosovo.

Frequency in Time and Space of Measurements

The monitoring frequency currently conducted at each station by the HMIK is usually once per month. In case of floods, monitoring is performed more frequently, i.e., every 24 hours. Support for much more frequent measurements is aimed to reach by the system, namely:

- Real-time measurements, say every 7 or 10 minutes, or on-demand.
- Different timing depending on:

Figure 3. Water quality monitoring stations in river Sitnica

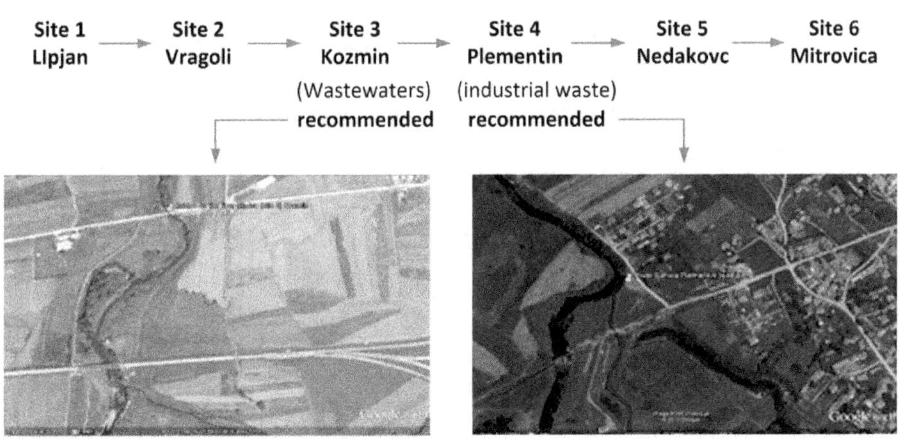

- The actual parameter(s) measured, e.g.:
  - Turbidity is usually not measured so frequently since no significant changes happen, except if rainy days. However, in Kosovo, more frequent (regular) measurements of turbidity shall provisionally be planned due to river meanders and unregulated riparian zones alongside Sitnica.
- Certain events, e.g.:
  - More frequent measurements when rain-events (due to presumable agricultural discharge).

Design and Implementation

Driven by the requirements specification presented in the previous section, the conceptual design of the WSN system in Plemetin (one of the recommended sites for deployment) for monitoring water quality in Sitnica is developed as depicted in Figure 4, and consists of three types of nodes:

Figure 4. Conceptual WSN design for water quality monitoring in Sitnica

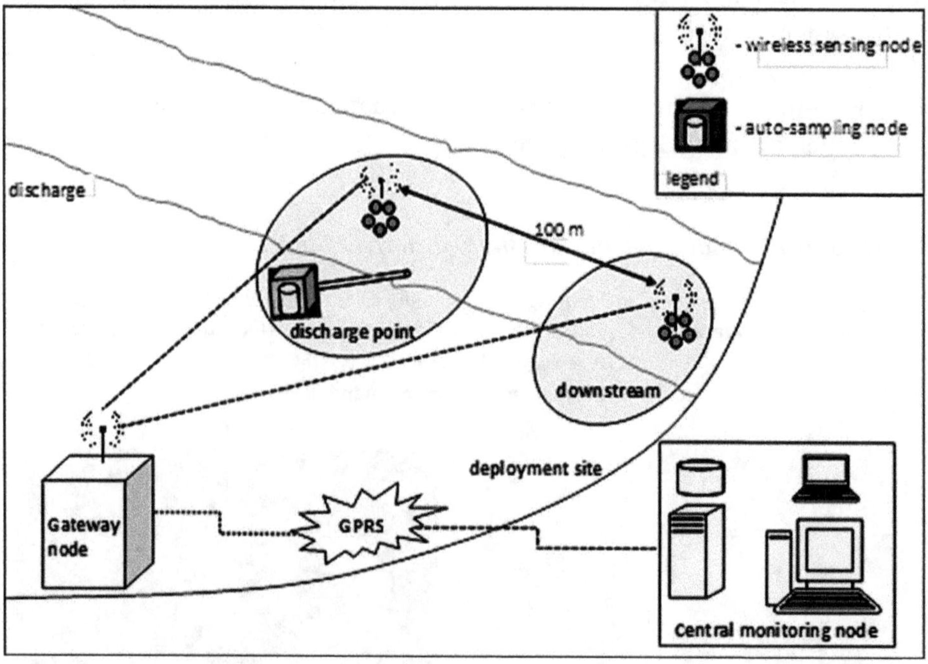

- Sensing nodes,
- A gateway node, and,
- A central monitoring node.

The description of each type of node and its implementation, as well as how they communicate to each other is next provided.

Sensing Nodes

Three types of sensing nodes are implemented as suggested for the Sitnica WSN system:

- **Static Wireless Sensor Nodes:** Comprising of certain number of sensors, each for a given parameter among dissolved oxygen (DO), temperature (T), pH, conductivity (EC), and turbidity.
- **Auto-Sampler Nodes:** To overcome the shortage (due to cost issue) in types of parameters that can be measured online. Triggering automatic sampling of additional arbitrary parameters ad-hoc or in case of certain events is supported, i.e., when certain values of given parameters measured by wireless sensors are reached. The samples could be used for laboratory analyses by the water quality monitoring institute (HMIK).
- **Mobile Wireless Sensor Nodes:** To overcome the shortage when measuring through static sensors in sensing area coverage, in lifetime of the sensing activity, as well as in portability.

In total, there are two static wireless sensor nodes, one auto-sampler node, and a mobile wireless sensors component implemented as part of the system. The physical dislocation of static sensor devices and of auto-sampler in Sitnica, the Plemetin village near the capital city Prishtina, is (1) in accordance with the conceptual design of the system (Figure 4), i.e., along two regions where measurements shall take place (dots in yellow in Figure 5), and also (2) taking into consideration the physical characteristics of the recommended deployment sites, like is the security issue of the sites prone to thefts, and the terrain specifics:

- On the discharge (i.e., point) source (see the discharge tube in Figure 4 which extends to the discharge point), the measurements are performed with wireless static sensors placed at the wireless sensing node, as well as with the auto-sampler, both maintained from within the housing (left node in Figure 5).

Figure 5. Physical design of WSN in Plementin: Two measuring regions (dots in yellow); housing and a manhole/pipe

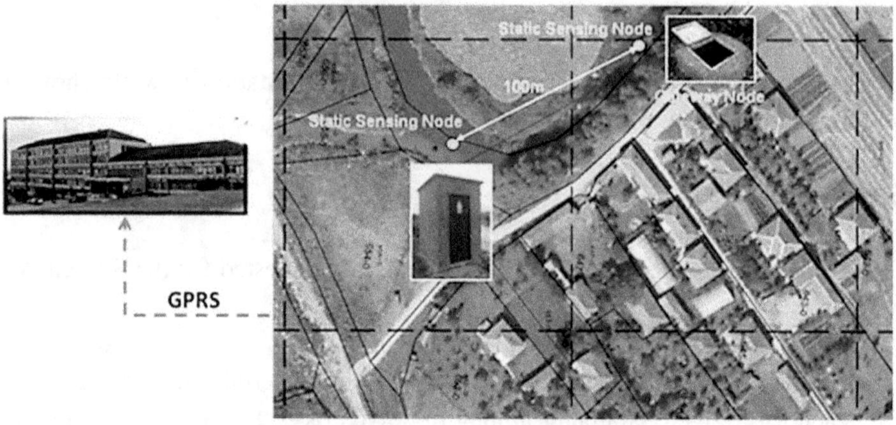

**For a more accurate representation see the electronic version.*

- Downstream the river in ca 100m distance from the point source, the measurements are performed with wireless static sensors placed at the wireless sensing node maintained from within a manhole (right node in Figure 5).

It is of interest to measure in both points (regions) for the sake of data gathering and data analyses.

Wireless static sensors support measuring in real-time of the following parameters: dissolved oxygen (DO), temperature (T), pH, conductivity (EC), and turbidity. Programming of sampling time intervals is supported, e.g. currently, sampling every 10 min at both points (housing and manhole) is being performed in both stations installed in Plemetin village along Sitnica river (Figure 6). Measured data are being transmitted via GPRS to the remote server at University of Prishtina (Figure 5).

The auto-sampler (Figure 6), on the other side, supports automatic sampling via tubes of ad-hoc arbitrary parameters whenever instructed in distance via SMS or in case of certain events (say, when static sensors send critical pH values in real-time), to afterwards analyse the measured parameter values at a lab. Moreover, certain predefined expert if-then rules may be applied to trigger auto-sampling. The number of subsequent samples depends on the number of bottles being supported with the auto-sampler, currently 12 in our case.

In addition to continuous non-stop monitoring of water quality through static sensors which are restricted to measuring only at selected station(s) at a time (Plemetin), measurements on demand at any arbitrary station along Sitnica river is supported by wireless *mobile sensors* (Figure 6) and supplement the system. Such

Figure 6. System implementation in Sitnica river – Components: wireless sensor network (also referred to as static sensors) in Plemetin village near the capital Prishtina (middle in the water), auto-sampler (left-bottom), and mobile sensors (right).

"in a bag" portable sensors represent an added value to the system, being easy to deploy given also their battery supply. Mobile sensors actually support measuring dissolved oxygen (DO), temperature (T), pH, and conductivity (EC) of the water, and have already been deployed to measure in five different stations along Sitnica and transmit the measured data to the remote server at University of Prishtina (Figure 7). The technology used is based on Waspmote Plug and Sense, an open source wireless sensor network platform[16].

Gateway Node

The gateway node's task is to receive monitoring data from static sensing devices, and transmit them further to the central monitoring node at university (see Figure 4 for design and Figure 5 for actual implementation). Gateway devices are actually housed within the manhole of the WSN system implementation as depicted in Figure 5.

Central Monitoring Node

The central monitoring node is a node physically residing remotely and autonomous of any of the deployment sites, and where actually the whole monitoring data is stored, managed and visualized. Thus, for the whole WSN system, a single common central monitoring node should be installed (design in Figure 4). In case that in the future, more deployment sites become active in parallel, that same single central node may as well serve and get shared among all deployment sites.

Therefore, a database server machine is installed as a central monitoring node in the laboratories at University of Prishtina, Faculty of Civil Engineering and

Figure 7. The mobile sensors component attached to the existing WSN system (static sensors) for portability and broader area coverage

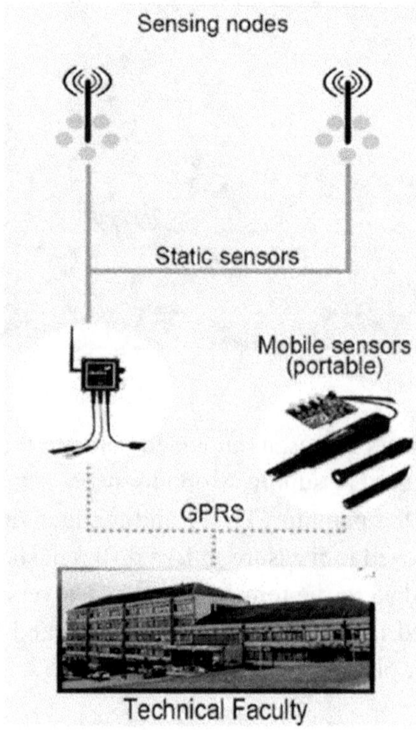

Architecture, Hydro-Technics Department, and connected to the gateway node at the Plemetin site as shown in Figure 5 and Figure 7.

Data format. Since the final objective is to transform the WSN for water quality monitoring into an intelligent platform to operate almost autonomously, the WSN is necessarily designed and implemented to support transmitting measurement data in easily exchangeable format between different software platforms, namely as a verbose XML data model, or a more compact structure in CSV. Location data and timestamps corresponding to the place and time the measurements occur are also transmitted in XML and CSV and synchronized to overcome different timing zones as well as satellite placements concerning the geolocation data.

Communication Between Nodes

For connecting sensing nodes and the gateway node, communication via the ZigBee protocol and cables has been implemented. The connectivity of ZigBee ranges close to 1 km which suits well for the monitoring system, reducing thus the need to only a single common gateway node shared among all sensing nodes at a given deployment site instead of deploying two or more gateway nodes at a site. Data packet format exchanged between nodes is in ASCII frames. These frames are supposed to facilitate the comprehension of data communicated. As the frame is composed of ASCII characters, it is easier to understand numerous fields included within the payload. In our scenario, the communication from the sensing nodes to the gateway is done in increments of time and the size of the packet is ~ 100 bytes. Then the gateway node transmits the aggregated data to the central monitoring node.

For communication between the gateway node and the central monitoring node, the GPRS network has been adopted which enables the system to operate remotely, and is a reliable method.

Initial Use of the System: Water Quality Monitoring in Sitnica River

The water quality monitoring in river Sitnica is performed through static sensors (WSN system) and mobile sensors.

Measurements through the WSN system (static sensors) have been conducted in Plemetin in two points, sensing node 1 (SH - housing) and sensing node 2 (P - manhole). Figure 8 shows the raw measurement data for temperature (T), pH, conductivity (EC), and turbidity. The measurement period covered is from 01.05.2015 to 12.07.2015. Furthermore, Figure 9 shows the comparison between those parameters, each measured at the same time (31.08.2015 at 10:15) and at the same monitoring station (in Plemetin).

Figure 8. Fluctuation in T, pH, EC, and turbidity over ca. two months May 1st to July 12th 2015, measured through the WSN system in Plemetin (raw data).

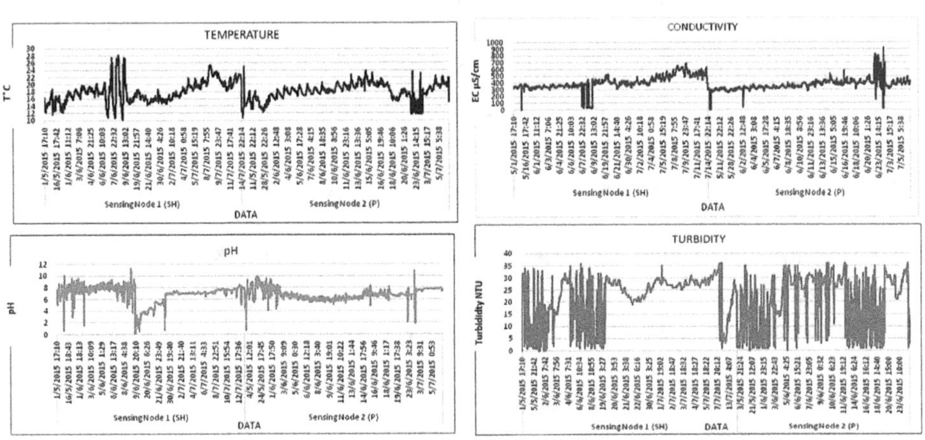

Figure 9. Comparison of four parameters measured at the same monitoring station in Plementin (measurements on 31.08.2015 at 10:15).

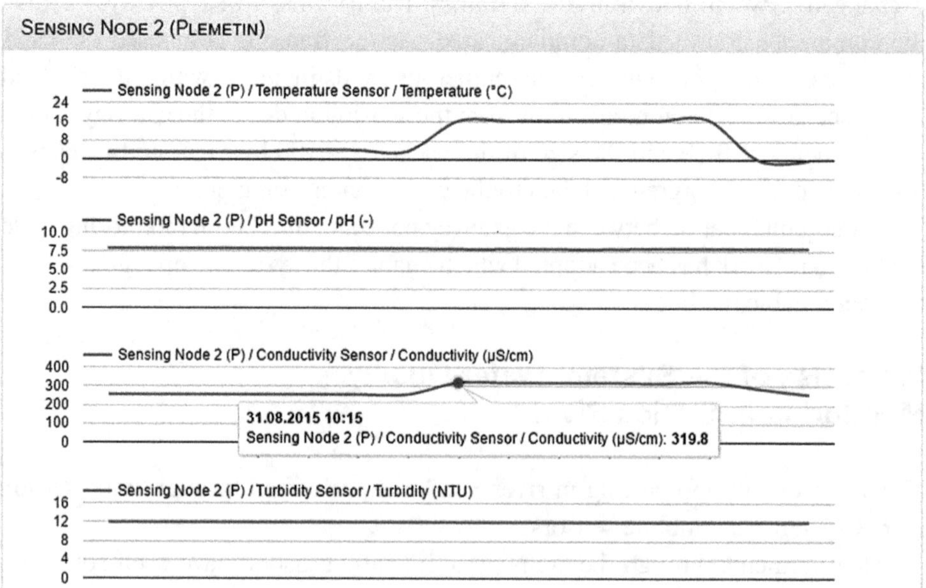

Measurements through the mobile component are performed over several consecutive days, covering five monitoring stations alongside Sitnica river (Lipjan, Vragoli, Plemetin, Nedakovc, and Mitrovicë). Hence, mobile sensors complement the static ones in terms of space coverage, i.e., the measurements by mobile sensors are conducted in locations not possible to cover by the deployed static sensors due to restrictions in measurement ranges. The mobile sensors' measurements at each station are performed by access through a bridge to the water at respective station. Figure 10 shows the results of measurements for temperature (T), and pH, conducted at five stations in days 26.05.2015 to 29.05.2015, as well as for conductivity (EC) at three stations on 26.05.2015 and 27.05.2015.

Sampling through the auto-sampler has been conducted in the same site where monitoring by the WSN system was performed, i.e., in Plemetin, such that more parameters of interest are measured through auto-sampler not supported by the actual WSN system. The selection of parameters is usually done based on activities being developed in the surrounding of the sampling station. i.e., the Kosovo Power Plant, agriculture activity, sewage, as well as waste disposal.

It is important to note that the analysis performed by static sensors and mobile devices only indicate the water quality data of quality parameters at the time they were sampled. Moreover, these data are preliminary raw data. Hence, to reflect the assurance data, following measures need to be taken:

Figure 10. Raw data for T and pH measured through mobile sensors over four days at five stations, as well as for EC over two days at three stations.

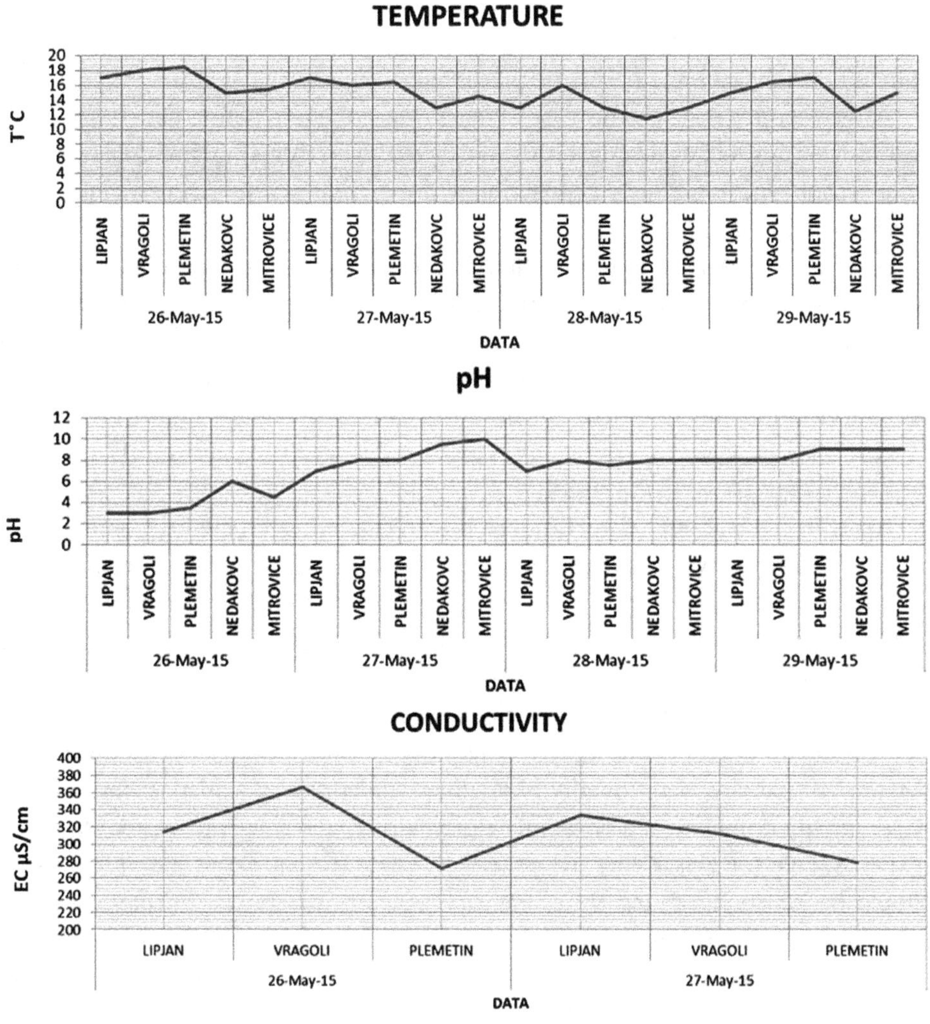

- Validate the reliability of data and their accuracy prior to analysing them.
- Analysis to be performed regularly in time and space to identify trends and changes of data.

Towards an Intelligent System Through the Portal

In a previous work (Ahmedi, 2015) within the InWaterSense project, a Web portal[17] has been introduced initially developed to work with simulated data. The portal has quickly become operational over real measurement data as demonstrated in this paper. Water quality data of Sitnica river monitored through WSN (static sensors) and mobile sensors are stored in the portal and made available in real-time to the wide public, water experts and WSN engineers. In the previous section, real measurement data inferred through the portal are presented.

A Web service has been developed to automate the population with transmitted data, as well as synchronization issues addressed when receiving time data remotely from sensors due to GPS and connection with different satellites covering different time zones (Figure 11).

For example, *Sensor Node 2* (with id 2) contains on-board sensors that measure water parameters (turbidity, temperature, conductivity and pH), as well as a GPS sensor to measure latitude, longitude and timestamp. This node transmits data to the gateway node through wireless link. The gateway node is further connected with Web service through GPRS to transmit sensor data to the database server. As Figure 11 illustrates, the database server stores also the date and time (as EntryDate) when the sensor data arrive to the server.

Apart from managing observational stream data on water quality coming from wireless sensors – dynamic data, the portal supports managing data from engineers describing the system itself, e.g., wireless sensor network or the mobile sensor component, their devices (e.g., sensor devices, node types) and the corresponding site allocation data (e.g., municipality, river) – static data (Figure 12).

In addition, an experts' module of the system has been developed which infers new data about water parameters given the experts' data, as well as the expert's rules (Figure 13) like are water quality regulatory standards and statistical formulas in water quality domain. As an example expert's rule, a Water Framework Directive (expert's) rule for classifying pH observations states that pH as individual value should be between 4.5 and 9.0 (Statutory Instruments, 2011). Further, working with data describing potential pollutant sources along the water flow is supported.

Thus, the portal provides already the infrastructure to turn the InWaterSense system into an intelligent platform able to reason over stored data emitted by sensors, data specifying the sensor system itself, as well as experts' data and experts' pre-defined rules of kind:

Figure 11. GPS data and sensor data transmitted by the WSN system merged into a single table

	Id	SensorNodeId	Parameter	Value	EntryDate	Timestamp	Longitude	Latitude
1	200184	2	Turbidity	35.00	2015-09-28 02:20:27	201509028020320000	21.0384311676025	42.7067031860352
2	200185	2	Temperature	14.69	2015-09-28 02:20:27	201509028020320000	21.0384311676025	42.7067031860352
3	200186	2	Conductivity	325.70	2015-09-28 02:20:27	201509028020320000	21.0384311676025	42.7067031860352
4	200187	2	pH	5.87	2015-09-28 02:20:27	201509028020320000	21.0384311676025	42.7067031860352

Figure 12. Raw data gathered through mobile sensor devices, and internals of the deployed system in terrain

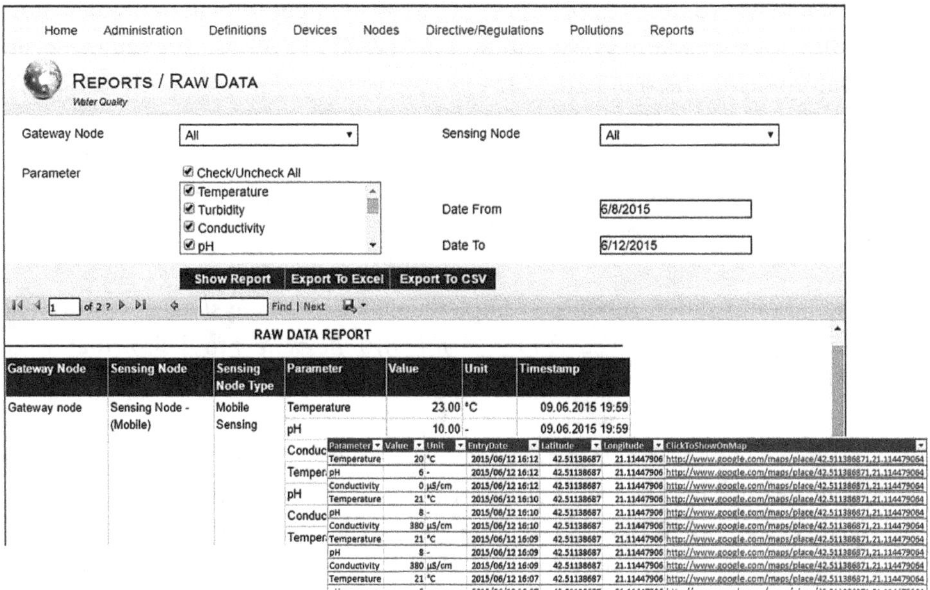

Figure 13. Experts' module of the intelligent system for administration of water directives/regulations, water status classifications, pollutants, water parameters, basins, rivers, etc.

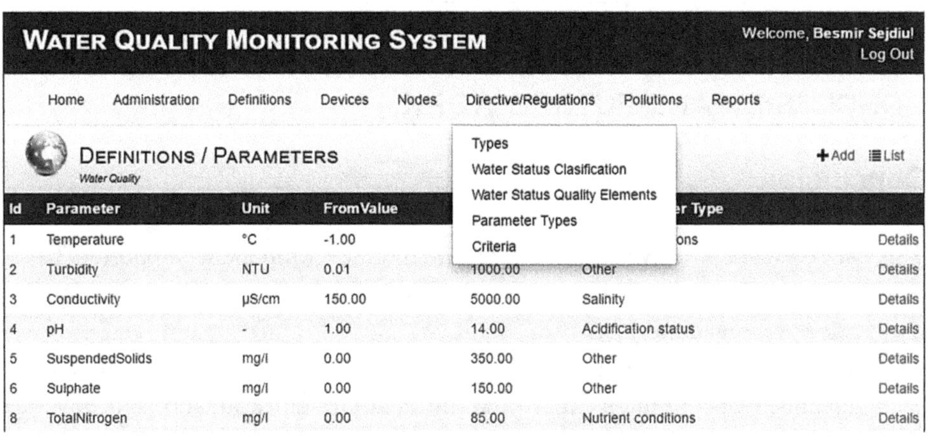

- Given a certain regulatory standards' data, classify the water status at certain points.
- Given data describing potential pollutant sources along the water flow, indicate actual sources of pollution for a certain water quality parameter not conforming to the standard.

The portal has practical value, and is already in use. It may behave as an intelligent system as discussed above, and further get expanded to any other scenario of water quality monitoring in distance and real-time:

- The administration module is implemented for both system engineers and water experts to easily *customize (configure) the system* to their actual needs and manage the system remotely for any other water quality system in the future elsewhere in Kosovo or abroad.
- It allows *adding with ease new components and functionalities in the future* when deemed necessary, e.g., add certain new regulatory documents for water quality, like are WFD (Water Framework Directive), EPA (Environmental Protection Agency), or other, and directly compare through the portal the data measured by sensors with the quality standard values as defined by those corresponding regulations. Insertion of arbitrary potential pollutant sources for consideration as responsible for polluted water is also supported.
- It is capable to reason over expert rules for a rich set of water quality parameters *not restricted to only a set of parameter types* supported by the actual WSN system deployed in Sitnica.

CONCLUSION AND FUTURE WORK

Conclusion

This work presents a WSN system and a mobile component supported by an intelligent portal system developed in a river in Kosovo to monitor its water quality in distance and real-time.

That readily provides with advantages such as enabling convenient monitoring arrangements, like continuous, real-time, and spatially-dense and broad in coverage measurements, remote monitoring, flexible disposition, and identification of long-term trends and fluctuations of parameters of interest, and similar. The auto-sampler attached to the system on the other hand is suggested to serve complementing the measurements performed by the WSN system in terrain with accurate lab analysis of samples gathered also in distance and triggered ad hoc as programmed by the expert. In addition, the mobile sensors compensate advantageous features not provided by the WSN system such as flexible disposition and arrangement of measurements in time and space due to portability of the sensors.

Although the implemented architecture bases on existing traditional systems, the novelty lies on:

- The unique combination, to the best of our knowledge not found elsewhere in the literature, of a triple of existing but separate technologies at one place, namely the WSN system, mobile sensors and auto-sampler, which complement each other to enable remote, portable, and reliable quality observations being characteristics of each single technology alone in this combination.
- The ease to expand the existing demonstrated system architecture to any surface water resources in the country or abroad.
- The level of scalability towards integration of other similar water quality data provision systems, not necessarily sensors, to the existing software (portal) system.
- Introduction of an innovative solutions – the intelligent basic infrastructure for water quality monitoring capable to (a) automate the work usually performed by humans, mainly for quality supervision and control procedures vis-à-vis the acting regulations, and even further, (b) outperform the performance a human expert might afford in the quality monitoring cycle due to his or her limited capacities in terms of availability, the amount of data able to reason with at a given time, and the processing power.
- Flexibility to allow adaptation of other similar expert views with their own data collections and rules for a certain domain of interest, like on quality standards, or activities along the water flow for identifying potential pollution sources.

The suggested surface water quality-monitoring framework with its intelligent and easy expandable features to other scenarios (surface waters) and target groups (authorities in the water sector at different levels: researchers, policy-makers, engineers, citizens) thus promises broad applicability prospects and effects in the future.

Future Work

The implemented exemplary remote sensor system for water quality monitoring is already being extended with more functionality.

As a proof of concept of the capacity of the system to behave intelligently, there is previous work but focusing on Semantic Web technologies within the InWaterSense project presented on modeling portal's data with semantics (Ahmedi, 2015), and also work introducing reasoning over such semantically-enriched data (Jajaga, 2015) (Jajaga, 2016a) (Jajaga, 2016b) (Jajaga, 2017). Furthermore, mining of context data like making distinction between day and night observations of water quality through the system modelled through ontologies is introduced in (Bytyçi, 2016).

Among other several extensions under way of the system to mention are:

- Data validation on accuracy of parameters measured through mobile devices and static sensors is being considered to be done through samples gathered by the auto-sampler, and then analysed in the laboratory.
- Expanding the existing system to other surface water resources in the country for demonstrating its broad coverage in terms of space and quality control.
- Augmenting the RFID tags and integration into the new Internet of Things (IoT) paradigm. Tagging the sensors or any other device (like microcontroller) in the monitoring system and their registration afterwards in the system inventory via NFC (Near Field Communication) service available in smart-phones is being considered. Each device (sensor and other arbitrary monitoring system devices) might get assigned an RFID tag, and hence uniquely identified and easily incorporated to the immense global Internet of Things infrastructure promising readily added values to the future.
- **Semantic water portal:**Moreover, the aim is to enrich the Web portal of the intelligent system with Semantic Web behaviour as a future portal model for water quality monitoring portals. The idea here is to merge the results of this work and of the previous published works already on the portal (Ahmedi, 2015) and the reasoning capabilities (Jajaga, 2015) (Jajaga, 2016a) (Jajaga, 2016b) (Jajaga, 2017) of the water quality monitoring system into a unique online portal able to behave intelligently, like identify potential pollutions in real-time given certain input quality data gathered by the system available through the portal.

ACKNOWLEDGMENT

The work described in this paper is supported by "InWaterSense: Intelligent Wireless Sensor Networks for Monitoring Surface Water Quality", an EU funded project managed by European Union Office in Kosovo, implemented by University of Prishtina.

REFERENCES

Adamo, F., Attivissimo, F., Carducci, C. G. C., & Lanzolla, A. M. L. (2015). A smart sensor network for sea water quality monitoring. *IEEE Sensors Journal, 15*(5), 2514–2522. doi:10.1109/JSEN.2014.2360816

Adu-Manu, K., Tapparello, K. S., Heinzelman, C., Katsriku, W., & Abdulai, J. D. (2017). Water Quality Monitoring Using Wireless Sensor Networks: Current Trends and Future Research Directions. *ACM Transactions on Sensor Networks, 13*(1), 4. doi:10.1145/3005719

Ahmedi, L., Jajaga, E., & Ahmedi, F. (2013). An Ontology Framework for Water Quality Management. In C. H. O. Corcho (Ed.), *Proceedings of the 6th International Workshop on Semantic Sensor Networks co-located with the 12th International Semantic Web Conference* (vol. 1063, pp. 35-50). Sydney, Australia: CEUR Workshop Proceedings.

Ahmedi, L., Sejdiu, B., Bytyçi, E., & Ahmedi, F. (2015). An Integrated Web Portal for Water Quality Monitoring through Wireless Sensor Networks. *International Journal of Web Portals, 7*(1), 28–46. doi:10.4018/IJWP.2015010102

Allan, I. J., Vrana, B., Greenwood, R., Mills, G. A., Knutsson, J., Holmberg, A., ... Laschi, S. (2006). Strategic monitoring for the European water framework directive. *TrAC Trends in Analytical Chemistry, 25*(7), 704–715. doi:10.1016/j.trac.2006.05.009

Bytyçi, E., Ahmedi, L., & Kurti, A. (2016). Association rule mining with context ontologies: An application to mobile sensing of water quality. *Communications in Computer and Information Science, 672*, 67–78. doi:10.1007/978-3-319-49157-8_6

Chaamwe, N. (2010). Wireless sensor networks for water quality monitoring: A case of Zambia. In *Proceedings of the 4th International Conference on Bioinformatics and Biomedical Engineering*. IEEE. 10.1109/ICBBE.2010.5515792

Chave, P. (2001). *The EU water framework directive*. IWA Publishing.

CIRCA. (2003). *Monitoring under the Water Framework Directive, Policy Summary to Guidance Document 7, Produced by Working Group 2.7 – Monitoring, Common Implementation Strategy for the Water Framework Directive (2000/60/EC), Published Guidance Documents*. CIRCA Library EPA Ireland - Water Framework Directive Monitoring Programme. Retrieved from http://www.epa.ie/pubs/reports/water/other/wfd/

Fleischmann, N., Staubmann, K., & Langergraber, G. (2002). Management of sensible water uses with real-time measurements. *Water Science and Technology*, *46*(3), 33–40. PMID:12227601

Goumopoulos, C., Kameas, A., & O'Flynn, B. (2007, July). Proactive agriculture: an integrated framework for developing distributed hybrid systems. In *International Conference on Ubiquitous Intelligence and Computing* (pp. 214-224). Springer Berlin Heidelberg. 10.1007/978-3-540-73549-6_22

Goumopoulos, C., O'Flynn, B., & Kameas, A. (2014). Automated zone-specific irrigation with wireless sensor/actuator network and adaptable decision support. *Computers and Electronics in Agriculture*, *105*, 20–33. doi:10.1016/j.compag.2014.03.012

Grath, J., Ward, R., Scheidleder, A., & Quevauviller, P. (2007). Report on EU guidance on groundwater monitoring developed under the common implementation strategy of the water framework directive. *Journal of Environmental Monitoring*, *9*(11), 1162–1175. doi:10.1039/b710665n PMID:17968442

Greenwood, R., Webster, J., & Regan, F. (2008). *Sustainable water: chemical science priorities*. RSC Report, 1-18.

IDS Monitoring. (n.d.). Retrieved from https://www.idsmonitoring.com

IPCC. (2001). *The Regional Impacts of Climate Change: An Assessment of Vulnerability*. Geneva: IPCC.

Irvine, K., Boelens, R., Fitzsimmons, J., Kemp, A., & Johnston, P. (2002). *Review of Monitoring and Research to meet the needs of the EU Water Framework Directive (200-DS-5-M1)*. Academic Press.

Jajaga, E., & Ahmedi, L. (2017). C-SWRL: SWRL for Reasoning over Stream Data information. *The 11th IEEE International Conference on Semantic Computing (ICSC 2017).*

Jajaga, E., Ahmedi, L., & Ahmedi, F. (2015). An Expert System for Water Quality Monitoring based on Ontology. *Communications in Computer and Information Science*, *544*, 89–100. doi:10.1007/978-3-319-24129-6_8

Jajaga, E., Ahmedi, L., & Ahmedi, F. (2016a). StreamJess: Enabling Jess for Stream Data Reasoning and the Water Domain Case. *20th International Conference on Knowledge Engineering and Knowledge Management*. Bologna, Italy: Springer Lecture Notes in Computer Science.

Jajaga, E., Ahmedi, L., & Ahmedi, F. (2016b). StreamJess: Stream Data Reasoning System for Water Quality Monitoring. *International Journal of Metadata, Semantics and Ontologies*, *11*(4), 207. doi:10.1504/IJMSO.2016.083507

Jiang, P., Xia, H., He, Z., & Wang, Z. (2009). Design of a Water Environment Monitoring System Based on Wireless Sensor Networks. *Sensors (Basel)*, *9*(8), 6411–6434. doi:10.339090806411 PMID:22454592

Johnson, K. S., Needoba, J. A., Riser, S. C., & Showers, W. J. (2007). Chemical sensor networks for the aquatic environment. *Chemical Reviews*, *107*(2), 623–640. doi:10.1021/cr050354e PMID:17249737

Könnölä, T. (2008). *Future outlook on water quality monitoring technologies*. VTT Technical Research Centre of Finland Report.

Lawlor, A., Torres, J., O'Flynn, B., Wallace, J., & Regan, F. (2012). DEPLOY: A long term deployment of a water quality sensor monitoring system. *Sensor Review*, *32*(1), 29–38. doi:10.1108/02602281211197125

Nasser, N., Ali, A., Karim, L., & Belhaouari, S. (2013). An efficient wireless sensor network-based water quality monitoring system. In *Proceedings of the 2013 ACS International Conference on Computer Systems and Applications*. IEEE. 10.1109/AICCSA.2013.6616432

O'Flynn, B., Martinez, R., Cleary, J., Slater, C., Regan, F., Diamond, D., & Murphy, H. (2007). SmartCoast: a wireless sensor network for water quality monitoring. In *Local Computer Networks, 2007. LCN 2007. 32nd IEEE Conference on* (pp. 815-816). IEEE. 10.1109/LCN.2007.34

O'Flynn, B., Regan, F., Lawlor, A., Wallace, J., Torres, J., & O'Mathuna, C. (2010). Experiences and recommendations in deploying a real-time, water quality monitoring system. *Measurement Science & Technology*, *21*(12), 124004. doi:10.1088/0957-0233/21/12/124004

Oliveira, L. M., & Rodrigues, J. J. (2011). Wireless Sensor Networks: A Survey on Environmental Monitoring. *JCM*, *6*(2), 143–151. doi:10.4304/jcm.6.2.143-151

Quevauviller, P. (2006). Chemical monitoring activity under the common implementation strategy of the WFD. *Journal of Soils and Sediments*, *6*(1), 2–3. doi:10.1065/jss2006.02.002

Rasin, Z., & Abdullah, M. R. (2009). Water quality monitoring system using ZigBee based wireless sensor network. *IACSIT International Journal of Engineering and Technology*, *9*(10), 24–28.

Ruggaber, T. P., Talley, J. W., & Montestruque, L. A. (2007). Using embedded sensor networks to monitor, control, and reduce CSO events: A pilot study. *Environmental Engineering Science*, *24*(2), 172–182. doi:10.1089/ees.2006.0041

Seders, L. A., Shea, C. A., Lemmon, M. D., Maurice, P. A., & Talley, J. W. (2007). LakeNet: An Integrated Sensor Network for Environmental Sensing in Lakes. *Environmental Engineering Science*, *24*(2), 183–191. doi:10.1089/ees.2006.0044

Silva, S., Nguyen, H. N., Tiporlini, V., & Alameh, K. (2011). Web based water quality monitoring with sensor network: Employing ZigBee and WiMax technologies. In *Proceedings of the Conference on High Capacity Optical Networks and Enabling Technologies*. IEEE. 10.1109/HONET.2011.6149804

Statutory Instruments. (2009). *European Communities Environmental Objectives (Surface Waters) Regulations 2015*. Retrieved from http://www.irishstatutebook.ie/eli/2015/si/386/made/en/pdf

Stemp-Morlock, G. (2007). Natural Resources: Australia's War on Drought. *Environmental Health Perspectives*, *115*(7), A348. doi:10.1289/ehp.115-a348

Tateson, J., Roadknight, C., Gonzalez, A., Khan, T., Fitz, S., Henning, I., ... Marshall, I. (2005). Real world issues in deploying a wireless sensor network for oceanography. *Workshop on Real-World Wireless Sensor Networks REALWSN'05*.

Voigt, T., Osterlind, F., Finne, N., Tsiftes, N., He, Z., Eriksson, J., . . . Hjort, K. (2007). Sensor networking in aquatic environments-experiences and new challenges. In *Local Computer Networks, 2007. LCN 2007. 32nd IEEE Conference on* (pp. 793-798). IEEE. 10.1109/LCN.2007.23

Wang, J., Ren, X., Shen, Y., & Liu, S. (2010). A Remote Wireless Sensor Networks for Water Quality Monitoring. In *Proceedings of the 2010 International Conference on Innovative Computing and Communication and 2010 Asia-Pacific Conference on Information Technology and Ocean Engineering*. IEEE Computer Society. 10.1109/CICC-ITOE.2010.9

Wang, Z., Wang, Q., & Hao, X. (2009). The design of the remote water quality monitoring system based on WSN. In *Proceedings of the 5th International Conference on Wireless Communications, Networking, and Mobile Computing*. IEEE. 10.1109/WICOM.2009.5303974

Yang, X., Ong, K. G., Dreschel, W. R., Zeng, K., Mungle, C. S., & Grimes, C. A. (2002). Design of a wireless sensor network for long-term, in-situ monitoring of an aqueous environment. *Sensors (Basel)*, *2*(11), 455–472. doi:10.339021100455

ENDNOTES

1 http://ec.europa.eu/environment/water/water-framework/index_en.html (accessed 01/01/2017)

2 http://www.irishstatutebook.ie/eli/2003/si/722/made/en/print (accessed 01/01/2017)

3 http://www.housing.gov.ie/sites/default/files/migrated-files/en/Legislation/Environment/Water/FileDownLoad%2C20824%2Cen.pdf (accessed 01/01/2017)

4 http://www.irishstatutebook.ie/eli/2010/si/610/made/en/pdf (accessed 01/01/2017)

5 http://www.irishstatutebook.ie/eli/2011/si/489/made/en/pdf (accessed 01/01/2017)

6 http://www.irishstatutebook.ie/eli/2014/si/350/made/en/pdf (accessed 01/01/2017)

7 http://ec.europa.eu/environment/nature/legislation/habitatsdirective/index_en.htm (accessed 01/01/2017)

8 http://eur-lex.europa.eu/LexUriServ/LexUriServ.do?uri=OJ:L:2006:064:0037:0051:EN:PDF (accessed 01/01/2017)

9 ec.europa.eu/environment/industry/stationary/ied/legislation.htm (accessed 01/01/2017)

10 ec.europa.eu/environment/water/water-nitrates/index_en.html (accessed 01/01/2017)

11 http://www.rcep.org.uk/reports/24-chemicals/documents/24-chemicals.pdf pg 221 (accessed 01/01/2017)

12 CSIRO http://www.csiro.au/en/Research/Environment/Water (accessed 01/01/2017)

13 WRON Brochure. https://www.seegrid.csiro.au/wiki/pub/Main/SeeGridIIIProgram/ACKLAND_SEEGRIDIII.pdf (accessed 01/01/2017)

14 http://www.smartbay.ca (accessed 01/01/2017)

15 https://teacheratsea.files.wordpress.com/2011/06/dsc_5884.jpg?w=150&h=99

16 http://www.libelium.com/development/plug-sense/documentation/waspmote-plug-sense-quick-overview/

17 http://inwatersense.uni-pr.edu

Chapter 4
Semantic Web-Based Agricultural Information Integration

Kaladevi Ramar
Jerusalem College of Engineering, India

Geetha Gurunathan
Jerusalem College of Engineering, India

ABSTRACT

A huge volume of information is available in the worldwide web. However, the demand is on relevant information rather than available information, which are often heterogeneous and distributed. Agriculture is one such domain, which includes information like soil, crops, weather, etc. under one roof. This information is in different representations and structures (e.g., weather). This scenario leads to a challenge of how to integrate the available and heterogeneous agricultural information to deliver better production. As the information on the web is syntactically structured, the need is to provide semantic linkage. The semantic web supports the existing web to easily process and interpret information. In this chapter, a semantic-based agricultural information system (AIS) is proposed that addresses heterogeneity issues among weather systems and integrates information like soil, weather, crop, and fertilizers. AIS helps the farmers regarding the type of crop/soil, crop/climate, fertilizer applications, diseases, and prevention methods using effective retrieval of information from integrated systems.

DOI: 10.4018/978-1-5225-5978-8.ch004

INTRODUCTION

Huge volume of information in agriculture and its related fields is being published on the internet in diverse representation, structure and formats. Agriculture is the main occupation of many countries in the Asia-Pacific region as nearly 60 percent of its people derive their livelihood from agriculture. Recent FAO 2016 (Food and Agriculture Organization of the United Nations) report estimates that to meet the projected demand, global agricultural yield will have to increase at least 60 percent above 2006 levels. At the global level, the yield rates for many crops have been decelerating in recent decades, while demand continues to increase in absolute terms. A holistic approach in agriculture production is the need of the hour. This can be achieved by providing relevant information to the farmer and using knowledge engineering methods to extract efficient agricultural information from integrated information sources.

Though the available information on the web is huge, most of them are heterogeneous due to the design and development autonomy of users and systems. For effective utilization of available information, the heterogeneity among the information systems must be redeemed and integration of related information systems has to be done. This assures enhanced information retrieval from the integrated information sources.

The motivation for information integration in agriculture system is to provide required knowledge to the farmers to make more informed and valuable decisions. Recently available systems support syntactic based service and complete integration of services are not adequately available. The existing approaches are time consuming, difficult and expensive. There is a need for a sophisticated system, which is more efficient. In order to provide a technological development for the growth of the agriculture, a semantic based agriculture framework is proposed that resolves heterogeneities among weather systems, integrates related information from disparate sources such as soil, crop, pesticides, weather and fertilizers thereby ensuring proper information retrieval.

In this paper, a semantic based agriculture framework is proposed which represents the information using ontologies. Ontology mapping technique is used to reduce heterogeneities among information systems thereby providing semantic interoperability between them. Then ASDOM algorithm (Asymmetric Source Driven Ontology Merging) is used to integrate mapped ontologies and INRON algorithm (INformation Retrieval from ONtology) is used for effective information retrieval.

This paper is organized as follows; in next section, some related works are discussed to assess the state of the art methods and techniques that are used to improve agriculture production using semantic web technologies. Third section narrates proposed AIS framework and its functionalities. Last section concludes the proposed work and gives directions for future use.

RELATED WORK

Agriculture is an essential domain, which is entrenched in our lives and economy. To fulfill the food demands of the increasing population and to improve the economy of the nation, existing agriculture systems need to cope up with emerging technological inventions. Various information systems based on semantic web techniques to improve the agriculture production are discussed below.

An application framework is proposed to support the agricultural knowledge integration for better production based on semantic web concepts (Sun, Zhu, Gu, Wu & Feng 2009). This framework is designed to overcome existing agricultural knowledge representation, which is a complex and difficult task due to the distributed and heterogeneous characteristics of the agricultural knowledge sources. Layered approach is suggested for structural and functional description. An Integrated Agriculture Information Framework (IAIF) is developed to enable knowledge extraction from many domain related repositories (Shoaib & Basharat 2010). IAIF has been designed with the core aim of representing, combining and aggregating the data available from existing knowledge repositories through the help of metadata and domain ontologies. This domain ontology is used to link the sources to the domain knowledge with the help of corresponding metadata. A model that uses ontology, semantic web technology into agriculture knowledge system in order to improve precision and recall rate of information retrieval is discussed in (Hongye 2010). To overcome the issues in keyword based knowledge retrieval service, content-based representation is used with the help of semantic web. A semantic data model is introduced for Indian agriculture domain (Swaran, Bhaskar, Ela, Somnath & Raghu 2013). This e-governance model is developed from raw unstructured government related data spread over various sites. Searching and querying RDF data from ontology of agriculture domain has been done with the help of semantic technologies. Domain specific semantic web based search engine is developed to address the difficulties with the keyword-based search engines (Mukhopadhyay, Banik, Mukherjee & Bhattacharya 2011).The search engine is designed for an agricultural domain that hosts agricultural information. This search engine retrieves the results directly from knowledgebase rather than the URLs (as returned by an existing search engine). The retrieval process is depends on class name or instance name or both.

An application is developed using FAO AGROVOC ontology along with Google AJAX API in (Yunpeng, Peng, Shihong & Sufen 2010). This AGROVOC ontology contains agricultural concept collection as well as the relationships between the concepts. Google API is used to retrieve results from Google search engine using keywords. These two combinations can help the users of agriculture domain to get more desired information and navigate the knowledge into some degree. AGROVOC is used as a main vocabulary to build the ontology (CROPont) to provide individual farmers with required level of information (Bansal & Malik 2011). For this, a framework is developed which makes available of relevant contextual and scientifically correct information with respect to crop production lifecycle. CROPont and AGROVOC can act as a knowledgebase when information (in the form of instances and inference rules) is appended by the domain experts via semantic web techniques. The system is capable of providing advice to individual farmers. The PAP (Pre-process Associate and Predict) system architecture is developed to support farmers in decision making on which mechanism is useful for a better productivity (Manikandan, Sakthiganesh, Kumar & SenthilKumar 2012). The focus of the system is to improve the profit of the farmers by enhancing the efficiency of agricultural input and reduce the risk and cost of production. This can be achieved by providing timely access and advice to the farmers like dynamic weather information and use of knowledge engineering to deduce best appropriate agricultural information from various sources.

Ontology based technique is used to enhance the identification, search and classification of agricultural information (Liu, Duan & Zhang 2012). Learning based algorithm is applied to extract topic knowledge of agricultural related information. The classification of agricultural information using ontology has high accuracy, which enhances the reliability and stability of information searching and retrieval of agricultural products. A variety of innovations that integrate Information and Communication Technologies (ICT) into the dissemination of agricultural information to farmers is analyzed in (Muriithi, Bett & Ogaleh 2009). Farmers can access recent and reliable information on prices and commodity offers at an affordable rate, right hybrid seeds to plant in their respective agro ecological fields using their mobile phones. It recommends adoption of ICT by institutions such as metrological department to enhance the system on climate for appropriate decision making. A framework is proposed to provide interoperability and integration among heterogeneous weather systems using ontological based techniques (Kaladevi, Geetha, & Narayanasamy 2016). The proposed mapping algorithm is best suited for domain with less lexical and structural similarities and leads to enhanced information retrieval from the integrated system.

Liao, Li & Liu (2015) proposed an integrated ontology based agriculture information system to enhance the agriculture process. The system can support users for effectively access the desired agricultural information using ontology techniques provided by other independent AIS platforms. This intelligent decision system uses a classification algorithm, which is based on neural network to determine the relevance of the collected results, and the retrieved results are arranged in a proper order. An ontology based agricultural knowledge fusion method is developed that can be used as a comprehensive basis to solve agricultural information inconsistencies, analyze data, and discover knowledge (Xie, Wang, Ma, Zhang, Sun & Guo 2015). This fusion method leverages semantic web and big data technologies, that will enhance the identification and fusion of new and existing data sets to make big data analytics more possible. Empirical results show that this knowledge fusion method is useful for knowledge discovery. Junfeng and Yong (2016) presented the way of computing the semantic similarity values based on the relationship between concept pairs of agricultural ontologies at different depth of ontology structure. Accuracy of semantic similarity calculation between four algorithms shows proper selection of parameters and comprehensive similarity computation measures can significantly reduce difficulty of distinguishing the concept of weak correlation.

The above mentioned systems are focused on providing technology based support for agriculture production and the ways to integrate various services using semantic web technology. However, the existing systems do not address the complete integration of related systems for a domain and retrieval of information from these systems. In this paper, a semantic web based framework is developed for Agriculture Information Systems for the integration of heterogeneous sources for effective information retrieval.

PROPOSED METHODOLOGY

Availability of information is increasing on one side and new technologies are emerging on other side. Still there exists a gap between the available and desired information. In such scenarios, there is a need of new techniques to bridge the gap between decelerating agriculture and accelerating information technologies. Semantic Web is one among such technology that provides a unique representation of knowledge in the form of ontology in which semantics or meaning of information is formally defined (Berners-Lee *et al.* 2001). This formal definition can easily make the machines to interpret and relate information content more effectively.

The main process areas of Agriculture Information Systems (AIS) are soil system, weather system, crop system, fertilizer and pesticides system. Some information is available from more than one source for the process area but information given by the existing single source may not be sufficient for certain cases. For example, weather information is provided by more than one source for a particular region. The objective is to reduce heterogeneities among weather systems and the integration of relevant information sources for agriculture. The proposed methodology constructs an Agriculture Information System (AIS) that resolves the integration issues using semantic web and improves information retrieval as shown in Figure 1.

AIS Ontology

Ontology is an explicit formal specification of a conceptualization (Gruber 1993). Normally ontologies are represented as taxonomies with domain concepts, attributes of concepts, properties, individuals, constraints on properties.

AIS ontology can be designed as a graph where each node denotes the related domain concept such as *Crops, Fertilizer, and Soil* etc and edges denotes the relations between two concepts. Each domain concept is the sub ontology of the core AIS ontology. The domain concepts need to be integrated for the effective retrieval of information. This integration is achieved by the relationship between concepts of ontologies. *Soil* and *Crop* concepts are related via '*growsIn*' relation. *Soil* and *Fertilizer* are related using '*isNutrientTo*' relation. Likewise, *Pesticides* and *Weather* are related through '*isInfluencedBy*' relation. In AIS system, main process areas are modeled as ontologies. For instance, *Crop, Soil, Pesticides, Fertilizer* and *Weather* are the ontologies that are considered as sub concepts of core concept AIS. The generic framework is shown in Figure 2.

Figure 1. Semantic web based integrated agriculture information system (AIS)

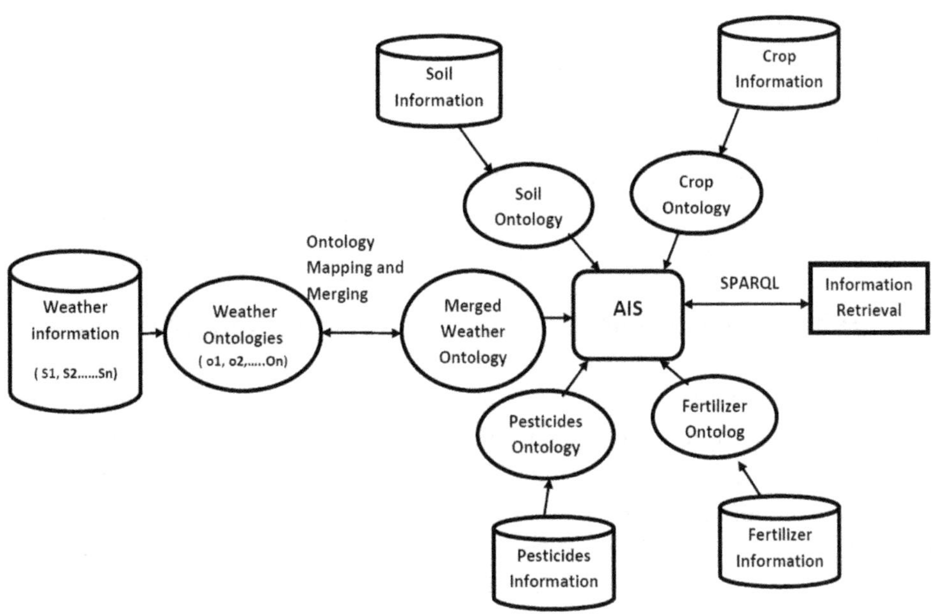

Figure 2. Integration of heterogeneous information sources

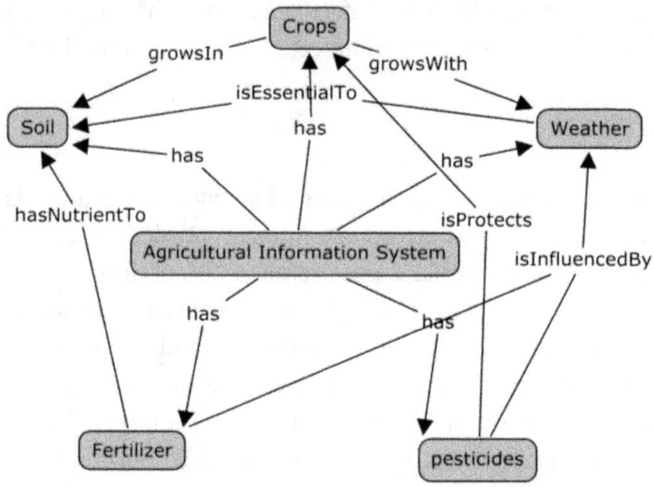

To specifically show the relationships between the different components of AIS, rice crop is taken. Rice is one of the most important food crop in the world and it is the staple food for over 2.7 billion people. Information about the rice crop is modeled as ontology based on experts suggestion and also refereed similar ontologies (Neha 2012). It describes the process areas and the relationship among the concepts of process area are shown in Figure 3. Rice ontology contains relevant information for soil and weather parameters, growing season and duration, disease affecting the crop and cures for those diseases. Similarly information on fertilizer requirements, pests which attack the crop, their precautions and cure are also given. The ontology also consists of knowledge on the various stages of farming like, nursery management, sowing, irrigation, ploughing, and fertilizing methods.

A domain expert must periodically review, append and check all knowledge sources, their relation, redundancy of entities, consistency and correctness between concepts, addition of new concepts with respect to the original domain.

Integrating Heterogeneous Sources Using Ontology Merging

Increasing the awareness of semantic information representation makes the availability of multiple ontologies for the same domain. The same information is interpreted differently in various ontologies. Therefore, single ontology is no longer enough to support the demands and multiple ontologies need to be accessed to fulfill the requirements of several applications. Ontology mapping aims to find the correspondence between the entities of heterogeneous ontologies, which provide

Figure 3. Snippet of rice ontology

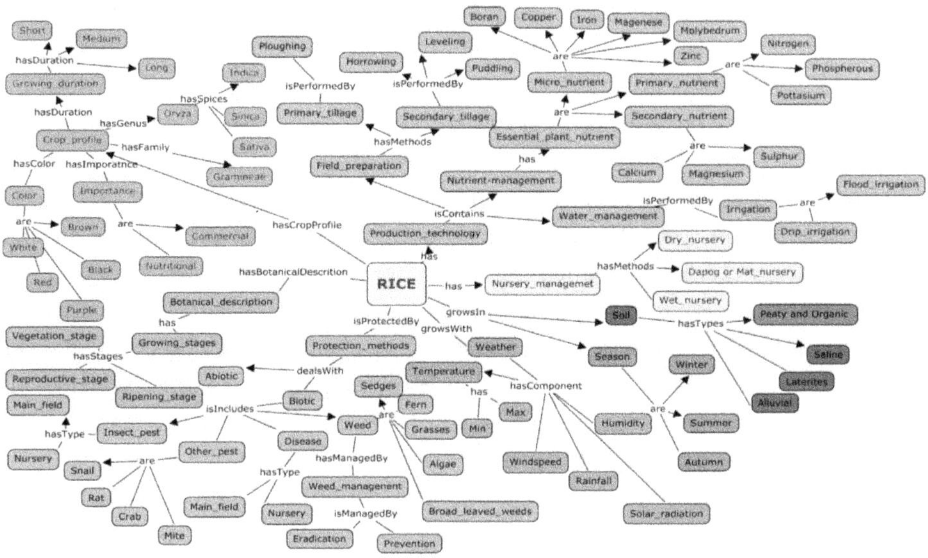

interoperability among various systems (Patel, Koch, Doerr & Sinaraki 2005). Ontology merging method is concerned with integrating related ontologies in order to form a single coherent ontology (Ghidini & Giunchiglia 2004). This is used to develop a common repository of knowledgebase for a domain.

Weather plays an important role in agricultural production. It has a profound influence on the growth, development, yields of a crop, incidence of pests and diseases, water and fertilizer needs in terms of nutrient and water requirements (Ogbuene 2010). Weather information is provided by many sources such as, KEA, Accuweather, and Worldweatheronline etc. An agricultural weather system should refer to all weather elements, which affect farm-planning operations. The elements can vary from place to place and from season to season. Normally an agriculture weather system includes temperature, rainfall, humidity, wind speed and direction, extreme weather events like heat and cold waves, frost, fog, thunderstorms, hail, low-pressure areas, cyclones, tornados.

Weather has more impact on rice crop in various stages. For example, increase in 1°C daily average temperature causes more than 7% decrease in rice grain yield. Paddy blast occurs when cloudy weather, drizzling, high relative humidity and low night temperature happens. Excess moisture promotes root diseases. Storm, hurricane, and wind direction changes are likely to alter the spread of both windborne pests and of bacteria and fungi that are agents of crop disease. Crops can die in a blizzard, because of the extreme cold.

The weather information provided by the weather system may not contain all the weather parameters required for agriculture. For example, Wunderground system does not include wind direction parameter. Dewpoint temperature is not present in Worldweatheronline system. KEA and Wunderground systems do not provide heat index, wind chill parameter information. Certain weather systems provide forecast report for four days. However, mid range forecasting (upto 10 days) is useful to take emergency warnings for farmers and government organizations. The systems such as weather network, Weather-forecast.com provides 10 days forecast. In such situations, there is a need for integration of more than one source.

Heterogeneities between weather systems are classified into entity level, attribute level, abstraction level and data level based on (Kashyap & Sheth 1996; Meenakshi, Verma, Sheth, & Miller 2007). Entity level incompatibilities arise when using different descriptions for semantically similar entities. It includes naming and schema isomorphism. Attribute level incompatibilities arises when semantically similar attributes are modeled using different descriptions. These include naming, data precision, data representation and data scaling conflicts. Abstraction level incompatibility arises when two semantically similar entities or attributes are represented at different levels of abstraction. These include type conflicts and generalization conflicts. Data level incompatibilities arise due to the values of the data present in different databases. Temporal, acceptance inconsistencies and noisy data are of this kind. To resolve the semantic conflicts between concepts of ontologies, ontology mapping is the suggested solution (Mao 2008).

Ontology mapping is the discovery of correspondences between ontologies. Most of the available ontology mapping algorithms relies on lexical or structural similarity of the entities and they do not perform well when less lexical or structural similarity between entities is available. For such scenarios, background knowledge is useful for mapping between ontologies. Proposed mapping approach OMFIM (Ontology Mapping For Information Management) is based on background knowledge. Existing background knowledge for weather systems is enhanced using domain ontologies, Wordnet and domain knowledge is captured from National Weather Service (NWS), METAR, National Oceanic and Atmospheric Administration (NOAA) and 40 other weather systems.

Ontology mapping is a statement of 4-tuple [ID_{ij}, $C_i(\boldsymbol{S})$, $C_j(\boldsymbol{T})$, $\boldsymbol{R}$], where ID_{ij} is a unique identifier of the given mapping pair, i=1...M; j =1,..., N, M is the number of concepts in source ontology, N is the number of concepts in target ontology. $C_i(\boldsymbol{S})$ is the i-th concept of the source ontology, $C_j(\boldsymbol{T})$ is the j-th concept of the target ontology and $\boldsymbol{R}$ specifies a semantic relation between the concepts of source and target ontologies. $\boldsymbol{R}$ refers to equivalent {=}, more general {>}, less general {<}, sibling {||} and mismatch {⊥}. OMFIM is used to identify the semantic relationships between the concepts of input ontologies, which are explained using OAEI (Ontology Alignment Evaluation Initiative) benchmark dataset in (Kaladevi & Mirnalinee 2015).

Ontology merging is a technique applied to integrate two or more ontologies into a single knowledge base. In general, merging methods can be symmetric or asymmetric with respect to the input ontologies. Symmetric methods (PROMPT (Noy & Musen 2000), SAMBO (Lambrix & Tan 2006), HCONE (Kotis & Vouros 2004)) are most common and it completely integrates all input ontologies with equal priority. Asymmetric methods take one of the input ontology as the target or source and merge the other input ontology into this target or source (ATOM (Raunich & Rahm 2011)) thereby giving preference to the target or source ontology. The proposed ontology merging algorithm ASDOM (Asymmetric Source Driven Ontology Merging) is based on mapping relations from OMFIM. ASDOM is automatic top down asymmetric source driven, i.e. merge target ontology concepts into the source ontology concepts. Asymmetric approaches reduce semantic overlap, redundant paths and allow source ontology extension by adding more target ontologies. ASDOM retains the meaning of structure (concepts, relation and attributes) and contents (instance) of the ontologies as much as possible. Proposed ontology merging algorithm is given below.

Algorithm

Input: Source ontology ***S***, target ontology ***T*** and mapping relation ***R*** between concepts of ***S*** and ***T*** are given as inputs.

Output: Merged ontology

1. If ***R***← '=',
    a. Find the relation between attributes of ***S*** and ***T*** concepts by using lexical matching. If attributes are same, that is termed as ***Att-Equal,*** and then moves the target attribute label to source attribute node.
    b. Else the attributes are termed as ***Att-NotEqual*** and create a slot for target attribute in ***S*** concept.
2. Repeat steps 1.1, 1.2 until all attributes are merged into ***S*** concept. Finally move ***S*** concept label to ***T*** concept label (node).
3. If ***R***← **'>'**, then create a node for ***T*** concept with its attributes as a child of ***S*** concept
4. If ***R***← **'<',** then existing parent of ***S*** concept becomes the grand parent of ***S*** concept and ***T*** concept is a new parent of ***S*** concept.
5. If ***R***← '||', then create a node for ***T*** concept with its attributes as a child of parent of ***S*** concept

6. If **S** and **T** concepts are not in above-mentioned relations then it is termed as R← '⊥'. In that case, create a node for **T** concept with its attributes from a root of **S**.
7. Repeat steps 3,4,5,6 until all concepts and attributes of **T** are merged into **S**

An example mapping and merging scenario of ASDOM algorithm for weather system is shown in Figure 4. (a) and (b). The relation between concepts of ontologies is identified using OMFIM. For example, *Min* and *Low* concepts are semantically similar then the two concepts are merged, concept *Temp* has more general than relation with *High* then it becomes the child of *Temp* in merged ontology. *Gust* has less general than relation with *Wind* and *Speed* has sibling relation with *Direction*, so it becomes the child of *Wind*. Finally, *Daily rain* is not matched with concepts of source ontology and then it becomes a new node from the root of merged ontology. Now the merged ontology has extended knowledge from the source ontologies.

The merged ontology is not only contains the integration of all source and target concepts but also contains lexically and semantically similar concepts in the same node from both the ontologies. When there is an attempt to retrieve information from individual ontology, only the specific information from the ontology is retrieved, for example *Barometricpressure* from first ontology and *SeaLevelPressure* from second ontology. But from the merged ontology, information is retrieved either by using *Barometricpressure* or *by SeaLevelPressure*. Likewise if more ontologies are merged, then more similar information are available with same node for easy retrieval and complex queries can be answered using the available parameters. Most of the ontology mapping algorithm deals with equal relation only (PROMPT, ATOM).

Figure 4. (a) Mapping between weather ontology1 and ontology2; (b). Merged ontology

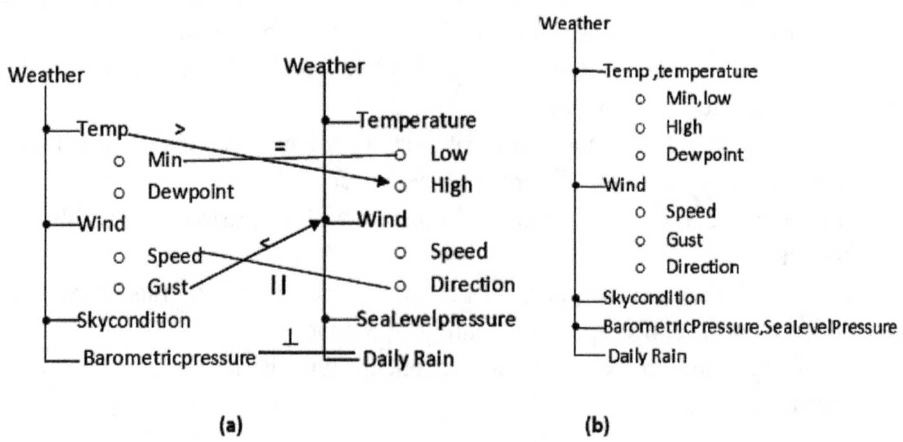

So while merging two or more such ontologies, other than equal relations between remaining concepts are difficult to handle. ASDOM finds correspondence between concepts based on all semantic relations such as equivalence (=), more general (>), less general (<), sibling (||), disjointness (⊥), which handles all the concepts in both the ontologies.

The evaluation of merged ontology is a tedious task. For mapping, numbers of metrics and benchmark datasets are available but for merging, no such facilities are available. Empirical comparison between existing and proposed algorithms is shown in Table 1. ASDOM algorithm considers terminological, structural and semantic matching for merging. From the comparison, existing merging algorithms needs more user support and that is not utilizing all the available information from ontologies. ASDOM exploits all possible information for matching and uses all semantic relations for merging.

Information Retrieval

A knowledgebase is a collection of Ontology Web Language (OWL) resources. Information retrieval from a single source is not sufficient for most of the applications

Table 1. Comparison between ontology merging systems

Sl. No.	Parameters	Chimaera (McGuinness, Fikes, Rice & Wilder 2000)	Onion (Mitra & Wiederhold 2002)	Prompt (Noy & Musen 2000)	ATOM (Raunich & Rahm 2011)	ASDOM
1	User interaction	More	More	More	Nil	Less
2	String matching	No	Yes	Yes	Yes	Yes
3	Semantic matching	Yes	Yes	No	No	Yes
4	Attribute matching	No	Yes	No	No	Yes
5	Structure matching	Yes	Yes	Yes	Yes	Yes
6	Ontology structure	Ontologua IDL	XML, OWL	OWL	OWL	OWL
7	Ontology selection	Based on concept	Based on concept	Based on concept	Based on concept	Based on concept
8	Type of merging	Symmetric	Symmetric	Symmetric	Asymmetric	Asymmetric
9	Mapping relations	Subsumption	Subsumption	Equivalent	Equivalent	Equivalent, Subsumption
10	Additional Resources	No	Wordnet	No	No	Background ontology

and the required information may be derived from more than one source. The user can retrieve required information from the merged ontology instead of searching each source separately. Simple Protocol and RDF Query Language (SPARQL) is a common query language for semantic web. Example SPARQL code are given below. First query is to find the crops that grow in alluvial soil with temperature above 30°C. Second query retrieves the favorable weather conditions to grow *stemborer pest* in rice.

Available semantic query languages completely rely on corresponding syntax and structure of the knowledgebase. Therefore, the end users need to have the knowledge of query syntax and structure of knowledgebase. Practically, it is not possible for most of the end users and they are only aware of their queries. The main objective in semantic based information retrieval is to process user queries over knowledgebase and retrieve relevant information. Existing systems have attempted to convert user queries to semantic queries or directly access the knowledgebase, but a definite solution is not yet attained. In order to focus on such information retrieval from knowledgebase, an algorithm named INRON (INformation Retrieval from ONtology) is proposed.

In INRON, the user or stakeholder of the system enters a domain related query. Input query is initially parsed by means of the parser to make the query simpler. The parsing is done to analyze the query syntactically, which determines the part of speech (POS) of every word in the query. Given query can be formulated into

Figure 5.

```
Query 1: PREFIX :<http://www.semanticweb.org/ontologies/2013/9/19/AIS-Ontology.owl#>
SELECT ?crop ?soil  ?temp
WHERE {
          ?crop :soil_type ?soil .
           ?crop :temperature ?temp .
FILTER ( str(?soil) = "Alluvial"&& ?temp >30)
        }
Query 2: PREFIX :<http://www.semanticweb.org/ontologies/2013/9/19/AIS-Ontology.owl#>
SELECT  ?Rice ?Stemborer ?Meantemperature ?Relative humidity
WHERE {
?Stemborer : Mean_Temp ?Meantemperature
?Stemborer : Relative_humidity ?Relative humidity
      }
```

a syntactic structure named Query Characterization Template (QCT) (Dasgupta, Patel, Padia & Shah 2013) which covers all parts of the query. Aim of using QCT is to identify the query desire and the relationship between desire and the query input. Normally in existing systems all tokens of the query are compared against knowledgebase in order to retrieve results. However, QCT forms the further step of accurate query formalization and it reduces the number of token comparisons. Due to the domain and time constraints, the scope of this paper is restricted to simple *wh* (what, which, where, when, who) queries.

A simple *wh* query can be characterized according to the following structure:

[Wh][[R_1]][[[Q]*][[M]*] [D]] [[R_2]][[[Q]*][[M]*][I]][?]

Where:

[[]]: second square bracket indicates optional component
[D]: Query desire class/instance - value restricted to {NN, NNP, JJ, RB, VBG}
[I]: Query input class/instance - value restricted to {NN, NNP, JJ, RB, VBG}
[R_1]: Auxiliary relation - includes variations of the set {is, is kind of, much, might be, does}
[R_2]: Relation that acts as (i) predicate of *D* as the subject and *I* as the object or (ii) action role of *I* as the actor - value restricted to {VB, PP, VB-PP}
[[Q]*]: Quantifier of *D* or I - values restricted to {DT}. The * indicates that *Q* can recur before *D* or *I*.
[[M]*]: Modifier of *D* or *I* - value restricted to set {NN, JJ, RB, VBG}. The * indicates that *M* can recur before *D* or *I*.

Parsing and formulation of QCT structure of two sample queries are given below.

Query: What are the crops that grow in alluvial soil with temperature above 30°C?

Parser output: What/WP are/VBP the/DT crops/NNS that/WDT grow/VBP in/IN alluvial/JJ soil/NN with/IN temperature/NN above/IN 30/CD °C/NNS? /.
QCT output: crop, alluvial, soil, temperature/ [D], crop, alluvial, soil, temperature/ [I], is /[R_1], above/[R_2] the/[Q], crop, alluvial, soil, temperature/ [M]
Query: What are the causes of stemborer in rice?
Parser output: What/WP are/VBP the/DT causes/NNS of/IN stemborer/NN in/IN rice/NN
QCT output: stemborer, rice/[D], stemborer, rice/[I]

Required information is accessed from knowledgebase using [D] and/ or [I]. If proper output is not retrieved, the parsed tokens are sent to the Wordnet to get the related synsets. Next is to form the query based on the information derived from Wordnet and retrieve information from the knowledgebase. Fifty simple wh queries are formed to evaluate the INRON algorithm and results are compared with SPARQL and GOOGLE search results. Google is a widely used search engine and SPARQL is a W3 recognized semantic query engine. The standard metrics such as Precision, Recall, and F -measure are used for evaluation as given below.

$$Precision = \frac{No.\ of\ correct\ found\ results}{No.\ of\ total\ found\ results}$$

$$Recall = \frac{No.\ of\ correct\ found\ results}{No.\ of\ total\ correct\ results}$$

$$F\text{-}measure = \frac{2 * Precision * Recall}{Precision + Recall}$$

Table 2 shows that the performance of INRON is comparatively high. From the table SPARQL's precision and recall rates are low and same respectively because it retrieves information only when the exact matching between the keyword and the knowledgebase has been done. Google lists more results, so its recall rate is high but precision is slighter low due to the failure in synonym match for certain cases. However, the INRON's retrieval is based on exact keyword as well as synonyms of the keywords from the merged knowledgebase, which enhances the retrieval performance.

Table 2. Performance comparison between information retrieval methods

	SPARQL	GOOGLE	INRON
Precision	53	76	84
Recall	53	100	94
F-measure	53	86.36	88.15

CONCLUSION

Semantic based Agricultural Information System (AIS) framework is designed using the semantic web technologies for the effective integration of heterogeneous information sources and retrieval of proper information from those systems. Heterogeneity between similar systems is resolved using ontology mapping algorithm OMFIM and semantic interoperability is ensured. Ontology merging approach ASDOM integrates heterogeneous sources using mapping relations and Information retrieval algorithm INRON retrieves information from knowledgebase based on user queries. AIS allows information to be integrated and shared across various related applications and disseminates essential information to the user and stakeholders. This system could be further enhanced by associating inference rules and complex queries for effective utilization of available information and visual tracking systems for easy understanding of the illiterate users.

REFERENCES

Bansal, N., & Malik, S. K. (2011). A framework for agriculture ontology development in semantic web. *Int. Conf on Communication Systems and Network Technologies*, 283 – 286. 10.1109/CSNT.2011.68

Berners-Lee, T., Hendler, J., & Lassila, O. (2001). The semantic web. *Scientific American*, *284*(5), 34–43. doi:10.1038cientificamerican0501-34 PMID:11396337

Dasgupta, S., Patel, R., Padia, A., & Shah, K. (2013). *Description Logics based formalization of Wh-Queries.* CoRR abs/1312.6948

Ghidini, C., & Giunchiglia, F. A. (2004). Semantics for abstraction. In *Proceedings of ECAI* (pp. 343-347). Academic Press.

Gruber, T. (1993). A translation approach to portable ontology specifications. *Knowledge Acquisition*, *5*(2), 199–220. doi:10.1006/knac.1993.1008

Hongye, C. (2010). Study of agricultural knowledge service system model based on semantic web. *International Forum on Information Technology and Applications*, 381 – 383.

Junfeng, S., & Yong, L. (2016). Semantic similarity algorithm based on agricultural ontology and its application on crop ontology. *Nongye Gongcheng Xuebao (Beijing)*, *32*(16), 175–182.

Kaladevi, R., Geetha, G., & Narayanasamy, P. (2016). Ontological based interoperability and integration framework for heterogeneous weather systems. *Rev. Téc. Ing. Univ. Zulia*, *39*(1), 185–192.

Kaladevi, R., & Mirnalinee, T. T. (2015). Heterogeneous information management using ontology mapping. *Journal of Engineering and Applied Sciences (Asian Research Publishing Network)*, *10*(5), 2078–2081.

Kashyap, V., & Sheth, A. (1996). Semantic and schematic similarities between database objects: A context-based approach. *The VLDB Journal*, *5*(4), 276–304. doi:10.1007007780050029

Kotis, K., & Vouros, G. A. (2004). The HCONE approach to ontology merging. In ESWS (pp. 137–151). Academic Press.

Lambrix, P., & Tan, H. (2006). SAMBO– A system for aligning and merging biomedical ontologies. *Journal of Web Semantics*, *4*(1), 196–206. doi:10.1016/j.websem.2006.05.003

Liao, J., Li, L., & Liu, X. (2015). An integrated, ontology based agricultural information system. *Information Development*, *31*(2), 150–163. doi:10.1177/0266666913510716

Liu, X., Duan, X., & Zhang, H. (2012). Application of ontology in classification of agricultural information. *IEEE Symposium on Robotics and Applications (ISRA)*, 451 – 454. 10.1109/ISRA.2012.6219222

Manikandan, N., Sakthiganesh, M., & Kumar, P. J., & SenthilKumar, M. (2012). Web based farmers bulletin for agricultural development using PAP Approach. *International Journal of Soft Computing and Engineering*, 2(1), 171–174.

Mao, M. (2008). *Ontology Mapping: Towards semantic interoperability in distributed and heterogeneous environments (Ph.D. dissertation)*. Pittsburgh, PA: Pittsburgh Univ.

McGuinness, D. L., Fikes, R., Rice, J., & Wilder, S. (2000). An environment for merging and testing large ontologies. Academic Press.

Meenakshi, N., Verma, K., Sheth, A., & Miller, J. A. (2007). Ontology driven data mediation in web service. *International Journal of Web Services Research*, *4*(4), 104–126. doi:10.4018/jwsr.2007100105

Mitra, P., & Wiederhold, G. (2002). Resolving terminological heterogeneity in ontologies. *Workshop on Ontologies and Semantic Interoperability*.

Mukhopadhyay, D., Banik, A., Mukherjee, S., & Bhattacharya, J. (2011). A domain specific ontology based semantic web search engine. *7th International Workshop on MSPT*, 1-9.

Muriithi, A. G., Bett, E., & Ogaleh, S. A. (2009). Information technology for agriculture and rural development in africa: experiences from Kenya. In *Conference on International Research on Food Security, Natural Resource Management and Rural Development*. University of Hamburg.

Neha. (2012). *Building crop ontology for farmers* (M.Tech thesis). Banasthali University, Rajasthan, India.

Noy, N. F., & Musen, M. A. (2000). PROMPT: Algorithm and Tool for Automated Ontology Merging andAlignment. In AAAI/IAAI (pp. 450–455). AAAI.

Ogbuene, E. B. (2010). Impact of meteorological parameters on rice yield: An approach for environmental resource sustainability in Ebonyi rice farmland, Nigeria. *Journal of Environmental Issues and Agriculture in Developing Countries*, *2*(2& 3), 103–116.

Patel, M., Koch, T., Doerr, M., & Sinaraki, C. (2005). *Semantic interoperability in digital library systems*. UKOLN, University of Bat.

Raunich, S., & Rahm, E. (2011). ATOM: Automatic Target-driven Ontology Merging. *Proc. 27th International Conference on Data Engineering (ICDE)*, 1276–1279.

Shoaib, M., & Basharat, A. (2010). Semantic web based integrated agriculture information framework. *IEEE Second International Conference on Computer Research and Development*, 285 – 289. 10.1109/ICCRD.2010.33

SPARQL Query Language. (n.d.). Available: http://www.w3.org/TR/rdf-sparql-query/

Sun, X., Zhu, H., Gu, J., Wu, H., & Feng, C. (2009). Research on the semantic web-based technology of knowledge integration for agricultural production. *Sixth International Conference on Fuzzy Systems and Knowledge Discovery*, 2, 361-366. 10.1109/FSKD.2009.58

Swaran, L., Bhaskar, S., Ela, K., Somnath, C., & Raghu, A. (2013). Semantic web query on e-Governance data and designing ontology for agriculture domain. *International Journal of Web & Semantic Technology*, *4*(3), 65–72. doi:10.5121/ijwest.2013.4307

Web Ontology Language (OWL). (n.d.). Available: http://www.w3.org/TR/owl-ref/

WordNet. (n.d.). Retrieved from www.w3.org/TR/wordnet-rdf/

Xie, N., Wang, W., Ma, B., Zhang, X., Sun, W., & Guo, F. (2015). Research on an agricultural knowledge fusion method for big data. *Data Science Journal*, *14*(7), 1–9.

Yunpeng, C., Peng, T., Shihong, L., & Sufen, S. (2010). Study of agricultural search engine based on FAO Agrovoc ontology and Google API. *Conf on World Automation Congress (WAC)*, 439 – 444.

Chapter 5
Processing and Visualizing Floating Car Data for Human-Centered Traffic and Environment Applications:
A Transdisciplinary Approach

Patrick Voland
University of Potsdam, Germany

Hartmut Asche
University of Potsdam, Germany

ABSTRACT

In the era of the internet of things and big data, modern cars have become mobile electronic systems or computers on wheels. Car sensors record a multitude of car- and traffic-related data as well as environmental parameters outside the vehicle. The data recorded are spatio-temporal by nature (floating car data) and can thus be classified as geodata. Their geospatial potential is, however, not fully exploited so far. In this chapter, the authors present an approach to collect, process, and visualize floating car data for traffic- and environment-related applications. It is demonstrated that cartographic visualization, in particular, is an effective means to make the enormous stocks of machine-recorded data available to human perception, exploration, and analysis.

DOI: 10.4018/978-1-5225-5978-8.ch005

INTRODUCTION

Motivation

In our wired world, vehicles have developed into complex electronic systems. As early as 2011 the Canadian Author Cory Doctorow characterized the current status as follows, "We don't have cars anymore; we have computers we ride in" (Doctorow 2011). Cars are equipped with a large number of sensor devices essential for smooth technical operation (monitoring, e.g., vehicle speed or engine revolutions) and environmental monitoring (measuring, e.g., barometric air pressure, ambient air temperature). Such constantly acquired data are indispensable for a growing number of so-called Advanced Driver Assistance Systems (ADAS). Automotive sensor data are a prerequisite for semi-autonomous and autonomous driving systems currently in development and testing.

Existing technical solutions in communications and bandwidth coupled with a constant drop of transmission costs have paved the way for using automotive sensor data in a variety of uses not or loosely linked to car-related applications. The range of applications includes communication models, such as vehicle-to-vehicle (v2v), vehicle-to-infrastructure (v2i), or vehicle-to-x (v2x) communication. In this context, so-called Extended Floating Car Data (XFCD) expand the existing concept of Floating Car Data (FCD). In fact, that XFCD data will play a pivotal role in future data-driven automobile developments. Currently cars are being turned into interconnected sensors that collect and provide a vast amount of mobile data on vehicles, vehicle drivers, routes, traffic, transport infrastructure and the environment. The enormous value of this massive data stock (big data) has only recently been recognized by the industry. Driving ongoing developments such as autonomous cars, (extended) floating car data will have a significant impact on our everyday life.

Objectives

In addition to semantic parameters, automobile sensor data record locational and temporal information. They can hence be considered geospatial data allowing for geographical analysis and visualization. Application of both methods make these machine data accessible to wider human use. Extending the original machine-to-machine data communication and processing, data-to-human communication and filtering opens up a wide and novel range of infrastructure, environmental and planning applications or analysis based on car sensor data. Moreover, this approach helps to extract and utilize relevant and meaningful information for traffic or environmental planning from the ever-growing massive stock of automobile sensor data.

The ongoing research1 discussed in this paper aims at identifying and utilizing the spatial and temporal components of automotive sensor data, in particular, to process, analyze and visualize them for applications in traffic monitoring, driver assistance and selected environmental issues.

Underlying this research is the basic assumption that a human vehicle operator is and remains an integral part of the automobile system and thus is the creator of his/her own spatial mobility. Using machine data for human information, planning and decision support, i.e. for car-to-man communication, we develop a new field of applications that complements the existing realm of machine-to-machine communication.

For that purpose, car sensor data were extracted via the On Board Diagnostics (OBD) interface, a mandatory component of a FCD system, by a smartphone-based procedure. In a preliminary study the data acquired have been processed and visualized for a defined use case (Voland, 2014). In the work presented here, we aim to make available automotive sensor data (XFCD in particular) as geodata facilitating spatio-temporal analysis and geovisualization to aid the analysis of individual driving styles, environmental behaviour, or environmental conditions in a traffic network. We develop a modular process to handle XFCD based on the visualization pipeline. We also investigate whether open-source software tools can be used developed to implement the process for geospatial applications.

In the research context outlined above there are a number of issues that require further investigation not covered here. How and what data can be recorded, processed, visualized, and displayed automatically? How can the data be combined, processed, visualized, analyzed, and interpreted in a logical and productive manner? Which visualization components and strategies of are most effective and how can they be defined, combined, and eventually applied?

Automotive Sensor Data

The recording of traffic data (specifically of automobiles) can be divided into the general recording of traffic and the recording of individual car routes or trajectories through incremental spatio-temporal tracking of a vehicle's position (e.g. by GPS Global Positioning System) in a given traffic network. The acquisition of data on general traffic is mainly effected by the monitoring of traffic flow (e.g. by induction loops, photoelectric barriers, camera surveillance systems) Constant traffic flow monitoring requires a network of localized and, in most cases, costly sensors. Individual FCD are taken by mobile sensors on random trajectories over defined time spans. Given the availability of multitudes of automotive sensor data one obvious option is to examine whether individual FCD can augment and enhance existing stationary traffic flow data. We assume that by merging stationary point and linear mobile data a broader and deeper database will be available promoting traffic-related research, applications and decision making. Of particular interest is which automotive sensor data of the individual vehicle need to be provided compared against the overall traffic system and beyond.

We have pointed you that both FCD and XFCD (Figure 1) allow for the analysis of the spatial and temporal attributes of each data set. Considering the technical implications, the integration of road traffic and car data is feasible.

Floating Car Data (FCD)

FCD record geopositions and timestamps of the semantic parameters incrementally recorded during car drives. The result is a geolocated dataset of individual vehicle operating states and selected environmental data for one individual space-time trajectory within the overall road traffic system. The more FCD are generated, the more the database of the overall data system will grow, the more trajectories can be analyzed, and the more relevant the analysis result will tend to be. In this massive database, single vehicles generating FCD can be identified as samples to access and evaluate the overall traffic situation. The analogy of "corks swimming in the river" adequately describes this principle (Pfoser, 2008a, p. 321).

From a technical perspective, the main component of the FCD system is the so-called On Board Unit (OBU). The OBU is located in the vehicle. It features data binding (e.g. via the mobile phone network GSM Global System for Mobile Communication) and a central system or server to which the information is sent through programmed algorithms. The FCD concept extensively draws on the existing "Floating Cellular Data" or "Floating Phone Data" (FPD) used to collect and transfer data. That is why special equipment or hardware is required. A mobile phone with GPS or mobile phone tracking and data connection is all that is needed for the recording and transmission of data.

Figure 1. Schematic representation of the composition of a FCD and XFCD information

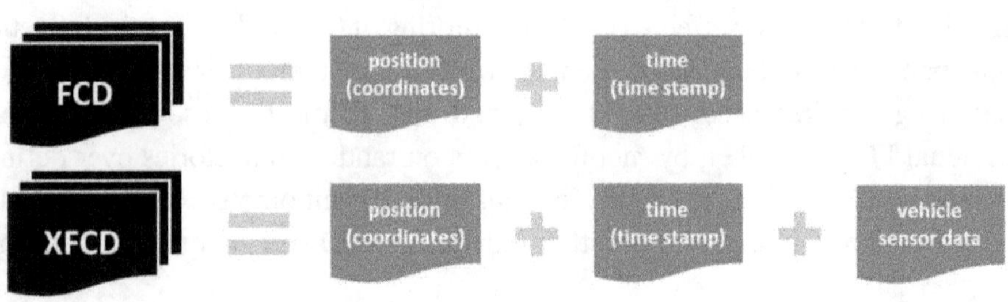

Both FCD or FPD data sets are point data determined by the acquisition at an incrementally defined position of the vehicle (geospatial reference) at an incrementally defined time (time reference). The spatial and temporal components of these data can be subjected to further processing. For instance, travel time and driving progress can be calculated, the velocity of an individual vehicle can be determined as well as individual driving style regarding acceleration or braking. Information derived from the bulk of individual car data can help to predict regional traffic status in an overall traffic system by use of statistical methods.

Based on FCD mass data, the range of applications is manifold (Lorkowski et al., 2003), including:

- Deviation of average velocity
- Current traffic status
- Automated reports of traffic jams
- Dynamic route planning with on- and off-board navigation systems
- Generation of digital road networks

It needs to be noted that a large number of vehicles equipped with FCD systems is mandatory for any reliable automotive sensor database compiled from individual recordings. Vehicle fleets, such as cab or bus fleets provide the required data and are hence often used for traffic data acquisition.

Extended Floating Car Data (XFCD)

As early as 1999 the German automotive company BMW investigated the extension of FCD by on-board sensor data from car electronics permanently recording car aggregates' conditions. "The location and timestamp information are enriched with vehicle status information derived from the in-vehicle bus system or additionally attached sensors" (Schneider et al., 2010, p. 164). These data are labelled XFCD. Along with other car manufacturers BMW produces cars equipped with specific OBUs. In line with the XFCD concept, the data recorded are eventually merged by means of telematics services (under the brand name "BMW Connected Drive"). Without endorsing the commercial specification of BMW it can be stated that XFCD amply demonstrates the potential generated by connecting FCD spatio-temporal parameter recordings with aggregate status monitoring. In fact, all information of the automotive electronics (such as ABS, ESP, rain sensors, etc.) is being used and analysed for different situations (condition of traffic, weather, condition of roads, etc.)" (Halbritter et al., 2008).

Today, modern automobiles are equipped with an ever-growing number of car sensors. It can rightly be said that cars have turned into mobile sensors. The event and status data constantly recorded cannot be used for traffic-related issues, such as traffic monitoring, but also for environmental issues, such as particulate emissions. Moreover, XFCD facilitate the development and implementation of traffic services, such as (Breitenberger et al., 2004). See Table 1.

The range of potential applications can be characterized as follows: "The emergence of real-time location data has created an entirely new set of location-based services from navigation to pricing property and casualty insurance based on where, and how, people drive their cars" (McKinsey, 2011, p. 6). XFCD applications include:

- Traffic monitoring (e.g. traffic intensity and quality)
- Environmental monitoring (e.g. weather events)
- Emissions (e.g. ecological footprint and air quality)
- Driving behaviour (e.g. routes, driving speeds and brake activation)
- Navigation (e.g. depending on traffic or environmental situation)
- Electro-mobility (e.g. route planning and range estimate)

It is obvious that data and application use cases like the ones mentioned do potentially interfere with individual privacy issues. At any rate, potential user groups or individual users can be identified by their respective user profiles. Users include:

- Private users or end-user
- Planners (e.g. traffic and environment planning)
- Managers (e.g. automobile fleet management)
- Decision makers (e.g. politics and administration)
- Industry players (e.g. automaker, insurer)

With the above user groups in mind, further investigations are required how to facilitate human use and analysis of the processed vehicle data. Humans perceive their environment largely with their eyes. They are literally used to read and process

Table 1. Content and services based on XFCD (selection)

XFCD content	XFCD services
• sensory data • data for vehicle operation • traffic situation • beginning and end of traffic jams • cruising speed in traffic jams	• traffic information • environment and weather information • dynamic guidance • travel times and routes • individual consumption analysis (emission and fuel)

space-related data, thereby generating new information for human use. It has been pointed out that both XFCD and FCD data can be classified as geospatial data. Visualization of geospatial data in map form is a common method to provide users with easy-to-read geographic information. Hence the generation of maps from car sensor data is instrumental in making these data available for further spatial analysis. The temporal component of XFCD and FCD data requires advanced visualization to communicate the spatio-temporal development of car sensor recordings on vehicle trajectories in a road network.

One possible use of spatio-temporal car sensor data can be the augmentation of environmental data, such as air pollutants, recorded by fixed stations at fixed time intervals. Merging environmental point data from random trajectories with environmental point data of fixed stations will increase the area coverage of the recorded data, allowing for the application of reliable interpolation methods to complete the area coverage. Another option would be to analyse the different acquisition levels (sensor height above ground) of fixed and mobile sensors. In cars, XFCD or FCD sensors, respectively, record environmental parameters at a height of 0.5 to 1.5 metres above ground, depending on car manufacturers and models. Using GPS positioning technology (GPS height/3D value) the detection height of automotive sensors can be determined. Comparing the data from different acquisition levels can help to study and analyse the three-dimensional change of environmental conditions of an area.

The bulk of permanently recorded automobile sensor data constitute the database for ongoing technology developments, such as semi-autonomous and autonomous driving (however, transmission of data from the system of the vehicle is not mandatory in this case). Potential uses of XFCD go well beyond driverless cars. Technologies, such as "vehicle-to-vehicle communication", enable car-to-car communication of traffic-related warnings, such as traffic jams, accidents or icy roads, based on data recorded by individual vehicles on their trajectories. „The key aspect of XFCD is that they have the potential to indicate hazardous conditions before they turn into real incidents" (Schneider et al., 2010, p. 164). Prerequisites are, i.e. (Stottan, 2013, p. 52):

- Low operating costs (especially communication costs)
- Applicable and scalable anonymization
- Reliable and suitable OBU
- Easy and usable standardisation for data interchange and application

Of these, transparency and data protection (effective data anonymization in particular) as well as standardisation of data models prove to be problematic. To date, the largely proprietary nature of these data is already seriously affecting data exchange and wider data use in industry and science contexts.

Data Acquisition and Processing

The increase of electronic components in automobiles is coupled with a growing number of electronic control units (ECU). In Mercedes Benz E Class models, for instance, we see an increase from 17 components in 1995 (Model W210) to 67 in 2011 (Model BR212) (cf. Schäffer, 2012, pp. 19-22). Each ECU is equipped with an intelligent, software-driven microcontroller designed for a specific task or activity area. The EBU receives signal of the assigned sensors and generates signals for the output device connected with it. The electronic interconnection within the automobile is realised through an on-board power supply. The whole setup can be considered a connected or allocated system of a specialised control device enabling the single control devices to exchange data with each other through a common data bus. Data bus systems can be seen as networks for data exchange, while ECUs can be considered network nodes.

Considerations how to use sensor data from car electronics, such as XFCD (which have been designated as geodata) technically, will have to be put into this context. In this paper, we investigate the complete procedure of data recording and handling, from acquisition in the vehicle to external processing and visualization. To do so, we develop and apply a technical process cycle to augment the data generated in the car with a spatial and time reference. It has to be noted that the technical approach developed can not only be applied to present-generation vehicles but also to older models. To record and capture the data no additional hardware had to be installed in the car. However, reasonably priced hardware is required to analyse the data. Data processing is completely based on free and open-source software components.

Automakers have developed a wide range of tools, such as communication protocols, diagnostic codes or connectors to analyse the vehicle systems and its components, particularly for error analysis in repair workshops. The range of variations results from the different specifications of automakers and proprietary technical transmissions on the one hand and the lack of standards on the other.

In the United States, the California Air Resource Board (CARB) has defined measurement criteria and technical specifications called “On Board Diagnostics” (OBD or OBD II). Technical details, such as form and layout of the diagnostic connector as well as defined diagnosis routines and “Diagnostic Trouble Codes” (DTC), have found international acceptance and were formalized in ISO Standard 15031 “road vehicles – communication between vehicle and external equipment for emission-related diagnostics.” The list of OBD (II) parameters is long, although not all of them are accessible (cf. Schäffer, 2012, pp. 205-224). Of more than 80 defined parameters, less than 32 parameters are regulated by legislature for supervision of exhaust relevant components and systems.

During communication, the OBD-supported information or parameters are called OBD PID (Parameter Identifier). The conclusive identification with PID enables the specific access to sensors or measured data and routines. The query of the respective supported parameter is possible (in blocks) here. Inquiries (always binary) and responses are expressed through the hexadecimal system and need to be converted for further use (cf. ib., 96f) The following table specifies the PID for further investigations as an example (Table 2).

Thus, the mandatory OBD interface is the defined technical interface for automotive electronics. Given that diagnostics hardware is available at reasonable cost, car sensor data can be collected and subjected to further analysis this way. Current mobile devices (smartphones) do provide the required technical features. They are hence suitable to communicate (wireless or wire-connected) with the OBU hardware, process, store and transmit the data via a combination of a web-based or server system. While acquiring the data, space-time attributes can be added to the data by using a GPS-supported smart phone, thus allowing for the generation of XFCD.

The technical solution outlined above is taken as the basis for the development and implementation of a technical process (Figure 2), the main components of which are (1) data acquisition and transmission, (2) server-side data processing and (3) web-based visualization and application (cf. Voland 2014).

Data Acquisition

Acquisition of car sensor data is made possible by equipping the OBD interface of a vehicle with an OBD adapter connected to a smartphone by wireless Bluetooth transmission. Different OBD adapters have been applied. On the smartphone, a mobile application is used for communication with the OBD hardware or interface to retrieve data (at defined time increments, e.g.). To facilitate this process, several software solutions can be resorted to. However, the availability of open-source

Table 2. Selected OBD PID (incl. range and und measuring units)

PID (Hex)	Initials	Description	Min	Max	Unit
0C	RPM	engine revolutions per minute	0	16.383,75	1/min
10	MAF	mass air flow	0	255	g/s
0D	VSS	vehicle speed	0	655,35	km/h
33	BARO	outside barometric air pressure	0	255	kPa
46	AAT	outside air temperature	-40	215	°C

Figure 2. Smartphone supported process for generating of XFCD from OBD data with subsequent web based process and visualization (schematic illustration of the setup)

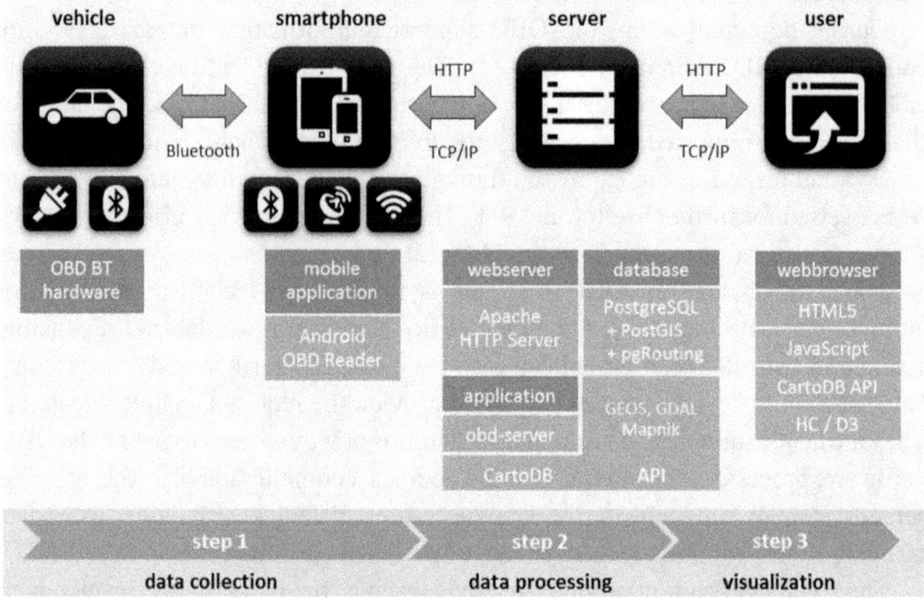

software meeting advanced requirements is very limited. For the Android open-source system, the Android OBD reader (under apache license 2.0, cf. AOBDR 2016) is one of the available software tools. In the framework of the research project discussed here a fruitful collaboration with the above software project has been established to make the recorded OBD data (XFCD) available through a mobile application. This smartphone-based mobile application augments the acquired OBD data with real-time GPS coordinates and a time stamp of the respective acquisition time. The XFCD generated are either cached or transmitted from the automobile (identifiable by its vehicle identification number, VIN) to an internet-connected server. The tasks of the mobile applications can be summarized as:

- Selection of OBD PID, configuration of the interval of inquiries
- Communication with the OBD via (ELM327-based) OBD Bluetooth hardware (sending requests and processing responses)
- Generation of datasets: analysis of values, augmentation of spatio-temporal reference (coordinates, time stamps, spec. through GPS)
- Data transmission to server via existing internet connection or data caching for subsequent transmission (in case of non-availability of internet connection), respectively

In our research, the following setup has been chosen and implemented (Figure 3). We coupled the selected OBD Bluetooth hardware (APOS BT OBD 327) with an OBD-Simulator board (Ozen Elektronik ECUSim mOByDic 1610) which is connected to a smartphone via Bluetooth. The mobile application (Android OBD reader) as well as a GPS simulation application (to simulate movement) is run on the smartphone. The OBD simulator is equipped with an OBD interface designed to simulate (one or more) different data bus systems (e.g. CAN-Bus). It also facilitates the manipulation of (one or more) different OBD PID. The latter can be helpful in the development of the mobile application and to perform tests (before or between using a real vehicle).

Data Processing

In the framework of the open-source project we developed a server-side software component "OBD server" (under apache license 2.0, cf. AOBDR 2016) as a service. This allows for the uploading of data from the mobile application to the server (or the server-side software component). We further use the CartoDB software system to connect the following open-source software components: PostgreSQL/ PostGIS as spatial database system, GEOS (libgeos) and GDAL (libgdal) to process vector

Figure 3. Experimental arrangement for development: OBD-Simulator with connected OBD-Bluetooth-Device and active connection to the smartphone with mobile application "android-OBD-reader"

and raster data, and Mapnik to generate map tiles as well as raster and vector data, and leaflet as a map viewer. CartoCSS is used for the generation of graphical presentations, based on CSS (Cascadian Style Sheets). CartoDB provides a specific API (Application Programming Interface), i.e. a REST API (Representational State Transfer), to access the underlying PostgreSQL/PostGIS database (cf. CartoDB 2016).

The OBD server component is constantly receiving data from the mobile application "Android-OBD reader. Data collected can be reviewed through a soft copy on the server. From the data transmitted to the server database tables are created, as well as a quick clearing of data or data bank. The interchange format we use is JSON. This format is supported by PostgreSQL/PostGIS as well as CartoDB. Data are now ready for further processing including visualization (transformation of vector to raster or image data, respectively) and provision of a web-based application.

Data Analysis

Processing of the collected car sensor data is significant to increase data quality and content. Among the procedures applied the following are briefly dealt with below.

Map Matching

Determining the location of a recorded automobile sensor dataset (by GPS, e.g.) might yield an inaccurate position for several reasons. Hence techniques need to be applied to establish the exact location of a vehicle or its potential location at a specific time (cf. Pfoser, 2008b). This can be achieved by using of Map Matching algorithms (in particular Kalman filtering). These can be applied either during data acquisition, data import or a subsequent processing step.

It goes without saying that traffic flow detection requires precise determination of the vehicle's position. For that purpose, we use the open-source extension pgRouting of PostgreSQL/PostGIS and augment the road network (with or without routing capability) with OpenStreetMap (OSM) Data. We finally implement a map-matching algorithm.

Generation of Trajectories

As has been pointed out above, XFCD are considered point data from a geometric perspective. This allows for the application of a technique to match the recorded point data to the trajectory line on which they are acquired. Point snapping to trajectory lines can be run during the map matching process. Only when this has been carried out, a sufficient number of geocoded data collected from a reasonable number of use cases is available to be subjected to further analysis. Processing is performed using specific algorithms. When it comes to simple processing or construction of trajectories, such as single tracks of a specific vehicle, data storage becomes an issue. In spatial DBS, such as PostgreSQL/ PostGIS, a data table contains data of predefined data types only, e.g. point, line or polygon. To derive trajectories, data are converted from point to line geometry. As the database table requires different types of geometry in this case, a modified setup of the database or the whole database scheme is required as a consequence. To derive areal data from the collected point data and constructed line data, respectively, advance spatial interpolation methods need to be applied which are not dealt with here.

Anonymization

To protect user privacy, data anonymization is an important requirement against the background of the relevant data protection laws. Hence privacy should be ensured in the process chain as early as possible. Depending on the technical execution of this task, this can already be done before data transmission in the vehicle observed or directly after transmission to the server. In any case, data anonymization both for vehicle and driver should be effected before data transmission. It has been shown that trajectories of a particular vehicle can be derived from the available data (in terms of evaluation of movement profiles and determination of the start and end point). Anonymization of trajectories by alteration of individual start and end coordinates or points constitutes a spatial approach to data generation. For this, three strategies can be identified in Figure 4.

One approach for the anonymization of trajectories is the construction of a buffer, i.e. the removal of a trajectory's segment, which resides within a circular buffer area around the start and end point, respectively. Another way is the removal of a

Figure 4. Anonymization strategies (modification of start and end points of a trajectory)

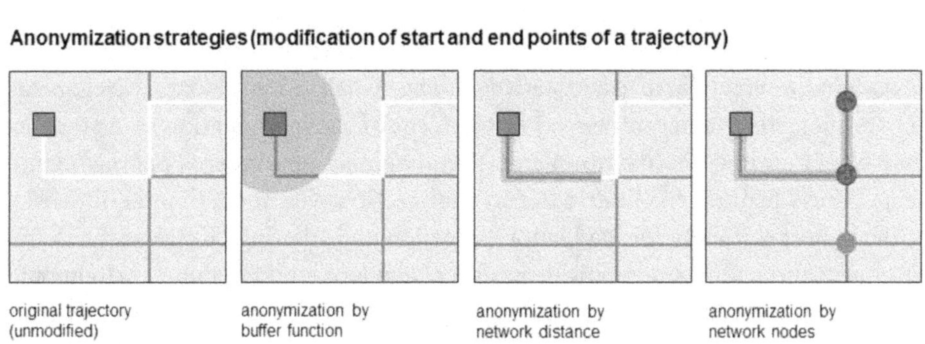

trajectory's segments within a defined distance between start and end point. Since movement and distance covered is determined from the road network, this method can be referred as ‚network distance'. A third approach is the removal of trajectory segments up to a defined number of nodes passed from start to end point. To increase the reliability of anonymization, all three methods should preferably use random values in a defined range for the defined variables. We feel that most applications can be dealt with this way. However, the strategy proposed might not be adequate for every potential use case. Nevertheless, the procedure described represents a practical, geometry-based approach strongly related to the structure of the XFCD. Presently, automobile manufacturers claim an exclusive right to the vehicle sensor data of their cars. A car holder is either unaware of this fact or because of a lack of alternatives is forced to sign an agreement of data transfer when buying a new car. To improve the situation for the car driver/holder, the General Automobile Club of Germany (ADAC) has developed a five-layer model to structure and explain the source and use of data from inside the vehicle system in the context of privacy aspects discussed above.

Visualization

The last stage of the process presented the user is able to utilize the server-based application via web browser (cross-platform). The ‚torque' library, provided by CartoDB, allows for the use of a map-based dynamic display for movement through JavaScript, CartoCSS and HTML5 (spec. HTML5 Canvas). See Figure 5. To facilitate efficient transfer of the massive bulk of data from server to client, a compressed ‚data cube'- format is used as a kind of multidimensional OLAP (online analytical processing) cube or dataset. The edge of this cube is equivalent to the data dimension.

In addition, a complete visualization concept has been developed to fit the specific needs and (visual) analysis requirements for defined application scenarios, such as monitoring of traffic and environment). Visualizations include cartographic presentations, user interface and interaction design. For its generation prototyping techniques, wireframes and mock-ups were used (Figure 6).

The example shown here comprises a one-window view (for map or statistics) and a two-window view (for map and statistics). The visualization features two elements: (1) an interactive temporal legend in the form of the combination of a so-called time wheel or time pie (daytime) and (2) a time diagram (records per day/month/ year). These features facilitate extensive interaction with the temporal dimension of the data specifically for navigation, selection and filtering (Kraak et. al. 2002). They also allow for „geographical, statistical and temporal brushing" (Monmonier 1990). Following Monmonier, this approach integrates the different visualization modes map, temporal legend and statistical chart/ diagram (e.g. scatter-plot) with

Figure 5. Output from the server (top), CartoDB User Interface: database table / “Table View” (middle), CartoDB User Interface: Visualization / “Map View” (bottom).

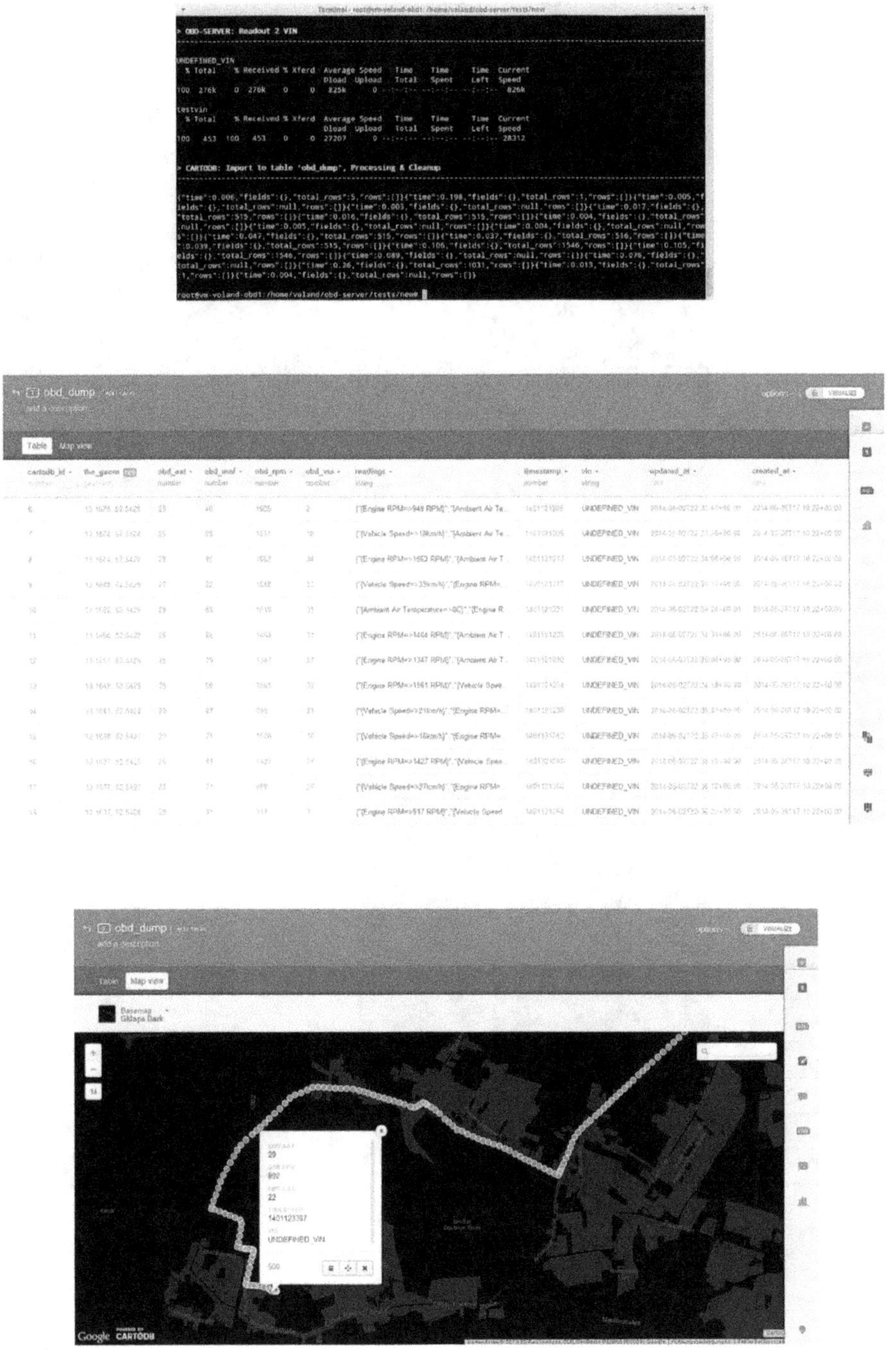

Figure 6. User interface mock-ups: one-window view (top), two-window view (bottom)

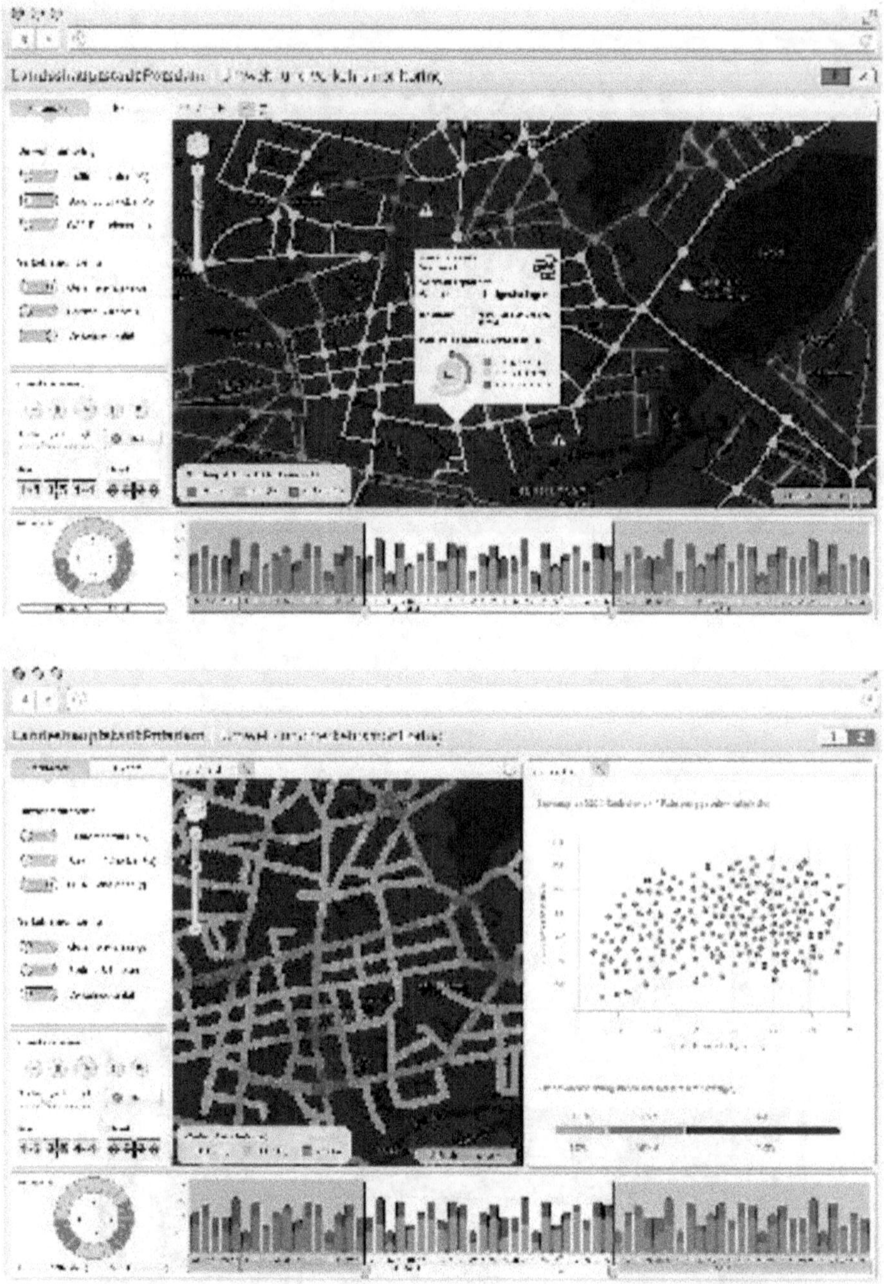

the interaction forms of brushing and linking (focusing). The user is thus put in a position to select, highlight and compare the data in different visualization forms (or data dimensions) through reciprocal interaction between the different interfaces.

Need for Integrated Data Handling

There are two general procedures to acquire car sensor data from a data stock, either by methods of data mining or by calculation of new data values from existing ones. One example of the latter method is the calculation of a correlation between fuel consumption and CO_2 emission. See Figure 7. Depending on the data available (automobile or fuel specific), the calculation is performed by using different formulas with different accuracies. The calculation presented here is based on parameters recorded by the air-mass sensor (MAF), the automobile's speed (VSS) as well as on information of the fuel used. In case the latter information is not available, it can be determined by means of OBD PID 51 FUEL_TYPE. In our calculation, we assume the fuel used is petrol. We can then apply fixed variables: the air volume required to combust 1 litre of fuel (14.7kg) and the average petrol density (740kg/m3) (cf. Schäffer, 2012, p. 183).

Figure 7. Example: Calculation of the current fuel consumption and CO_2 emissions

Sample calculation: current fuel consumption

$$\text{current fuel consumption} = \frac{\frac{\left(\frac{\text{air mess}}{\text{air volume for combustion}}\right)}{\text{density of fuel}} * 3600}{\text{speed}} * 1000 \, \frac{l}{100\text{ km}}$$

Required:
air mass (MAF, PID 10, e.g. 0,2 g/s),
speed (VSS, PID 0D, e.g. 50 km/h),
air volume to combust 1 litre of petrol (14,7 kg),
average density of fuel (740 kg/m³)

$$\frac{\frac{\left(\frac{0{,}2\frac{g}{s}}{14{,}7\text{ kg}}\right)}{740\frac{kg}{m^3}} * 3600}{50\frac{km}{h}} * 1000 \, \frac{l}{100\text{ km}} = 7\frac{l}{h}$$

Sample calculation: current CO_2 emissions

$$CO_2 \text{ emissions (kg)} = \text{fuel cosumption (l)} * \text{emission factor } \left(\frac{kg}{l}\right)$$

Required:
fuel consumption (FUEL_RATE, PID 5E or from calculation, e.g. 7 litres),
vehicle speed (VSS, PID 0D, e.g. 50 km/h),
emission factor (for petrol: 2,874 kg/l)

$$7l * 2{,}7874\frac{kg}{l} = 19{,}51 \text{ kg } CO_2 \text{ emissions (over 7l fuel consumption)}$$

From these parameters, the fuel consumption (so-called current fuel consumption) and the resulting CO_2 emissions of a vehicle at any recorded position and time can be calculated. For that purpose, it is essential to know the specific emission factor, i.e. the relation between the amount of fuel used and the amount of the subsequent emission (CO_2 and equivalents). The adjusted emission factor (CO_3 equivalents included) is specified as 2,874 kg/l for petrol which equals the production of about 2.87kg of CO_2 emission per litre of petrol. The procedure outlined above can be applied to various other calculations, such as average consumption value (based on kilometres covered, trajectories or route, defined time, or aggregated road segments). In addition, all values calculated can be used for statistical analysis. The approach presented was tested successfully with different automobiles, smartphones, and OBD Bluetooth hardware

To be able to produce effective visualizations out of vehicle sensor data for traffic-related and environmental applications, demands and requirements of the targeted user groups have been discussed with informed representatives of the respective user groups. Utilizing the expertise gathered, a user interface and potential cartographic visualizations have been generated in the form of wireframes and mock-ups. These were then assessed by the experts talked to. Their positive response indicates that visualizations based on spatio-temporal XFCD or on geospatial databases augmented with these car sensor data are expedient and desirable for the presentation and analysis of car and traffic-related topics as well as for environmental issues. It has been stated that effective cartographic visualization of these data facilitates decision making processes in the respective application fields.

For technical implementation of the visualization process we adapt the well-known visualization pipeline (Haber & McNabb 1990) which controls the processing of raw geospatial input data to generate a map graphic output through a sequence of processes, namely filtering, mapping and rendering. It is obvious that the graphic representation of data is largely determined by the filtering component (included in XFCD). The mapping and rendering components are driven by rule bases from data-to-graphic conversion (vector data) and graphic-to-image conversion (raster data), respectively. Both rule sets are based on the well-established map modelling principles of thematic cartography.

Applying the visualization pipeline to generate effective map graphics of traffic data can result in recurrent filtering, transformation and visualization tasks. These can be aggregated to patterns comprising a sequence of tasks and managed in a pattern library. Subject to the respective visualization goal, the visualization rule base determines the application and combination of patterns and additional single tasks to be executed automatically. The concept of design or GeoViz patterns (Heidmann 2013) can be adapted for the development of car sensor data visualization patterns. A related approach is suggested by Asche & Engemaier (2012). In order to generate spatio-temporal visualizations based on cartographic methods and principles in a systematic way, the visualization process can be formalized and automated (Figure 8).

We consider the application of rule-based systems an appropriate way to automatically generate effective visualizations. Decision diagrams (as well as their connection and combination) help to formulate and display a set of decision rules required for the visualization process. The combination of selected methods of software engineering, such as design patterns or libraries, service oriented architecture, and model view controller (MVC), and a rule-based approach may contribute to solve the issue. However, such approach is rather complex and will require additional research work. Some relevant aspects are briefly outlined below.

On Conceptual Level

On a conceptual level, it is necessary to identify the tasks and operations to be performed in the visualization process. In loose analogy to the visualization pipeline these can be grouped into functional components, namely data storage (database engine), data filtering (data querying engine), mapping (generation of vector data) and rendering (generation of raster data). The components mentioned can be used as generic building blocks in the visualization process. The component-based approach a can be taken further by developing rule-based visualization patterns. These patterns represent sample solutions for identical or similar visualization tasks. Visualization pattern are stored and managed in a pattern library.

Figure 8. Formalization of map production environment

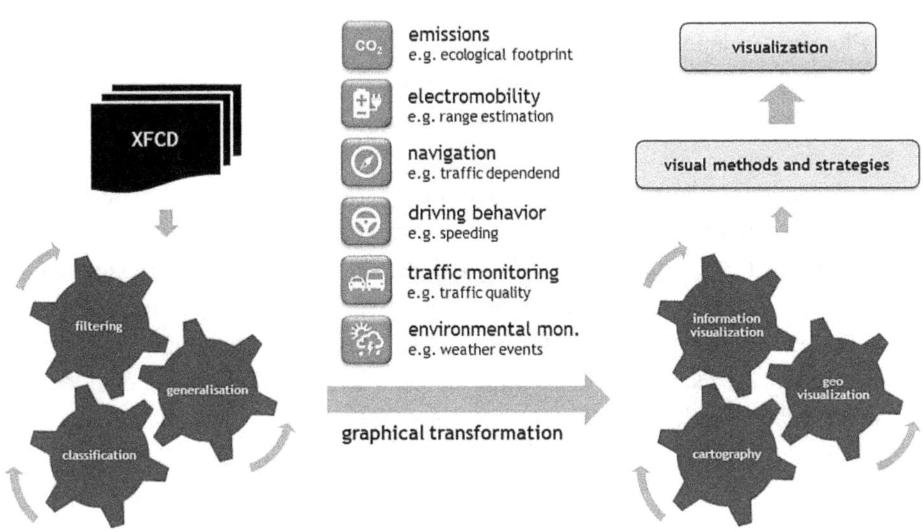

Depending on the visualization solution targeted at, the necessary patterns are combined in an executable process. Such pattern set can be characterized as a specific rule set applicable to a range of similar map visualizations. Different combinations of patterns from the pattern library allow for visualization of a variety of geospatial data in various contexts or use cases.

On Implementational Level

Additional considerations are required to transfer the visualization pattern approach into a technical implementation. The main task is to separate the technical structure into a 'data core' (a raw database) and a topic-specific 'application shell' (analysis and visualization). This facilitates the adaptation of the above process to different application contexts. Due to the functional component structure adapted from the visualization pipeline it is possible to select and integrate any software application (open-source or proprietary) complying to the component requirements (e.g. for raster generation), by encapsulating it into individual, connected services (or modularized architecture, cf. Nadareishvili et al., 2016). This concept draws on the recent software development paradigm of microservices (cf. Newman 2015), an alternative concept to monolithic software architecture. In this context, the conceptual design can directly be used as a blueprint for transformation into a technical implementation.

In line with the selected configuration and under the use of innovative software technologies, such as application virtualization (with application containers, e.g. environments like Docker, LXC), substantial flexibility is possible when it comes to, e.g.:

- Highly distributable operating
- Platform-independent operation
- Vertical and horizontal scalability
- Exchangeability of components involved

Interchange between these (application-container based) microservices is facilitated by standardized interfaces and data formats (like REST, JSON). To connect, control and manage individual services across the complete visualization process workflow, two strategies can be followed: service orchestration (centralized) and service choreography (decentralized). To achieve the advantage in flexibility mentioned above, a service choreography (e.g. with event stream) is considered to be the adequate solution (cf. Exposito & Diop 2014).

CONCLUSION

In this paper, we aimed to present an approach to gather, process and visualize car sensor data (XFCD, FCD) automatically acquired by a vehicle's onboard electronics while driving. We have shown that these data have an implicit spatio-temporal reference in addition to their semantic attributes, thus allowing for geospatial analysis and visualization. XFCD, in particular, provide information on the current environmental situation in urban areas, either recorded directly or calculated during data processing. In this research, the OBD diagnose interface was used to access the data bus system of an automobile. This way, a large number of car sensor data could be collected via a smartphone application and made available to further server- or client-side processing and subsequent analysis. FCD are by now in use for traffic monitoring. XFCD which extend the FCD concept, allow for a wide range of applications, both onboard and offboard. The environmental data recorded by XFCD outside a vehicle can be considered an added value, allowing for spatio-temporal environmental analysis of road networks and related topics, such as various emissions, status of surrounding air quality, temperature of pressure.

Based on these findings we intend to integrate environment-related XFCD into environmental databases compiled from stationary monitoring stations at defined heights above ground. Doing so will augment existing 2D point databases with linear data acquired on trajectories at different, generally lower heights above ground. First, augmentation of the database yields aggregated data with a substantially larger areal coverage, thus allowing for various kinds of geospatial interpolation facilitating better analysis of the (horizontal) areal distribution of an environmental phenomenon. Second, spatio-temporal data recorded at different acquisition heights might allow for 3D analysis of the vertical distribution of an environmental phenomenon in a so-called space-time cube. Both ways will also improve traffic and environmental planning issues and decision making. It has been shown that visualization of floating vehicle sensor data provides access to these data for various audiences by making them visible and thus perceptible to humans.

That is why visualization literally puts man into a hitherto machine-to-machine as an active player. In this way data acquired without our involvement and/or knowledge are returned to the use by the generators (car drivers) or a wider public. As has been mentioned, an important prerequisite is the standardization of data recording and output. For the time being, proprietary solutions exist from different car manufacturers who are thereby enforcing an exclusive access to the XFCD generated by their cars without considering privacy issues. To open the already existing enormous data stocks to wider if not general use, we feel that a car driver, the de-facto generator of the recorded data.

As a minimum, the vehicle or driver specific identifying features should be removed to protect privacy. Further approaches need to be evaluated in this context. It is crucial that data is not privileged to interested industry players, e.g. automobile manufacturer or insurance companies. Instead it is necessary that the driver, as the generator of the data, is not only put in a position to determine the utilization of data but also is able to gain insight into "his" own data. As has been shown, geovisualization can play a helpful role in communicating the information and making it available for further investigation.

ACKNOWLEDGMENT

This research work is part of a PhD project funded by the German Research Foundation (DFG) in the research training group 1539 "Visibility and Visualisation - Hybrid Forms of Pictorial Knowledge" at the University of Potsdam. This support is gratefully acknowledged. The PhD of Patrick Voland project is supervised by Hartmut Asche (University of Potsdam) and Frank Heidmann (Potsdam University of Applied Sciences).

REFERENCES

AOBDR. (2016). *android-obd-reader and obd-server, Github Repository Android OBD-II Reader Project*. Retrieved from https://github.com/pires/

Asche, H., & Engemaier, R. (2012). From Concept to Implementation: Web-based Cartographic Visualisation with CartoService. In *Computational Science and its Application – ICCSA 2012* (Vol. 2, pp. 414–424). Springer International Publishing. doi:10.1007/978-3-642-31075-1_31

Breitenberger, S. (2004). Extended Floating Car Data – Potenziale für die Verkehrsinformation und notwendige Durchdringungsraten. In M. Rohleder (Ed.), *Straßenverkehrstechnik, 10/2014* (pp. 522–531). Bonn: Kirschbaum Verlag.

Carto, D. B. (2016). *CartoDB Github Repository*. Vizzuality Inc. Retrieved from https://github. com/CartoDB/cartodb/

Doctorow, C. (2011). *The Coming War on General Computation*. Presented at 28C3 conference in Berlin, Germany. Retrieved from https://github.com/jwise/28c3-doctorow/blob/master/transcript.md

Exposito, E., & Diop, C. (2014). *Smart SOA Platforms in Cloud Computing Architectures*. Hoboken, NJ: John Wiley and Sons Inc.; doi:10.1002/9781118761489

Haber, R. B., & McNabb, D. (1990). Visualization Idioms: A Conceptual Model for Scientific Visualization Systems. In G. Nielson & B. Shriver (Eds.), *Visualization in Scientific Computing*. Los Alamitos, CA: IEEE Computer Society Press.

Halbritter, G. (2008). Strategien für Verkehrsinnovationen: Umsetzungsbedingungen, Verkehrstelematik, internationale Erfahrungen. Edition Sigma.

Heidmann, F. (2013). Interaktive Karten und Geovisualisierungen. In W. Weber, H. Burmeister, & R. Tille (Eds.), *Interaktive Infografiken* (pp. 66–67). Berlin: Springer-Verlag; doi:10.1007/978-3-642-15453-9_3

Kraak, M.-J. (2002). Cartographic Animation and legends for temporal maps: exploration and interaction. In *Multimedia Cartography* (2nd ed.). Berlin: Springer-Verlag.

Lorkowski, S. (2003). Erste Mobilitätsdienste auf Basis von "Floating Car Data". In *Stadt, Region, Land – 4. Aachener Kolloqium "Mobilität und Stadt" (AMUS)*. Aachen: Institut für Stadtbauwesen der RWTH.

McKinsey. (2011). *Big data: The next frontier for innovation, competition, and productivity*. New York: McKinsey & Company Inc. Retrieved from http://www.mckinsey.com/business-functions/ business-technology/our-insights/big-data-the-next-frontier-for-innovation

Monmonier, M. (1990). Strategies for the visualization of geographic time-series data. *Cartographica, 27*(1), 30–45. doi:10.3138/U558-H737-6577-8U31

Nadareishvili, I. (2016). *Microservice Architecture – Aligning Principles, Practices, and Culture*. Sebastopol, CA: O'Reilly Media Inc.

Newman, S. (2015). *Building Microservices – Designing fine-grained systems*. Sebastopol, CA: O'Reilly Media Inc.

Pfoser, D. (2008a). Floating Car Data. In S. Shekhar & H. Xiong (Eds.), *Encyclopedia of Geographical Information Sciences* (p. 321). New York: Springer-Verlag.

Pfoser, D. (2008b). Map-Matching. In S. Shekhar & H. Xiong (Eds.), *Encyclopedia of Geographical Information Sciences* (pp. 632–633). New York: Springer-Verlag.

Schäffer, F. (2012). *OBD-Fahrzeugdiagnose in der Praxis*. Franzis Verlag, Haar.

Schneider, S. (2010). Extended Floating Car Data in Co-operative Traffic Management. In J. Barceló & M. Kuwahara (Eds.), *Traffic Data Collection and its Standardization* (pp. 161–170). New York: Springer-Verlag. doi:10.1007/978-1-4419-6070-2_11

Stottan, T. (2013). XFCD als Basistechnologie für die Mobilität 3.0 - Entstehung, Entwicklung, Zukunftsanwendungen, Marktentwicklung. In *Schritte in die künftige Mobilität* (pp. 47–59). Wiesbaden: Springer Fachmedien. doi:10.1007/978-3-8349-4308-8_4

Voland, P. (2014). *Webbasierte Visualisierung von Extended Floating Car Data (XFCD) - Ein Ansatz zur raumzeitlichen Visualisierung und technischen Implementierung mit Open Source Software unter spezieller Betrachtung des Umwelt- und Verkehrsmonitoring* (Master Thesis). University of Potsdam.

ENDNOTE

1 Conducted within and funded by the DFG research training group 1539, Visibility and Visualisation - Hybrid Forms of Pictorial Knowledge Retrieved from http://www.sichtbarkeit-sichtbarmachung.de/

Chapter 6
New Design Approach to Handle Spatial Vagueness in Spatial OLAP Datacubes:
Application to Agri-Environmental Data

Elodie Edoh-Alove
Irstea Centre de Clermont-Ferrand, France & Université Laval, Canada

Sandro Bimonte
Irstea Centre de Clermont-Ferrand, France

François Pinet
Irstea Centre de Clermont-Ferrand, France

Yvan Bédard
Université Laval, Canada

ABSTRACT

Spatial OLAP (SOLAP) technologies are dedicated to multidimensional analysis of large volumes of (spatial) data. Spatal data are subject to different types of uncertainty, in particular spatial vagueness. Although several researches propose new models to cope with spatial vagueness, their integration in SOLAP systems is still in an embryonic state. Also, analyzing multidimensional data with metadata brought by the exploitation of the new models can be too complex and demanding for decision makers. To help reduce spatial vagueness consequences on the exactness of SOLAP analysis queries, the authors present a new approach for designing SOLAP datacubes based on end-users' tolerance to the risks of misinterpretation of fact data. An experimentation of the new approach on agri-environmental data is also proposed.

DOI: 10.4018/978-1-5225-5978-8.ch006

INTRODUCTION

Spatial Online Analytical Processing (SOLAP) technologies are dedicated to multidimensional analysis of large volumes of (spatial) data (Bimonte, Boulil, Pradel, & Chanet, 2012). This type of OLAP systems includes spatial measures or spatial dimensions.

Spatial data always suffer from some levels of uncertainty. Not dealing with uncertainty when making high-level decisions based on SOLAP aggregated data increases the risks of data misinterpretations (Gervais, Bédard, Levesque, Bernier, & Devillers, 2009). This leads to faulty trend analysis, missed problems and inexact comparisons between regions or periods. The uncertainty is the result of semantic imprecision, logical inconsistency, temporal incoherence, etc. and/or *spatial vagueness* (Lotfi Bejaoui, 2009). In particular, spatial vagueness refers to a frequent imperfection on boundaries or spatial location of represented geographical objects (e.g. forest, fire, lake). To deal with uncertainty in SOLAP systems, two main approaches are investigated. The first one tries to reduce uncertainty (overabundance of observations to increase spatial precision for example) from the data or to provide decision-makers with visual feedbacks about the uncertainty (Bimonte, Nazih, Kang, Edoh-Alove, & Rizzi, 2013; Lévesque, 2008; Worboys, 1998). The second one proposes to handle uncertainty issues by using new uncertainty-aware spatio-multidimensional models and operators {Jadidi, 2012 #750;Siqueira, 2012 #652} (Jadidi, Mostafavi, Bédard, & Long, 2012; Perez, Somodevilla, & Pineda, 2007; Siqueira, Aguiar Ciferri, Times, & Ciferri, 2012), that are based on the representation of the vague objects with fuzzy or exact models. Nonetheless their implementation is still in an embryonic state.

Motivated by the desire to offer a solution that presents a symbiotic trade-off between the theoretical accuracy on spatial vagueness, the implementation feasibility in current technologies and the usability by intended end-users, we come up with a third approach: instead of dealing with the complexity of manipulating complex vague objects models in SOLAP systems, we propose to manage the risks of SOLAP datacubes misinterpretations, related to spatial vagueness, that the end-users incur. In particular, we are interested in two types of risks of misinterpretation which are the Risk-Geometry (related to the vagueness of the geometric members of the level) and the Risk-Aggregation (risk related to the aggregation formula used to compute measures for a given level).

To do so, we define a new SOLAP datacubes design approach that can take those risks into account during the datacubes modeling process. Such approach leads to the development of a classical SOLAP datacube which not only fits the end-users' usage, but can also be implemented in existing (commercial) SOLAP tools and explored with classical SOLAP operators.

It extends existing methodologies with three main elements. First, our new approach takes simultaneously into account available data sources, end-users' needs and end-users' tolerance levels to "well-identified risks of SOLAP datacubes misinterpretations due to spatial vagueness issues". Second, it delivers to end-users, different versions of SOLAP datacubes (according to their tolerance levels) where the possibility of making erroneous SOLAP analyses is minimized. Third, it enriches the SOLAP datacubes elements with visualization policies to properly communicate risks of misinterpretations to end-users if necessary.

The paper is organized as follows: Section 2 presents a the state-of-the-art on spatial vagueness management in SOLAP systems; motivation of our work using an agicultural case study is presented in Section 3; in section 4, we define and classify the risk of misinterpretation before moving on to defining our new risk-aware design approach requirements as well as the whole new design process proposed in section 5; in section 6 we detail our contributions regarding the risk of misinterpretation assessment and management in the new approach; finally the approach is tested on the case study in section 7.

RELATED WORK

Concepts of Spatial Vagueness

Spatial vagueness is an uncertainty on spatial data that can be categorized into shape vagueness and location vagueness.

Shape vagueness designates an uncertainty on the shape of an object that represents a phenomenon which has parts with broad boundaries (e.g. forests, lakes with islets, flood zones) (L. Bejaoui, Pinet, Bedard, & Schneider, 2009). Location vagueness is either an uncertainty on the position of the spatial object, resulting from a lack of knowledge about the position of an object with an existing sharp boundary, or an uncertainty on the measurement, resulting from the inability to measure such an object precisely (Hazarika & Cohn, 2001; Schneider, 1999).

To address spatial vagueness in spatial data modeling, researchers have investigated the use of vague objects models to represent real world phenomena. The *fuzzy model* (Burrough & Frank, 1996; Dilo, de By, & Stein, 2007; Lagacherie, Andrieux, & Bouzigues, 1996; Schneider, 1999; Wang, Brent Hall, & Subaryono, 1990) is based on the fuzzy set theory, which describes the possibility (indicated by the "membership degree ") that an individual is a member of a set or that a statement is true. *Exact models* were also proposed with a representation of the geographical object by a complex geometry consisting of at least two crisp geometries: one represents the minimum extent (Lotfi Bejaoui, 2009) of the phenomenon (area where the phenomenon is certainly present) also called core (Pauly & Schneider, 2010) or egg yolk (Cohn & Gotts, 1996), and the other represents its maximum extent (Lotfi Bejaoui, 2009) (area where the phenomenon is probably present) (Cohn & Gotts, 1996; Pauly & Schneider, 2010).

Spatial Vagueness Management in SOLAP Systems

Very little researches focus on the integration of vague objects models in SOLAP systems. The authors (Siqueira et al., 2012) have recently proposed an extension of the multidimensional model for taking into account the exacts models of (Lotfi Bejaoui, 2009; Pauly & Schneider, 2010) with specific techniques for storing and querying (vague window query) the vague SOLAP datacubes (e.g. storing separately the core and the conjecture region of a vague object). Yet, they do not offer tools based on a practical implementation of their new definitions and techniques. (Jadidi et al., 2012) propose an algorithmic approach based on fuzzy set theory to integrate vagueness in the decision making process. Concretely, they work out a fuzzy spatial model (fuzzy spatial dimension, spatial fact, spatial measure etc.), a fuzzy spatial aggregation model (fuzzy union, intersection, overlay, etc.) and fuzzy indicators to manage and analyze areas at risk of coastal erosion in SOLAP datacubes. The implementation of the fuzzy spatial datacube is identical to classic datacubes except that we have to add an attribute to support the membership values with regards to the spatial members. Nevertheless, the measures should be calculated with the fuzzy operators instead of classic ones. The implementation of the defined operators in commercial SOLAP clients is yet to be done, and also, the computation of the membership functions requires specific practical tools that are not necessarily available.

As far as we know, research on spatial vagueness introduction in SOLAP datacube visualization is non-existent for now, although SOLAP has been used to visualize spatial vagueness in transactional geospatial data (Devillers, Bédard, & Jeansoulin, 2005; Devillers, Bedard, Jeansoulin, & Moulin, 2007).

In the literature related to the quality of OLAP systems, there have also been proposals dealing with data fuzziness in particular. Some of the authors in this field advocated the use of fuzzy set theory (Delgado, Molina, Sánchez, Ariza, & Vila, 2004; Fasel & Zumstein, 2009; Laurent, 2002) or rough set theory (Naouali & Missaoui, 2005, 2006) to handle the data incompleteness and imprecision. The first ones proposed fuzzy multidimensional database models and fuzzy slices, dices and aggregations while (Naouali & Missaoui, 2005, 2006) contributions involve data approximation techniques by means of rough sets (upper and lower approximations). Also, the authors Pitarch, Favre, Laurent, and Poncelet (2012) exploited the fuzzy logic in proposing new contextual hierarchies to allow the consideration of data experts knowledge for the definition of the hierarchies.

Although it is interesting to integrate vague objects models in SOLAP datacubes, there is still much to do before we can design, implement and exploit these vague datacubes in practice. Existing SOLAP tools (Extract Transform Load tools -ETL, Servers, Clients etc.) and databases (Pauly & Schneider, 2010) only allow the storage and querying of spatial data modeled as crisp entities (point, line, and polygon).

(S)OLAP Datacubes Design Methods

OLAP and Spatial OLAP systems are based on complex multi-tiers architectures where (spatial) data should be extracted from heterogeneous data sources and transformed before being loaded into the datacubes. Since SOLAP datacube multidimensional model represents required SOLAP analyses, the success of the whole SOLAP application project depends upon the correctness of the designed SOLAP datacube(s). In the previous years, several multidimensional modeling methodologies where presented in the literature (Di Tria, Lefons, & Tangorra, 2012; Giorgini, Rizzi, & Garzetti, 2005; Guimond, 2005; Malinowski & Zimányi, 2008; Pardillo, Mazon, & Trujillo, 2010; Prat, Akoka, & Comyn-Wattiau, 2006; Romero & Abelló, 2009, 2011). They offer guidelines, processes and/or tools to designers to help them draw (manually or automatically) a multidimensional intelligence from end-users' requirements (requirement-driven approach) or data sources (sources-driven approach) or a combination of both (hybrid approach; de-facto used for SOLAP systems). It is implemented in a Model Driven Architecture (MDA) method (Glorio & Trujillo, 2008) and/or a rapid prototyping one (Guimond, 2005). The analysis of these methodologies shows that (spatial) data uncertainty issues do not explicitely influence the resulting datacube multidimensional elements definition.

Indeed, on one hand, user-driven methods focus on determining and exploiting users' needs in defining the multidimensional elements. They are not designed to identify and/or consider the data types or uncertainty issues regarding the data to be exploited in the datacubes. On the other hand, the sources-driven methods allow the inventory of attributes present in the sources and their types so that the measures and dimensions can be extracted from the sources; but the methods do not allow considering the attributes content. However, knowing that an attribute holds a spatial vague object is needed to consider spatial vagueness issues soon enough during the design process. Ultimately, most hybrid methods present the same limits since they usually focus on resolving the merging of user-driven and sources-driven results (Bonifati, Cattaneo, Ceri, Fuggetta, & Paraboschi, 2001; Mazón, Trujillo, & Lechtenbörger, 2007). The methods used in practice in Geomatics, advocate the analysis of the sources and their quality according to the needs but in those methods, the quality issues (spatial accuracy, incompleteness, generalization issues (Bédard, Proulx, & Rivest, 2005) etc.) are whether resolved during the ETL process, or during the datacube visualization and exploration. Recently, approaches have been proposed to integrate a risk of misuse management method to the design methods, allowing the producers to identify and assess those risks early on during the design. It is for example the proposal of a collaborative platform for the risks identification and analysis by (Grira, Bédard, & Roche, 2013; Grira, 2014); or the work of (Lévesque, 2008) leading to the proposal of risks of misuse classification and identification formal tools, as well as contextual alerts definition allowing the display of the risks to the end-users during the datacubes exploration, to reduce the risks. (Grira et al., 2013) work is set on risk definition and assessment and can be seen as complementary to our proposal. (Lévesque, 2008) do not address spatial vagueness in particular and also the risks are not exploited in defining the final SOLAP datacubes elements.

Motivation

We consider a stripped down agri-environmental case study on pesticide spreading activities data. We consider the case where the intention is to build a SOLAP application for decision-makers to support their decisional process in the context of the control of surface water contamination by pesticide. The control activity implies the monitoring of the quantity of pesticide that can contaminate surface waters.

To do so, we have on one hand (1) decision-makers analysis needs and on the other hand (2) the available source data.

Regarding Decision-Makers Analysis Needs

Decision-makers expect a SOLAP datacube that will help them visualize and interrogate data related to all pesticide spreading activities in an easy and interactive way. The datacube should allow answering queries such as:

Q1: *"What is the total quantity of pesticide applied per year that can be found within flood risk areas?"*

Q2: *"What is the greatest amount of pesticide applied per year that can be found within flood risk areas?"*

To help decision-makers monitor the quantity of pesticide in spread zones and flood risk areas, the required datacube will need to have a spatial hierarchy composed of at least a Spread Zones level (holding areas where pesticide have been applied), a Flood Zones level (holding flood risk areas), and also a measure that holds the quantity of pesticide spread values (QuantityAsub).

Regarding the Source Data

First, we note that spread zones (defined as areas where pesticide has been applied) and flood risk areas (defined as regions where surface water can be present during an inundation) are vague objects (see sample on Figure 1). More specifically, both spatial objects have vague shapes.

In fact, the spreading activities should be conducted on plots parts defined beforehand as suitable areas. Those areas are surely spread but due to the spreading equipements and technics, pesticide can be found outside those suitable areas. Thus, while spread zones maximal extents are the limits of farming plots (i.e., no pesticide are spread outside the farming plots boundaries), their minimal extents are the suitable areas. Even if the quantity of applied pesticide is exactly recorded for a whole plot, it is not possible to determine with accuracy where that quantity has been applied inside the limits of farming plots. However, in the sources, this vagueness has been neglected and spread zones are represented by polygons covering the whole farming plot.

Concerning the flood risk areas, official data sources provide well calculated geometries representing flood-prone areas. However, during an inundation, surface

Figure 1. Map showing the vagueness on flood risk areas and spread zones

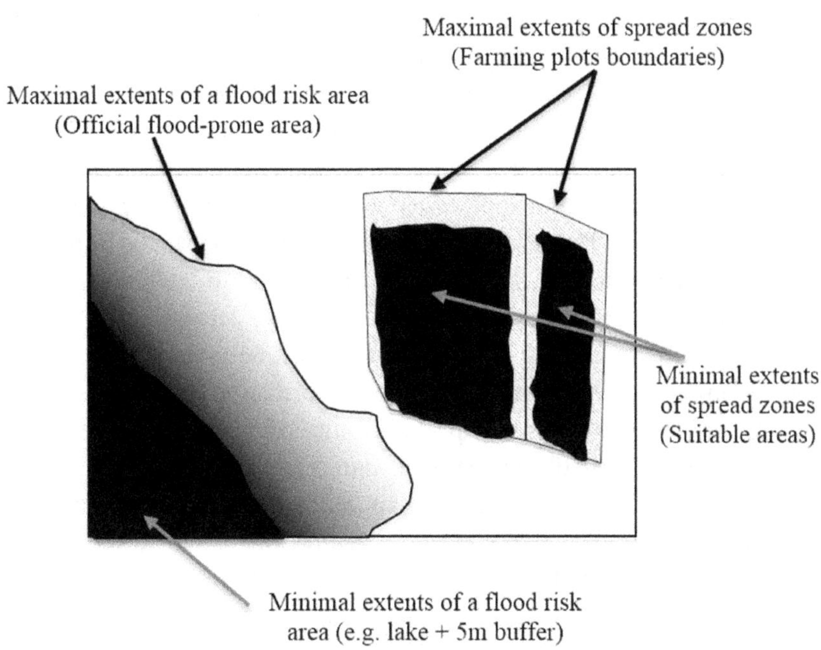

water does not always cover a whole flood-prone area. Therefore the limits of those areas are rather maximal extents of flood risk areas. Meanwhile, areas that are certainly covered by water during an inundation, meaning minimal extents of flood risk areas, correspond to the actual limits of waterbodies such as rivers, lakes etc. plus a little buffer of 5 meters. The latter are stored in spreading activities database as unsuitable areas for pesticide spreading.

Farming plots are areas where agricultural operations are conducted; they are delimited by well-defined cadastral limits. Farms are polygons representing the union of farming plots that belong to the same farm operator. Both geographic objects are spatial vagueness free.

Exploiting the available source data (where spatial vagueness were neglected) to feed the expected datacube, decision-makers will be provided with a datacube where the spatial vagueness is also neglected. Therefore, the results of the queries will be uncertain (e.g. the analyzed value, as well as the members' geometry shown, for a level composed of spread zones), yet end-users may think they are accurate (risk of misinterpretation).

Risks of Misinterpretation: Definition and Classification

In this section which launches our proposal, we answer two key questions related to our approach about the risk of misinterpretation: What does one call risk of misinterpretation? What type of risks of misinterpretation can one identify?

Starting from the risk of misuse definition proposed by (Lévesque, 2008) following the International Organization for Standardization (ISO) risk analysis standards (ISO/IEC-51, 1999), the risk of datacube (measures and level attributes) misinterpretation can be defined as being *the result of combining the probability of occurrence of a datacube misinterpretation with the severity of the potential impact of the misinterpretation*. The risk of misinterpretation is represented by a qualitative value (e.g. low, medium, high), which varies according to the end-users' context of use since both the occurrence probability and impact depends upon this context.

A datacube (measures and level attributes) misinterpretation occurs when a decision-maker makes decisions from faulty information leading to unexpected results. Our interest is towards unexpected results when spatial vague objects are involved. Also, we focus in this paper on the numerical measures misinterpretation.

We recall that in our symbiotic trade-off approach, datacubes geometric attributes are represented with simple polygons to allow an implementation in existing SOLAP systems. Accordingly, to still take into account the vagueness on spatial data, the geometric attributes will contain either the minimal extent or the maximal extent of the vague objects (not both as it is done for the complex vague objects models in (Siqueira et al., 2012)). For example, a flood zone can be considered only in its maximal extent (official flood-prone area), which will be represented by a single polygon, instead of being represented by a combination of two polygons (one corresponding to the official flood-prone area and the other to the lake+ 5m of buffer extent) .

We classify the risks of measures misinterpretation in two main categories, which are intrinsic risks and extrinsic risks:

- Intrinsic risks are usage-independent risks. They are specific to the geometric data exploited in the datacube. They are first and for all potential risks, since a datacube end-user can still decide that their impact is negligible and that they do not represent a real risk for them depending on their usage. For these intrinsic risks, we will use the term *risks of poor measure evaluation*. The intrinsic risks can be:
  - *Over evaluation;*
  - *Under evaluation*;
  - *Non-significant*. There is nothing to report in this case; it happens for example when the vagueness is absorbed by the datacube aggregation structure.
- Extrinsic risks are usage-dependent risks. They are specific to the intended usage and depend upon the user context of application.

In this paper we deal with the intrinsic risks. More specifically, we focus on the intrinsic risks related to the exploitation of spatial level vague geometric members. In other words, our interest is towards numerical measures poor evaluation that can occur when spatial levels of a spatial dimension contain vague geometric members (minimal extents or maximal extents, not both).

Intrinsic risks are concretely induced either by the vagueness on the spatial level geometric members (we call it "*Risk-Geometry*"), or by the aggregation algorithm (we call it "*Risk-Aggregation*").

1. **Risk-Geometry:** When the geometric members of a given SOLAP datacube spatial dimension level are considered in their minimal (i.e. probably smaller than in reality) or maximal extents (i.e. probably larger than in reality), certain types of measures, involving a geospatial parameter related to the members' geometries, are over or under evaluated on that level (e.g. measure "number of kg of pesticide per hectare" which involves the surface of the members' geometries). In that case, the impact is noticeable on the measures values themselves since they vary according to the geospatial parameter. Other measures

not involving geospatial parameters may also be poorly evaluated on that level (e.g. number of kg of pesticide per spread zone). In this case, even if the value is computed with the highest accuracy, end-users can still misinterpret that.

For a detailed example, let's say that we record the exact quantity of spread pesticide as being 68kg in a given spread zone. The zone, which is geometrically vague as explained in section 3.(2), is considered in its maximal extents in the datacube. In such case, we do not know exactly where those 68kg have been spread inside the limits considered. An end-user presented only with the maximal limits and the quantity recorded may think that the quantity has been spread uniformly in the whole extent while it is only in sub-parts of that extent. Some decisions he will make based on such pretty sensitive information may lead him to unexpected results, which means he is exposed to risks of misinterpretation. It is important to qualify that risk to take it into consideration when designing the datacube, especially the spatial data visualization policies. For this example in particular, we can qualify that as a risk of under evaluation of the quantity: in fact, the quantity would have been greater for the maximal extents if the spreading where uniform on the whole geometry as unaware users may think.

2. **Risk-Aggregation:** It is a risk of misinterpreting the measures' values, related to a given level, risk that is not induced by a potential vagueness on the members' geometries themselves. Instead, it can be induced by:
    a. **The Aggregation Formula and/or the Geometries Involved in the Aggregation:** For the pesticide case above (see section 3), let us consider the low level Spread Zones (maximal extents) and the level Flood Zones (minimal extents). Only the level Spread Zones members that intersect a Flood Zone member are participating in the aggregations for that member. For example, to compute the total quantity of pesticide to be found a flood zone FZ1, one option would be to "sum" the weighted values related to the results of the "intersection" between FZ1 geometry and level Spread Zones members' geometries (see Figure 2). Since the flood zone is considered in its minimal extents, spread zones that only intersect the maximal extents of flood zones are totally left out of the aggregation. Also, the portion of spread zones that actually intersect the flood zone geometry is the smallest. Since the real limits are somewhere between the minimal extent and the maximal extent, the aggregated values present an uncertainty that is a source for risks of poor evaluation (see Table 1). In general, the uncertainty importance depends on the spatial predicates

Figure 2. Schema illustrating the aggregation on level Flood Zones

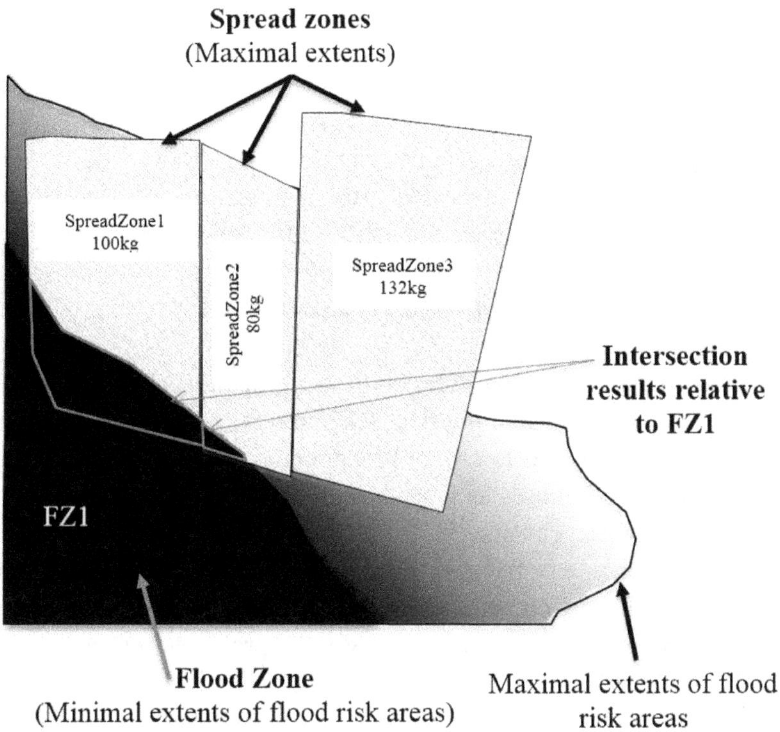

Table 1. Example of aggregation results for flood zones

Flood risk Areas	Spread Zones Considered	Aggregation Results	Risks of Misinterpretation
Minimal extents (FZ1)	SpreadZone1 SpreadZone2	Q1: 23kg Q2: 20kg	Under evaluation
Maximal extents	SpreadZone1, Spreadzone2 SpreadZone3	Q1: 240kg Q2: 98kg	Over evaluation

(e.g. intersect, touch, contain) used to select the lower level members that participate in the aggregation for a given higher level member, and the aggregator (e.g. sum, max, average).

b. *Existing uncertainty on the measures' values to be aggregated*: In fact, aggregations can simply propagate an already existing uncertainty on the measure's values for a lower level to a given level leading to a risk of misinterpretation on that level.

We designate by "*Risk-Level*" the risk created for a given level. It is a Risk-Geometry, a Risk-Aggregation, or a combination of both.

The Risk-Aware Design Approach

In this section, we present the requirements of the new risk-aware design approach, followed by the design process advocated by the approach, and then the new concepts and techniques that are needed to accomplish the different steps of the design process.

The Risk-Aware Design Approach Requirements

The design approach must not only render the resulting SOLAP datacubes Platform Independent Model (PIMs) and physical schemas, but also the aggregation rules, and the visualization policies to use for proper communication regarding the risks of misinterpretation. All those elements must be implementable with existing tools (Spatial Data Base Management System-DBMS, SOLAP server and client). Ideally, the design process should be based on an agile method. In fact, the design process, especially the transformation part, needs to be iterative to allow returns to the key steps of the process at any time during the SOLAP project to refine the design.

The Risk-Aware Design Process

To address the above described requirements, our main idea is replacing vague objects models handling with induced risks of datacube (data and analysis) misinterpretation management, i.e. delivering datacubes (PIMs) compliant with decision-makers' tolerances to risks.

To come up with the risk-aware design process, we extend the classic SOLAP datacube design process with the risk management method advocated by (Gervais et al., 2009; Lévesque, 2008). By classic SOLAP datacube design process we mean the following steps: Requirements specification (Computation Independent Model-CIM design in MDA method), conceptual design (Platform Independent Model-PIM design in MDA method), and finally the logical and physical design (Platform Specific Model-PSM design in MDA method). More specifically, we will present here new steps for the requirements specification and conceptual design phases. The obtained PIMs will be used to design the PSMs in a classic way.

In the rest of this section, we will first make an inventory of the type of actors involved in our risk-aware design process, followed by the description of all the steps advocated by our new process.

Actors Involved in the Risk-Aware Design Process

We consider two main profiles of actors involved in this design process: (Actor profile 1) geospatial systems and data users that are the end-users of the SOLAP application (decision-makers) and systems and data sources end-users such as application domain experts (farmers for example) who actually know, consciously or not, about the data sources quality; (Actor profile 2) the SOLAP experts who have the ability to design and implement the SOLAP application.

Delegates of those actors are chosen to form a project committee. They work together through the majority of the steps of our risk-aware design process. However, the datacubes schemas elaboration and tailoring to tolerance levels are done by the SOLAP experts (profile 2), while the tolerance levels assessment is solely done by the decision-makers (profile 1).

The Risk-Aware Design Process Steps

Our Risk-Aware Design Process has two phases: The requirements specification phase (see Figure 3) and the conceptual design phase (see Figure 4). Each phase includes new steps (in white) that did not exist in the classic design process.

Concerning the requirements specification phase (Figure 3), it starts with an end-users' identification (*Identify datacube end-users*), followed by an available data sources identification (*Identify available data sources*), all done by the profile

Figure 3. Risk-aware design process: Requirements specification phase

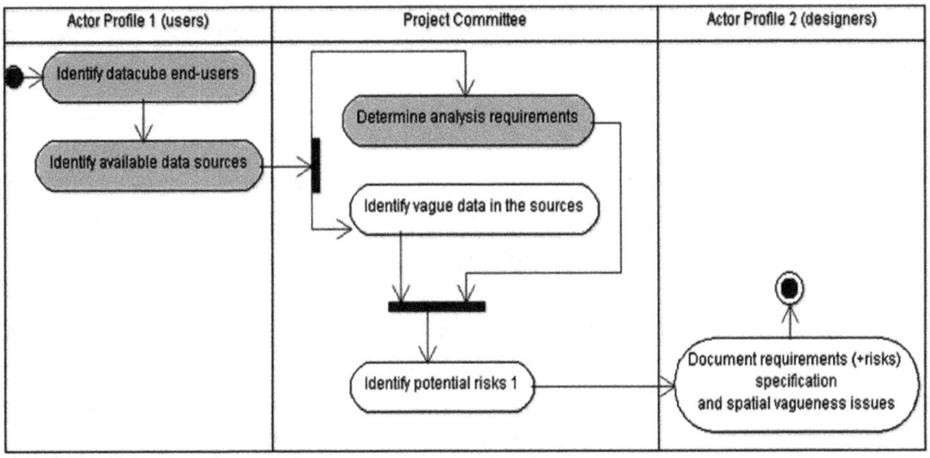

Figure 4. Risk-aware design process: Conceptual design phase

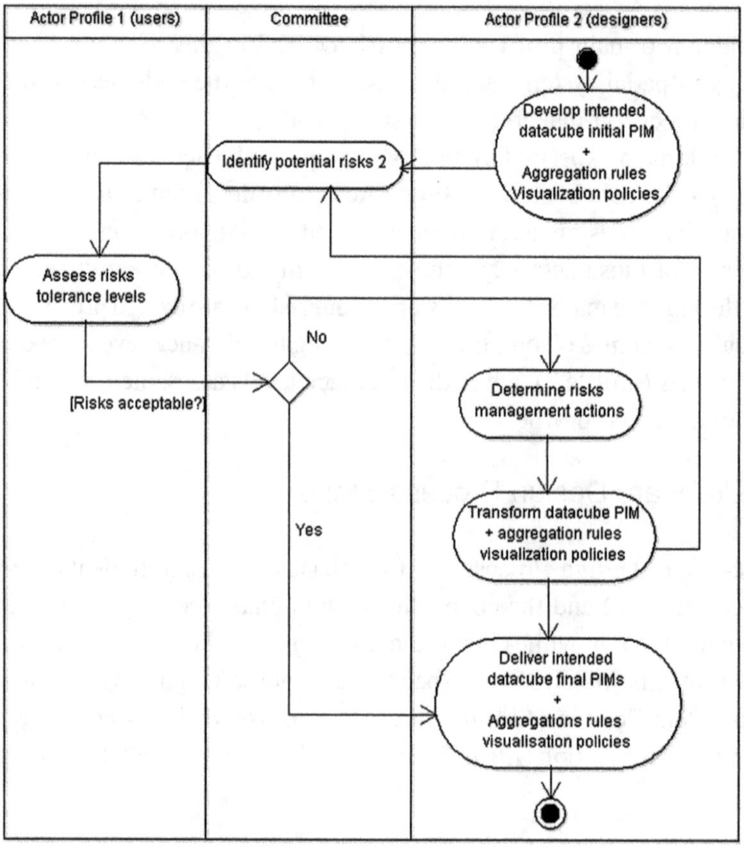

1 actors. With the results of those two steps, the project committee will work out the analysis needs specification (*Determine analysis requirements*). Then we have the new steps:

- **Identify Vague Data in the Sources:** This new step has to be done after the classic data sources identification (*Identify available data sources*). It consists of determining whether sources spatial objects present spatial vagueness issues and monitoring which objects are vague. This will be useful for the tagging of spatial vagueness on the SOLAP datacube multidimensional schema and therefore for intrinsic risks identification later.
- **Identify Potential Risks 1:** A first risks identification is done here by the project committee members. It is done by exploiting the analysis requirements and the spatial vagueness issues identified ("Risk of wrong interpretation of measures associated to flood zones" for example). It gives the committee a

succinct idea of the potential risks of misinterpretation (intrinsic or extrinsic ones). After that, the requirements specification is documented, as well as spatial vagueness issues on the sources and risks identified (*Document requirements (+risks) specification and spatial vagueness issues*), on a CIM for example. That closes the requirement specification phase. One can then move on to the actual conceptual design stage (Figure 4), beginning with the elaboration of an initial intended SOLAP datacube PIM (*Develop intended datacube initial PIM + Aggregation rules Visualization policies*). After that, the actual risk-aware design activities start (new steps).

- **Identify Potential Risks 2:** In this step, identified risks are updated by the project committee, knowing exactly the elements defined in the multidimensional model (PIM design results) and the vague spatial data SOLAP datacube end-users will have to deal with in the intended datacube analysis. For example, the "risk of under evaluation of QuantityAsub on level Flood Zones" can be added to the list.
- **Assess Risks Tolerance Levels:** Here, the project committee members representing the SOLAP application end-users (decision-makers) are asked to express their tolerance level to each identified risk (e.g. Risk of under evaluation of QuantityAsub on level Flood Zones: Totally Unacceptable i.e. tolerance level = 0). The possible tolerance levels are detailed in the following section 6. If all the identified risks are acceptable by them, the current SOLAP datacube PIM is provided (*Deliver intended datacube final PIMs + Aggregation rules Visualization policies*) and it is the end of the design process. If not, we proceed to the following steps.
- **Determine Risk Management Actions:** According to end-users' tolerance levels, different strategies, and thus actions, are defined to reduce the risks of misinterpretation. One action would be for example to remove a level with vague members to avoid the risk of under evaluation of measures values for that level (case of level Spread Zones).
- **Transform Datacube PIM + Aggregation Rules Visualization Policies:** Here risks management actions are applied to obtain appropriate SOLAP datacube's PIMs and corresponding aggregations rules and visualization policies. SOLAP experts can decide to only apply some of the identified actions. For example, they can only apply actions that have an impact on the multidimensional structure and/or on other identified risks at first. Then, the process goes back to the *Identify potential risks 2* step. If the updated risks regarding the produced datacube PIMs are acceptable for the end-users, the design process is over and the current datacube PIMs are supplied (*Deliver intended datacube final PIMs + Aggregation rules Visualization policies*); if not, the process goes on again.

The most important question left unanswered now is how end-users can actually assess tolerance levels and how designers can choose risks management strategies and actions. The following section presents the proposals in this regard.

Risks of Misinterpretation Assessment and Management

Before determining their tolerance to an identified risk, datacube end-users representatives need to evaluate that risk first. In a standard manner, risks evaluation is done by classifying the risks using the Kerzner (2006) risk degrees matrix where the axes are respectively a risk probability of occurrence (High, Medium or Low) and its gravity (High, Medium or Low). According to Kerzner (2006), the risk evaluation step is usually the longest, complex, but also most important one in a risk management process. Here, we have defined a risk degree scale composed of five main levels which should be enough to evaluate the risks for a conceptual design purpose. This risk degree scale is a simplification of the Kerzner risk degrees matrix.

Those five levels are:

- **Very High Risk:** Probability of occurrence of the risk is High and the severity is High (HH),
- **High Risk:** The risk probability of occurrence is Medium and its severity High (MH) or the probability of occurrence is High and Severity is medium (HM),
- **Medium Risk:** The probability of occurrence is Medium and the severity is Medium (MM),
- **Low Risk:** The probability of occurrence is Low and Severity is Medium (LM) or Probability of occurrence is Medium and Severity is Low (ML),
- **Towards Zero Risk:** Probability of occurrence is Low and Severity is Low also (LL).

Logically, the end-user evaluates the risk level according to his application context. Depending on the risk degree, he can then decide whether he can tolerate a risk or not. He can also decide if he can tolerate that risk directly based on the application context. What matters for our approach is the tolerance level (a qualitative parameter) that the user expresses relative to the risk, no matter at which degree he evaluates the risk beforehand (it's his in-house arrangements).

We define 4 tolerance degrees: 0 for totally unacceptable, 1 for preferably unacceptable, 2 for somewhat acceptable, 3 for totally acceptable (see Table 2).

Following risks evaluation, different mechanisms should be executed to manage risks according to the levels of tolerance expressed. According to Gervais et al.

Table 2. Risk tolerance levels and risks management actions

Risks level (Probability X Severity or Vice Versa)	Tolerance Level	Risk Management Strategies	Possible Actions	
			Impact on Multidimensional Data Structure	No Impact on Multidimensional Data Structure
Very High Risk (HH)	0: Totally Unacceptable Risk	Avoidance	Delete the risk source: dimensions, levels, members	Delete the risk source: aggregation rules
High Risk (HM)	1: Preferably Unacceptable Risk	Control	Modify members, levels, dimensions;	Add/Modify aggregation rules Prohibit some combinations, define access policies; List the recommended and non-recommended usages in a metadata support; communicate the risk if considered necessary.
Medium Risk (MM)	2: Somewhat Acceptable Risk	Transfer		Communicate the risk
Low Risk (LM)	3: Totally acceptable Risk	Indifference	No action	
Zero Risk (LL)				

(2009), there are different ways to cope with identified risks: avoidance, control, transfer or indifference. The *risk avoidance* aims at reducing an unacceptable risk by eliminating the source from which it emerges, the *risk control* aims at reducing the risk by taking preventive actions, the *risk transfer* aims at reducing the risk by transmitting it to a third party such as the end-users or an insurance company and in the *indifference* strategy, the existence of the risk is acknowledged without taking any specific action to reduce it.

Moreover, the ISO/IEC-51 (1999) guidelines advocate a risk reduction process that one should follow when choosing a risk reduction strategy. What is noted is that the risk reduction must be assumed by the SOLAP datacube producer first before transferring it to end-users if necessary. That means avoidance and control must be prioritized by the datacube producers. Also, datacube producers must do internal prevention first, then additional protection activities and finally security related information communication.

In our approach, datacube producers are the members of the project committee (the SOLAP experts and delegates of users); delegates of decision-makers in particular are the one deciding if a risk is acceptable or not, using our proposed tolerance level scale.

With all that in mind, we establish a one-to-one relationship between each tolerance level and a risk management strategy by issuing the following hypothesis:

- **Tolerance 0 (Totally Unacceptable Risk): Avoidance.** Eliminating a risk source implies doing more or less important internal changes on the SOLAP datacube, so we recommend it for unacceptable risk. Reciprocally, since the risk is totally unacceptable, we find it appropriate to completely avoid the source.
- **Tolerance 1 (Preferably Unacceptable Risk): Control.** The control is suggested here not only because it is the appropriate internal prevention strategy to reduce a risk to an acceptable level, but also because as suggested in the ISO/IEC-51 (1999) guidelines, datacube producers must prioritize control over transfer.
- **Tolerance 2 (Somewhat Acceptable): Transfer.** It is the preferred strategy here because since the risk is somewhat acceptable, we find it appropriate to leave the decision to use the data or not up to the actual end-user. The transfer can be done by properly communicating the risk itself to all end-users so they will make the decision depending upon their particular use case and tolerance.
- **Tolerance 3 (Totally Acceptable): Indifference.** Since the risk is totally acceptable, there is no need to make a particular effort to reduce it.

Even though for each risk reduction strategy, different mechanisms can be applied either before (during data collection, spatial ETL or datacube design) or after the datacube utilization to cope with the risks, in our work we are interested in actions that can take place during the datacube design stage. We distinguish two categories of such actions:

- Actions that change the multidimensional data structure: they are actions that can be taken on the datacube dimensions, hierarchies, aggregation levels, members and choice of measures;
- Actions that do not change the multidimensional data structure: they are actions to apply on the aggregation rules, the visualization policies, etc. without changing the multidimensional structure of the datacube.

Those mechanisms are classified according to the risk management strategies as presented in Table 2. The list is not exhaustive, and furthermore, this classification can be used in the future to identify what type of actions one can define to handle the different cases of vagueness presence in datacubes (in measures only, in measures and spatial dimensions etc.)

Application of the Risk-Aware Design Approach to Our Case Study

In this section, we propose to perform the design of the SOLAP datacube of our case study by adopting our risk-aware approach. This is done to illustrate the new approach.

Requirements Specification Phase

We recall that the datacube end-users are decision-makers in the environmental field (*Identify datacube end-users* step) and the available sources are the pesticide spread activities GIS containing spatial data listed in section 3 (*Identify available data sources* step). Also, end-users analysis requirements were presented in the same section (*Determine analysis requirements*).The vague data in the sources are: the area of applied pesticide and flood risk areas (*Identify vague data in the sources* step). The result of the step "*Identify potential risks1*" is shown on Table 3.

Conceptual Design Phase

Develop intended datacube initial PIM + Aggregation rules Visualization policies step: Analysis requirements are expressed on the initial datacube multidimensional model (knowing the available sources) showed on Figure 5, and the aggregation rules are shown on *Figure 6*. This multidimensional model is defined using the UML profile for SOLAP datacubes modeling presented in (Boulil, Bimonte, & Pinet, 2011).

This model has: (1) a spatial dimension "Zones" containing regions (represented by simple polygons) on which the quantity of applied pesticides is monitored; the regions are organized according to the hierarchy "Spread Zones → Farms → Flood Zones"; (2) a temporal dimension "Date" with the aggregation levels "Day→Month"

Table 3. Results of « Identify potential risks 1 »

Spatial Object	Definition	Identified risk
Spread zone	Region of applied pesticide (Represented by simple polygons)	End-users may think that every part of the polygons has been spread uniformly with pesticide, which is not the case (spread zones have vague shapes in reality).
Flood risk areas	Region where water can be present in case of inundation. (Represented by simple polygons).	End-users may think that the limits recorded are exact but actually the flood risk areas have broad boundaries.

Figure 5. Intended datacube multidimensional model

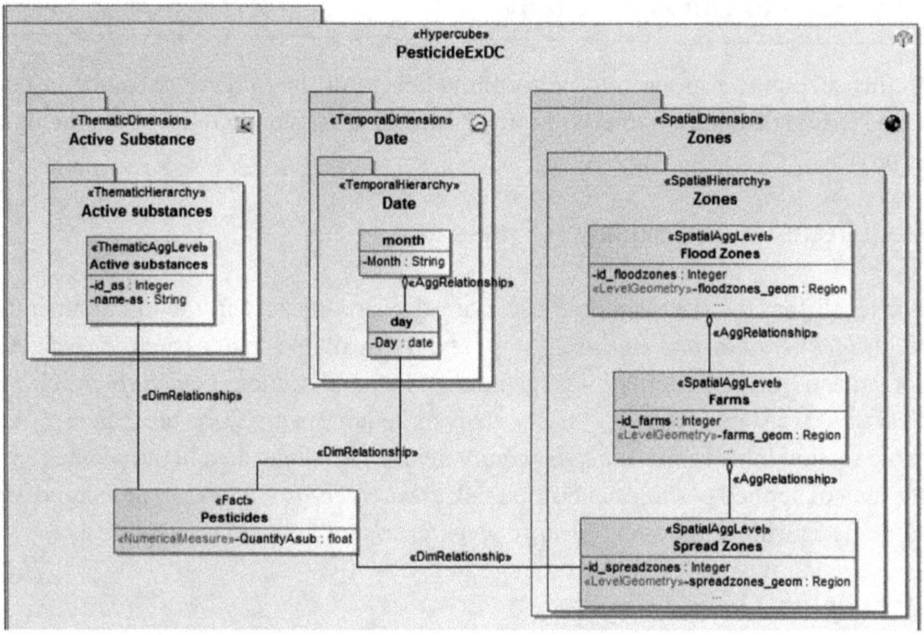

and (3) a thematic dimension "Active Substances" (contained in the pesticides) with the unique level "Active substances".

The spatial data to be used in the spatial hierarchies are the spread zones (for level Spread Zones), farms (for level Farms) and flood risk areas (for level Flood Zones). As explained in section 3.(2), spread zones and flood risk areas have vague shapes in reality. However in our symbiotic trade-off, only simple polygons are used to represent the spatial objects, vague or not. Thus, for the vague spatial objects, the actual geometries that will be stored in the implemented datacube later would be their minimal or maximal extents. In summary, for this initial datacube multidimensional model, we have identified for each spatial level, the corresponding spatial data and the corresponding geometry that is to be stored later in the SOLAP system (see Table 4).

The expected fact (Pesticides) is described by the measure QuantityAsub that represents the quantity (in Kg) of applied active substances. QuantityAsub will be aggregated along the hierarchies with the aggregation operation Sum to calculate the total quantity of applied pesticide (Top image of *Figure 6*), or with the aggregation operation Max to calculate the greatest quantity spread (Bottom image of *Figure 6*).

With this multidimensional model, it is actually possible to answer the SOLAP queries Q1 and Q2 (cf. section 3.(1)). Note that Q1 and Q2 expressions become:

Table 4. Dimension "Zones" spatial levels description

Spatial Level	Spatial Data	Geometry to Be Stored in the Datacube
Spread Zones	Spread zones as defined in section 3.(2).	Simple crisp polygons corresponding to the Maximal extents of spread zones as shown on Figure 1 (farming plots boundaries)
	They have vague shapes.	
	Ideally represented by vague complex geometries as shown on Figure 1	
Farms	Farms as defined in section 3.(2)	Polygons corresponding to Farms boundaries
	They have well-defined cadastral limits.	
	Represented by simple crisp polygons.	
Flood Zones	Flood risk areas as defined in section 3.(2)	Simple crisp polygons corresponding to Minimal extents of flood risk areas shown on Figure 1 (water bodies boundaries + buffer)
	They have broad boundaries	
	Ideally represented by vague complex geometries as shown on Figure 1	

Figure 6. Intended Aggregation rules

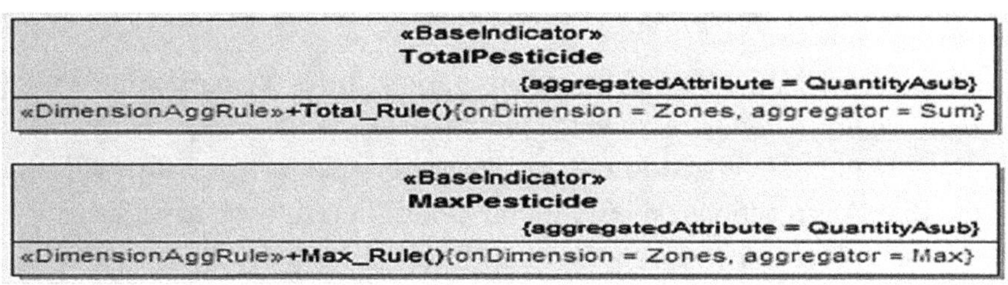

Q1: *"What is the **TotalPesticide** of **Active Substance** per **Year** for **Flood Zones**?"*
Q2: *"What is the **MaxPesticide** of **Active Substance** per **Year** for **Flood Zones**?"*

The visualization policies are at this point the classic ones (plain map and plain cells in pivot table).

- **Identify Potential Risks 2:** The Table 5 presents a summary of all the identified risks for the pesticide intended datacube.
- **Assess Risks Tolerance Levels:** Using our tolerance level scale, end-users delegates have expressed their tolerance levels to the identified risks as shown in Table 6.
- **Determine Risks Management Actions:** According to the tolerance levels, SOLAP experts have chosen appropriate actions among the possible actions presented in Table 2. See Table 6 for the actions.

Table 5. Risks identified for the Pesticide intended datacube

Levels	Risks	Risks Descriptions
Spread Zones (Maximal Extents)	Risk Geometry (*rgSpreadZones*)	Under evaluation
Flood Zones (Minimal Extents)	Risk Aggregation 1 on query Q1 (*raggFZ1*)	Under evaluation
	Risk Aggregation 2 on query Q2 (*raggFZ2*)	Under evaluation

Table 6. Risks + Tolerance + Actions for the intended Pesticide datacube

Risks	Tolerance Level	Actions
rgSpreadZones	0: totally unacceptable risk	Delete Spread Zones level (see Figure 7)
raggFZ1	1:preferably unacceptable risk	Modify the members geometries (use the maximal extents)
raggFZ2	1: preferably unacceptable risk	Modify the members geometries (use the maximal extents)

- **Transform Datacube PIM + Aggregation Rules Visualization Policies:** *Between the actions* chosen in the previous step, SOLAP experts have chosen to apply the one related to rgSpreadZones first, knowing that it will impact the Risk-Aggregations on level Flood Zones. In result, we have the new PIM shown on Figure 7 (the Spread Zones level has been deleted).
- **Return to "Identify Potential Risks 2" Step:** The risks are updated (see first column of Table 7 and then reassessed (see column New Tolerance Level).

For this datacube, the new actions will simply be risks communication through a visualization policy: color the pivot table cells in red for Flood Zone level. After applying these actions, SOLAP experts go back to the risk identification then the risk assessment steps. The risks are raggFZ1 and raggFZ2 still, and the tolerance levels are still 2 for both.

Since all the risks are acceptable now, the datacube PIM shown on Figure 7 is delivered plus the visualization policy defined previously.

CONCLUSION AND FUTURE WORK

In this paper, we have proposed a new SOLAP datacube design approach that takes into account SOLAP datacubes risks of misinterpretation and end-users

Figure 7. Example of final datacube schema: Tolerance 0 to Risk-Geometry on Spread Zones (under evaluation)

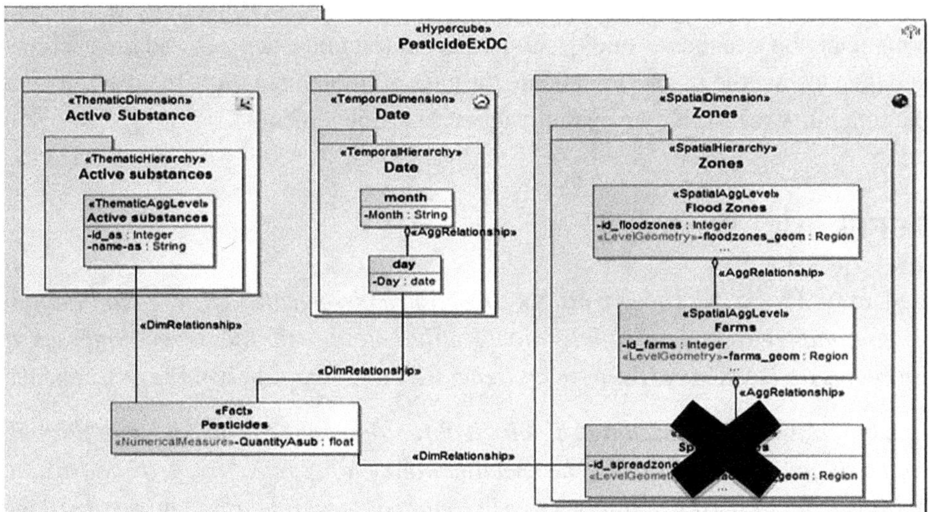

Table 7. Risks update and reassessment results

Risks	Old Tolerance Level	New Tolerance Level	Remarks
rgSpreadZones	0: Totally unacceptable risk		Does not exist anymore (level deleted)
raggFZ1	1: Preferably unacceptable risk	2: Somewhat acceptable risk	It is now a risk of over evaluation since the aggregation is supposed to take into account all the farms (new lowest level) that intersect the flood risk areas in their maximal extents.
raggFZ2	1: Preferably unacceptable risk	2: Somewhat acceptable risk	Same as for raggFZ1

tolerance levels to those risks. First we have defined and classified datacubes risk of misinterpretation. Then, steps of the new design process integrating a risk management method have been detailed. Then we have proposed a risk tolerance levels scale to help the risks assessment by end-users, as well as a classification of the risk reduction strategies that can be adopted to manage the identified risks. More specifically, we have established a one-to-one relationship between each tolerance level and one of the possible strategies to help SOLAP experts narrow the right actions for a given tolerance later. We have presented a small case study to illustrate the interest of our method.

We are working on the elaboration of a risk-aware prototyping method and tool, implementing this approach. Our future work concerns the evaluation of our proposal in a real SOLAP project. The evaluation will consist of experimenting the approach completely on a real case study, using different tolerance degrees and thus actions, and then testing the results to validate the ease of implementation, the usability and the reliability regarding the spatial vagueness management.

REFERENCES

Bédard, Y., Proulx, M. J., & Rivest, S. (2005). *Enrichissement du OLAP pour l'analyse géographique: exemples de réalisation et différentes possibilités technologiques. In Entrepôts de Données et Analyse en ligne, RNTI B_1* (pp. 1–20). Paris: Cépaduès.

Bejaoui, L. (2009). *Qualitative topological relationships for objects with possibly vague shapes: implications on the specification of topological integrity constraints in transactional spatial databases and in spatial data warehouses*. Université Blaise Pascal.

Bejaoui, L., Pinet, F., Bedard, Y., & Schneider, M. (2009). Qualified topological relations between spatial objects with possible vague shape. *International Journal of Geographical Information Science, 23*(7), 877–921. doi:10.1080/13658810802022814

Bimonte, S., Nazih, H., Kang, M.-A., Edoh-Alove, E., & Rizzi, S. (2013). *ProtOLAP: Rapid OLAP Prototyping with On-Demand Data Supply*. Paper presented at the DOLAP'13, San Francisco, CA, USA. 10.1145/2513190.2513199

Bonifati, A., Cattaneo, F., Ceri, S., Fuggetta, A., & Paraboschi, S. (2001). Designing data marts for data warehouses. *ACM Transactions on Software Engineering and Methodology, 10*(4), 452–483. doi:10.1145/384189.384190

Boulil, K., Bimonte, S., & Pinet, F. (2011). Un modèle UML et des contraintes OCL pour les entrepôts de données spatiales. De la représentation conceptuelle à l'implémentation. *Ingénierie des Systèmes d'Information, 16*(6), 11–39. doi:10.3166/isi.16.6.11-39

Burrough, P. A., & Frank, A. U. (1996). *Geographic Objects with Indeterminate Boundaries*. Academic Press.

Cohn, A. G., & Gotts, N. M. (1996). *Geographic objects with indeterminate boundaries*. Academic Press.

Delgado, M., Molina, C., Sánchez, D., Ariza, L. R., & Vila, M. A. (2004). *A flexible approach to the multidimensional model: The fuzzy datacube. In Current Topics in Artificial Intelligence* (pp. 26–36). Springer. doi:10.1007/978-3-540-25945-9_3

Devillers, R., Bédard, Y., & Jeansoulin, R. (2005). Multidimensional Management of Geospatial Data Quality Information for its Dynamic Use Within Geographical Information Systems. *American Society for Photogrammetry and Remote Sensing (PE&RS), 71*(2), 205–215. doi:10.14358/PERS.71.2.205

Devillers, R., Bedard, Y., Jeansoulin, R., & Moulin, B. (2007). Towards spatial data quality information analysis tools for experts assessing the fitness for use of spatial data. *International Journal of Geographical Information Science, 21*(3), 261–282. doi:10.1080/13658810600911879

Di Tria, F., Lefons, E., & Tangorra, F. (2012). Hybrid methodology for data warehouse conceptual design by UML schemas. *Information and Software Technology, 54*(4), 360–379. doi:10.1016/j.infsof.2011.11.004

Dilo, A., de By, R. A., & Stein, A. (2007). A system of types and operators for handling vague spatial objects. *International Journal of Geographical Information Science, 21*(4), 397–426. doi:10.1080/13658810601037096

Fasel, D., & Zumstein, D. (2009). *A fuzzy data warehouse approach for web analytics*. Springer. doi:10.1007/978-3-642-04754-1_29

Gervais, M., Bédard, Y., Levesque, M.-A., Bernier, E., & Devillers, R. (2009). Data Quality Issues and Geographic Knowledge Discovery. *Geographic Data Mining and Knowledge Discovery*, 99-115.

Giorgini, P., Rizzi, S., & Garzetti, M. (2005). Goal-oriented requirement analysis for data warehouse design. *Proceedings of the 8th ACM international workshop on Data warehousing and OLAP*. 10.1145/1097002.1097011

Glorio, O., & Trujillo, J. (2008). An MDA Approach for the Development of Spatial Data Warehouses. In I.-Y. Song, J. Eder & T. Nguyen (Eds.), Data Warehousing and Knowledge Discovery (Vol. 5182, pp. 23-32). Springer Berlin / Heidelberg. doi:10.1007/978-3-540-85836-2_3

Grira, J., Bédard, Y., & Roche, S. (2013). *Revisiting the Concept of Risk Analysis within the Context of Geospatial Database Design: A Collaborative Framework*. Paper presented at the World Academy of Science, Engineering and Technology 75, Madrid, Spain.

Guimond, L.-E. (2005). *Conception d'un environnement de découvertes des besoins pour le développement de solutions SOLAP*. Québec, Canada: Master, Université Laval.

Hazarika, S., & Cohn, A. (2001). *Qualitative Spatio-Temporal Continuity Spatial Information Theory*. Springer Berlin/Heidelberg.

ISO/IEC-51. (1999). ISO/IEC 51 Safety aspects - Guidelines for their inclusion in standards.

Jadidi, A., Mostafavi, M. A., Bédard, Y., & Long, B. (2012). *Towards an Integrated Spatial Decision Support System to Improve Coastal Erosion Risk Assessment: Modeling and Representation of Risk Zones*. Paper presented at the FIG Working Week 2012, Rome, Italy.

Kerzner, H. (2006). *Project Management: a systems approach to planning, scheduling, and controlling* (9th ed.). Academic Press.

Lagacherie, P., Andrieux, P., & Bouzigues, R. (1996). Fuzzyness and uncertainty of soil boundaries: from reality to coding in GIS. In P. A. Burrough & A. U. Frank (Eds.), *Geographic Objects with Indeterminate Boundaries, GISDATA series* (Vol. 2, pp. 275–286). London: Taylor & Francis.

Laurent, A. (2002). Bases de données multidimensionnelles floues et leur utilisation pour la fouille de données. Paris: Academic Press.

Lévesque, M.-A. (2008). *Formal Approach for a better identification and management of risks of inappropriate use of geodecisional data (Master of Science)*. Laval University.

Malinowski, E., & Zimányi, E. (2008). *Designing Spatial and Temporal Data Warehouses. In Advanced Data Warehouse Design* (pp. 307–343). Berlin: Springer - Verlag Berlin Heidelberg.

Mazón, J.-N., Trujillo, J., & Lechtenbörger, J. (2007). Reconciling requirement-driven data warehouses with data sources via multidimensional normal forms. *Data & Knowledge Engineering*, *63*(3), 725–751. doi:10.1016/j.datak.2007.04.004

Naouali, S., & Missaoui, R. (2005). *Flexible query answering in data cubes. In Data Warehousing and Knowledge Discovery* (pp. 221–232). Springer. doi:10.1007/11546849_22

Naouali, S., & Missaoui, R. (2006). *Approximation des cubes OLAP et génération de règles dans les entrepôts de données*. Academic Press.

Pardillo, J., Mazon, J.-N., & Trujillo, J. (2010). Designing OLAP schemata for data warehouses from conceptual models with MDA. *Decision Support Systems*, 1–13.

Pauly, A., & Schneider, M. (2010). VASA: An algebra for vague spatial data in databases. *Information Systems*, *35*(1), 111–138. doi:10.1016/j.is.2009.05.003

Perez, D., Somodevilla, M. J., & Pineda, I. H. (2007). Fuzzy Spatial Data Warehouse: A Multidimensional Model. *Proceedings of the Eighth Mexican International Conference on Current Trends in Computer Science*. 10.1109/ENC.2007.15

Pitarch, Y., Favre, C., Laurent, A., & Poncelet, P. (2012). *Enhancing flexibility and expressivity of contextual hierarchies*. Paper presented at the Fuzzy Systems (FUZZ-IEEE), 2012 IEEE International Conference on. 10.1109/FUZZ-IEEE.2012.6251176

Prat, N., Akoka, J., & Comyn-Wattiau, I. (2006). A UML-based data warehouse design method. *Decision Support Systems*, *42*(3), 1449–1473. doi:10.1016/j.dss.2005.12.001

Romero, O., & Abelló, A. (2009). *A Survey of Multidimensional Modeling Methodologies*. IGI Global.

Romero, O., & Abelló, A. (2011). Data-Driven Multidimensional Design for OLAP. In J. Bayard Cushing, J. French, & S. Bowers (Eds.), *Scientific and Statistical Database Management* (Vol. 6809, pp. 594–595). Springer Berlin Heidelberg. doi:10.1007/978-3-642-22351-8_51

Schneider, M. (1999). Uncertainty Management for Spatial Datain Databases: Fuzzy Spatial Data Types Advances in Spatial Databases. Springer Berlin/Heidelberg.

Siqueira, T. L. L., Aguiar Ciferri, C. D., Times, V. C., & Ciferri, R. R. (2012). Towards Vague Geographic Data Warehouses. In N. Xiao, M.-P. Kwan, M. Goodchild, & S. Shekhar (Eds.), *Geographic Information Science* (Vol. 7478, pp. 173–186). Springer Berlin Heidelberg.

Wang, F., Brent Hall, G., & Subaryono. (1990). Fuzzy information representation and processing in conventional GIS software: Database design and application. *International Journal of Geographical Information Systems*, *4*(3), 261–283. doi:10.1080/02693799008941546

Worboys, M. (1998). Computation with imprecise geospatial data. *Computers, Environment and Urban Systems*, *22*(2), 85–106. doi:10.1016/S0198-9715(98)00023-4

Chapter 7
Discriminating Biomass and Nitrogen Status in Wheat Crop by Spectral Reflectance Using ANN Algorithms

Claudio Kapp Jr.
State University of Ponta Grossa, Brazil

Eduardo Fávero Caires
State University of Ponta Grossa, Brazil

Alaine Margarete Guimarães
State University of Ponta Grossa, Brazil

ABSTRACT

Precision agriculture has the goal of reducing cost which is difficult when it is related to fertilizer application. Nitrogen (N) is the nutrient absorbed in greater amounts by crops and the N fertilizer application presents significant costs. The use of spectral reflectance sensors has been studied to identify the nutritional status of crops and prescribe varying N rates. This study aimed to contribute to the determination of a model to discriminating biomass and nitrogen status in wheat through two sensors, GreenSeeker and Crop Circle, using the resilient propagation and backpropagation artificial neural networks algorithms. As a result, a strong correlation to the sensor readings with the aboveground biomass production and N extraction by plants was detected. For both algorithms a satisfactory model for estimating wheat dry biomass production was established. The best backpropagation and resilient propagation models defined showed better performance for the GreenSeeker and Crop Circle sensors, respectively.

DOI: 10.4018/978-1-5225-5978-8.ch007

INTRODUCTION

Precision Agriculture has the goal of reducing cost which is difficult when it is related to fertilizers application. Currently, research has been carried out aiming to estimate the minimum fertilizers rates need to be applied involving aggregate technology equipment, such as the use of spectral reflectance sensors, which consist of equipment able to absorb bands of light reflected by plants and measuring them so that they can be interpreted and associated with different levels of cultures attributes. Among the key plants attributes being correlated with reading spectral reflectance sensors are potential productivity (Lofton et al., 2012), dry biomass accumulation (Hansen & Schjoerring, 2003), N accumulation in the leaves (Feng et al., 2008a; Yao et al., 2010).

Nitrogen (N) is a crucial element for plants development (Pathak et al., 2011), being essential for vegetative growth and the photosynthetic process. Nitrogen fertilizers account for a significant portion of the agricultural production costs. Due to the N dynamics in the soil, loss of considerable element may occur and may even lead to environmental damage (Shanahan et al., 2008).

Since there may be demands variables for N in the same agricultural area (Koch et al., 2004), it would be important have the N fertilizer distribution following the principles of Precision Agriculture, which consists of techniques that evaluate several features found along the same area in the agricultural production cycle, thus allowing analysis and interventions on the crops at variable rates (Inman et al., 2005; Kitchen, 2008).

Grohs et al. (2009) found a quadratic relationship between the Normalized Difference Vegetation Index (NDVI) and aboveground biomass production of wheat and barley cultivars. This led to the determination of the NDVI saturation point occurring when the curve reached its inflection point, without reducing the maintenance of NDVI response depending on the biomass production. Some papers in the literature reinforce these results looking for a positive correlation between reading spectral reflectance attributes of wheat plants increasing N rates during the crop development (Erdle, Schmidhalter & Mistele, 2011; Hansen & Schjoerring, 2003).

The correlation between reading plant reflectance attributes is also observed in studies with other crops such as cotton (Rossato et al, 2012), sugar cane (Portz et al., 2012), maize (Solari et al., 2008), and apple (Perry & Davenport, 2007).

All these studies support the use potential of this technology to develop algorithms to manage variable N rate application. For this, Raun et al. (2002) proposed the identification of the crop productive potential without fertilizer with NDVI (forming an index YP0) which multiplied by an response index to fertilizer application (RI - response index) can determine the crop yield potential with applying additional N rate, assisting in the management of nitrogen fertilizer. If the N response is expected, then management strategies can be changed to N application based on crop response (Mullen et al., 2003).

In this way, Bredemeier et al. (2013) suggest a unique model for N rate prescription based only on wheat crop yield potential. So for low yield potential sites, detected by lower NDVI value, the N amount to be applied would be lower than for sites with high yield potential, because in these places, the high yield dilutes the N content in plants. Raun et al. (2002) found an increase in N fertilization efficiency by 15% using the model proposed by Bredemeier et al. (2013).

In general, to perform correlations between crops attributes and the readings spectral reflectance sensors typically statistical techniques are used (Grohs et al., 2011; Feng et al., 2008b). Alternative methods have been proposed based on the concepts of Machine Learning (ML) (Jiang et al., 2010; Mishra et al., 2011). One technique that has proved to be workable is using Artificial Neural Network (ANN) through the Backpropagation algorithm (Yang et al., 2011, Liu et al. 2010). To improve the performance of the backpropagation algorithm the Resilient Propagation algorithm was proposed (Riedmiller & Braun, 1993).

This study was carried out to contribute to the generation of models in agriculture through the use of ANN, specifically studying their performance discriminating biomass and nitrogen status in wheat crop from indices obtained by reading spectral reflectance sensors. The objectives were (i) to study the amount of instances sufficient to obtain learning and correlating from the dataset, (ii) to compare the correlation between the readings spectral reflectance of the GreenSeeker and Crop Circle sensors and wheat crop attributes as affected by N rates application, and (iii) to compare the performance of the Backpropagation and Resilient Propagation algorithms related to the respective models generated.

SPECTRAL REFLECTANCE SENSORS

GreenSeeker Sensor

GreenSeeker is a sensor developed by the University of Oklahoma (USA), in 1990 (currently produced by nTech Industries, Inc., Ukiah, CA). There is the passive version of the sensor that receives reflected radiation from natural sources of light, and the active version that uses light emitting diodes (LEDs) as a source of radiation and detects the reflection in the spectral regions of the visible (VIS) and near infrared (NIR) (Erdle, Schmidhalter & Mistele, 2011). This sensor has been used for the management of N crops such as corn and wheat.

The sensor is designed to work at a distance of approximately 81-121 cm from the top plant. The active GreenSeeker operates by emitting radiation in the VIS red (650 ± 10 nm) or green VIS (550 ± 10 nm), as well as radiation in the NIR (770 ± 15 nm) on the plant canopy (Sheiver, Khosla & Westfall, 2011). The amount of visible and NIR light reflected by the canopy of plants is measured and transmitted to a portable computer adapted to the sensor. The files are obtained in ".CSV" to become the desired shape and then calculate the required ratios.

Crop Circle Sensor

Crop Circle ACS-210 (Holland Cientific, Lincoln, NE) is a spectral sensor similar to GreenSeeker. There are also the active and passive versions, but with a different electronic and optical design, resulting in different operational characteristics, such as the radiation emission, this sensor emits light from a single LED source in the VIS range (590 ± 5, 5 nm) and NIR (880 ± 10 nm). It is used a dual photo detector, which eliminates the chance of interference of the sample between the reflectance VIS and IR signals, also reducing the hysteresis of the detector.

The sensor is designed to work at a distance of approximately 25-213 cm from the top plant. The Crop Circle provides a number of classic vegetation index, as well as basic reflectance information of plant canopies.

Crop Circle ACS-470 (Holland Cientific, Lincoln, NE) is a more recent sensor model, has six spectral band filters each with a diameter of 12.5 millimeters detecting spectral ranges from 440 nm to 800 nm. In this version act simultaneously three filters to photodetection. It is recommended to allocate the filter NIR on channel 2 of the sensor, and in channels 1 and 3 put filters of VIS bands, because it does combinations of calculation between channels 1 and 2, and between the channels 2 and 3, and calculations of the traditional indices combine the VIS range with the infrared range.

Reflectance Indices

The reflectance indices are constructed by means of mathematical operations with the values resulting from the measurement of reflectance in the VIS and NIR range. The literature contains a number of indices applied in agriculture and the more important are shown in Table 1.

For this study it was used the NDVI and SR indices. Trotter et al. (2008) reported the NDVI is related to biomass, leaf area, vegetation cover, chlorophyll and N. NDVI saturates when the leaf area index is greater than 1.5.So the SR has also been investigated to predict biomass area with high vegetation, since it is less susceptible to saturation.

Table 1. Samples of indices that can be constructed by reading Crop Circle ACS-470 and GreenSeeker sensor.

Name	Acronym	Formula	Reference
Normalized Difference Vegetation Index	NDVI	(NIR - VIS)/(NIR + VIS)	Rouse et al. (1974)
Simple Ratio	SR	NIR/VIS	Jordan (1969)
Soil Adjusted Vegetation Index	SAVI	[(NIR – VIS) / (NIR + VIS+L)] *(1 + L)	Huete (1988)
Wide Dynamic Range Vegetation Index	WDR-NDVI	(a*NIR-VIS)/(a*NIR+VIS)	Gitelson (2004)

DATA ANALYSIS USING ARTIFICIAL NEURAL NETWORK

ANN consists of a methodology for solving problems from Artificial Intelligence (AI). They are compound of circuits that simulate the human brain, including their behavior, that is, learning errors and making findings. For that it is presented a behavior pattern related to the input variables that corresponds to a desired output. The training process consists of adjusting the synaptic weights to achieve a desired design goal (Singh & Verma, 2012). Synaptic weights are shown on the interconnections among neurons, and these may conveniently be arranged in layers. When a neural network is trained and comes up to a satisfactory error, she is able to respond to new inputs connected to the same situation training.

ANN methodology has contributed to the applied studies development in agriculture as relative humidity forecast (Białobrzewski, 2008), ethanol production prediction (Oliveira et al., 2010), identification of maize cultivar (Chen et al ., 2010) ; difference correction of NDVI predicted by the Kriging interpolation method (Yang et al., 2011).

Also in agriculture, ANN has been shown as a potential tool for the study of the relationship between readings spectral reflectance and cultures attributes. Panda et al.(2010) used spectral vegetation indices for crop yields prediction; Noh et al. (2012) constructed an artificial neural network to predict N stress levels in corn in the output layer (in SPAD units), using reading spectral bands in the red, infrared and green used in the input layer ; Liu et al. (2010) used RNAs to relate spectral reflectance indices in the input layer of the network with the chlorophyll content in the output layer in rice crops, obtaining satisfactory results.

One of the most used learning algorithms for neural networks is the Backpropagation. In this algorithm, training occurs in two phases, called forward and backward, and in each stage traverses the network in one direction. The forward phase is used to set the network output for a given input pattern. The backward phase uses the desired output and the output provided by the network to update the weights of their connections until a minimum error is obtained (Braga et al. 2000).

Selecting the adequate learning rate has a significant effect on the time required to obtain the convergence rate. If the rate is small it takes several steps to achieve an acceptable solution (Riedmiller & Braun, 1993).

Backpropagation algorithm has undergone some changes in order to obtain a better performance (Jacobs, 1988; Tollenaere, 1990) .

Among these changes there is the Resilient Propagation (RPROP) algorithm that makes convergence to the optimal solution occurs faster (Riedmiller & Braun, 1993). The most significant change in this algorithm compared to other heuristics based on the backpropagation variations, is the magnitude of the neuron weights correction and the learning rate does not influence their adjustment, which depends only on the sign of the error function gradient (Freitas et al., 2002).

RPROP algorithm try to eliminate the negative influence of the size of the partial derivative with respect to weight, and thus only the sign of the derivative is considered to indicate the direction of the weight update (Neyamadpour et al.,2009). To accomplish this, a value individual update is provided for each weight, and this only determines the size of the weight update. This introduces a second learning rule, which is to determine the evolution of the value update. This estimative is based on the behavior of the partial derivative for two successive interactions.

Freitas et al (2002) showed that each weight w_{ji} has its own rate of change (D_{ji}), which varies as a function of the time *t* as follows:

$$\Delta_{ij} = \begin{cases} \eta^{+}\Delta_{ij}(t-1) & if \frac{\partial E}{\partial w_{ij}}(t)\frac{\partial E}{\partial w_{ij}}(t-1) > 0 \\ \eta^{-}\Delta_{ij}(t-1) & if \frac{\partial E}{\partial w_{ij}}(t)\frac{\partial E}{\partial w_{ij}}(t-1) < 0 \\ \Delta_{ij}(t-1) & else \end{cases} \quad (1)$$

where:

(η) = learning rate and $0<n^{-}<1<n^{+}$.

(E) = error function;

(w_{ij}) = weight of the neuron (ij);

(t) = time;

(Δ_{ij}) = variation rate.

A change in the sign of the partial derivatives relative to the weight ω_{ji} indicates that the last change was large enough for the system to jump over a minimum point of the function $E(\boldsymbol{\omega})$, which then implies a decrease of the value of Δ_{ji} proportional to the factor η^{-}. Already the derived row with the same sign indicate that the system is continuously moving in a single direction, which thus implies a substantial increase of Δ_{ji} proportional to the factor η^{+} (Freitas et al., 2002).

Neyamadpour et al. (2009) explain that after the update of each weight is adapted, the weight is updated by following a simple rule: if the derivative is positive weight is reduced and if the derivative is negative the weight is increased by a value update, according to the following equations:

$$\Delta_{ij} = \begin{cases} -\Delta_{ij}(t) & if \dfrac{\partial E}{\partial w_{ij}}(t) > 0 \\ \Delta_{ij}(t) & if \dfrac{\partial E}{\partial w_{ij}}(t) < 0 \\ 0 & else \end{cases} \tag{2}$$

$$w_{ij}(t+1) = w_{ij}(t) + \Delta w_{ij}(t) \tag{3}$$

The total derivative of the partial error is given by (Neyamadpour et al., 2009):

$$\frac{\partial E}{\partial w_{ij}}(t) = \frac{1}{2}\sum_{p=1}^{P}\frac{\partial E_p}{\partial w_{ij}}(t) \tag{4}$$

Thus, the partial derivatives of the errors should be accumulated for all “P” training standards. This means that the weights are updated only after all training patterns are presented. A supplementary explanation regarding Resiliet Propagation algorithm can be achieved in the literature (Anastasiais et al, 2005; Riedmiller & Braun, 1993).

To build the network architecture we used the constructive growth method. According to Haykin (2001) this method starts with a simple structure and then adds up connections and synaptic weights until the network is able to meet the project specifications. The advantage of this method is that the structure tends to be the smallest possible, thereby using less computational processing.

Following this method, the network was trained with one or two hidden layers and in each case we used different nodes combinations (10, 20, 40, 50, 80, and 100) in the hidden layers, as found in the literature (Anastasiadis et al., 2005). The best combination following the constructive method found by Haykin (2001) is composed by one hidden layer containing 20 neurons.

For training the network it was used 5 cross validation folds. Was also used the tangential sigmoid activation function (output value comprised within -1 and 1) in the output layer, and a sigmoidal logarithmic function (output value comprised within 0 and 1) in the hidden layer.

FIELD EXPERIMENT

A field experiment was undertaken at Ponta Grossa, Parana State, Brazil (25°8′ S, 50°15′ W), on a Typic Hapludox under a no-till system. The climate in the region, according to Köppen classification is Cfb, with mild summer and frequent frosts in the winter. The average altitude is 970 m and annual precipitation is about 1.550 mm.

Wheat (*Triticum aestivum* L.), cv. Quartzo, was sown on June 13, 2011, after the maize crop (*Zea mays* L.), using 170 kg ha^{-1} of seeds and spacing of 0.17 m between rows. Fertilizers were applied at rates of 12 kg ha^{-1} N, 42 kg ha^{-1} P_2O_5, and 24 kg ha^{-1} K_2O at sowing.

A randomized complete block was used and four treatments were replicated 12 times. Plot size was 7 by 8 m. The treatments consisted of N application at the rates of 0, 50, 100, and 150 kg ha^{-1}, as urea, during the tillering of the wheat crop.

Data Collection

In each plot three points represented by lines of wheat plants were randomly selected. For purposes of leaf analysis, leaf samples were collected at flowering wheat crop, removing the flag leaf in 45 plants in each plot, being 15 plants in each row.

On the same selected wheat plants lines, by means of the sensor GreenSeeker using the spectral bands of 650 $\pm$ 10 nm and 770 $\pm$ 15 nm and Crop Circle ACS-470 using the 730 nm spectral bands on channel 1, 760 nm on channel 2 and 670 nm on channel 3, were held readings for a total of three readings per plot. To evaluate the wheat dry biomass production, the aerial part was collected in two continuous rows of 1.5 m (0.51 m^2) from each plot point, in a total of three points per plot . All these samples were taken on September 12, 2011.

Leaves and plants samples were washed in deionized water, dried in a forced-air oven at 60°C until a constant weight was achieved. After the dry biomass production evaluation, leaf and plants samples were ground, and N content was analyzed using sulfuric acid digestion and determination by the Kjeldahl method (Malavolta et al., 1997). The N extraction by wheat plants was calculated by multiplying dry biomass production by the N content in the biomass.

Database and Neurons Set

Two different databases were used in this work composed of:

- Crop Circle Database: indices generated from Crop Circle sensor readings, N rates, aboveground biomass production, N content in leaves and N extraction by the aerial part of the plant.
- Greenseeker Database: indices generated from GreenSeeker sensor readings, N rates, aboveground biomass production, N content in leaves and N extraction by the aerial part of the plant.
- For each database it was studied the performance of two sets of input neurons:
- Crop Circle 1 Set (SC1): NDVI (670 and 760 nm bands), NDVI (730 and 760 nm bands).
- Crop Circle 2 Set (SC2): NDVI (670 and 760 nm bands), NDVI (730 and 760 nm bands), SR (670 and 760 nm bands), SR(730 and 760 nm bands); 730 nm band, 760 nm band, 670 nm bands e N rates.
    - GreenSeeker 1 Set (SG1): NDVI and N rates.
    - GreenSeeker 2 Set (SG2): NDVI.

Each data set has been analyzed with three different numbers of instances:

- Total instances with 144 samples;
- 144 instances composed by the average of total instances;
- 48 instances composed by averaging the values of three samples from each plot.

The attributes of wheat crop used as output neuron were: aboveground biomass production, N content in leaves, and N extraction by the aerial part of the plant.

EVALUATING SENSORS AND ARTIFICIAL NEURAL NETWORK ALGORITHMS

N content in leaves was the first wheat crop attribute correlated with readings sensor and applied fertilizer rates. The correlation showed poor performance with similar behavior of the sensors and algorithms, regardless of the number of instances used.

Even when added N rates as input attribute (SC2 and SG1), the network failed to improve performance to the point of reaching a satisfactory result (Table 2).

The highest correlation, r = 0.45 and 1.62 error was obtained using SC2 dataset, total instances and the Backpropagation algorithm.

Considering the total instances, datasets showed higher performance with than without the input attribute "N rates" (SC2 and SG1). When we decreased the number of instances, the Crop Circle - SC1 dataset corrected this difference better than the GreenSeeker - SG2 dataset.

After the leaf-N analysis, the readings sensor and applied fertilizer rates were correlated to wheat aboveground biomass production, using ANN algorithms. The results of the correlation and mean absolute error are presented in Table 3.

The correlation in the databases to the wheat aboveground biomass production (Table 3) showed better performance compared to the attribute N-Leaf (Table 2). Reducing the number of instances did not impair learning network (Table 3) and the data linear behavior can explain these results.

Table 2. ANNs results to estimate N content in leaves according to Crop Circle and GreenSeeker sensor readings.

Algorithm	Total Instances				144 Instances				48 Instances			
	Backprop.		RProp.		Backprop.		RProp.		Backprop.		RProp.	
Result	r	Error	r	Error	r	Error	r	Error	r	Error	R	Error
SC1	0,34	1,74	0,33	1,74	0,40	1,77	0,37	1,88	0,42	1,24	0,40	1,77
SC2	0,45	1,62	0,42	1,67	0,38	1,78	0,38	1,80	0,42	1,54	0,39	1,28
SG1	0,41	1,76	0,41	1,70	0,36	1,65	0,39	1,82	0,42	1,54	0,37	1,94
SG2	0,20	2,67	0,22	2,50	0,30	1,84	0,30	1,84	0,36	1,67	0,38	1,56

Table 3. Artificial Neural Networks results to estimate the wheat aboveground biomass production according to Crop Circle and GreenSeeker sensor readings.

Algorithm	Total Instances				144 Instances				48 Instances			
	Backprop.		RProp.		Backprop.		RProp.		Backprop.		RProp.	
Result	r	Error	r	Error	r	Error	r	Error	r	Error	r	Error
SC1	0,52	543	0,52	541	0,60	496	0,58	536	0,64	487	0,74	363
SC2	0,58	524	0,59	520	0,57	533	0,58	535	0,67	406	0,68	408
SG1	0,55	531	0,57	518	0,63	596	0,60	513	0,64	428	0,67	403
SG2	0,53	532	0,54	534	0,62	515	0,63	494	0,73	386	0,66	441

The best correlation was 0.74 for SC1 dataset, with an error of 363. It is also observed that for total instances SC2 and SG1 sets containing N rate as the input attribute, showed more satisfactory performance for both algorithms. The same did not happen when 48 and 144 instances were used, showing that the learning gain cannot be related to the use or not of the N rate as the input attribute.

Finally, readings sensors and applied fertilizer rates were correlated the N extraction by plant, using ANN algorithms. The results of the correlation and mean absolute error are presented in Table 4.

As the results of aboveground biomass production (Table 3), the N extraction by plants presents the best correlation considering the dataset with for 48 instances (Table 4). In addition, the best mean absolute error values were also obtained with 48 instances. Because of the data linear behavior, decreasing in the instances quantity do not harms the network learning.

The best correlation t was obtained with the Resilient Propagation algorithm, SG2 set with 48 instances, presenting r = 0.76 and with an error of 12.50. Considering the total instances, the SC2 and SG1 sets, which had N rate as the input attribute, showed better performance for both algorithms. Already, with 144 and 48 instances, this relationship is reversed with the GreenSeeker sensor set, presenting correlation coefficients greater when N rate was not used as input neuron. The behavior of all instances remained similar to the Crop Circle sensor set, with the exception of 48 instances with the Backpropagation algorithm that performed best when N rate was not used as input neuron. The best coefficients obtained by the target attribute wheat aboveground biomass production (Table 3) were lower in all the numbers of instances, than those obtained by the target attribute N extraction by wheat plants (Table 4).

The results analysis, presented in Table 5, was made using the best results obtained with 48 instances for N- Leaf, aboveground biomass production, and N extraction by wheat plants. It is noteworthy that statistical correlation tests were performed for data presented and in all of them the correlation was significant at $P < 0.01$).

Table 4. ANNs to estimate N extraction by wheat plants according to Crop Circle and GreenSeeker sensor readings.

Algorithm	Total Instances				144 Instances				48 Instances			
	Backprop.		RProp.		Backprop.		RProp.		Backprop.		RProp.	
Result	r	Error	r	Error	r	Error	r	Error	r	Error	r	Error
SC1	0,59	16,19	0,55	16,14	0,62	15,54	0,63	15,19	0,71	12,00	0,70	11,63
SC2	0,64	14,82	0,65	14,80	0,64	15,10	0,67	13,56	0,67	11,65	0,74	11,15
SG1	0,65	15,27	0,66	14,82	0,58	15,27	0,65	15,52	0,69	12,61	0,64	13,40
SG2	0,59	15,60	0,57	15,08	0,67	15,06	0,66	15,59	0,71	14,80	0,76	12,50

Table 5. General results of the Backpropagation and Resilient Propagation algorithms performance.

Attribute	Crop Circle				Greenseeker			
	Backpropagation		Resilient Propagation		Backpropagation		Resilient Propagation	
	r	Error	R	Error	r	Error	r	Error
N-Leaf	0,42	1,54	0,40	1,77	0,42	1,54	0,38	1,56
Aboveground biomass production	0,67	406,00	0,74	363,00	0,73	386,00	0,67	403,00
N extraction by plants	0,71	12,00	0,74	11,15	0,71	14,80	0,76	12,50

The sensors showed similar performance, even with different input neurons. Whereas the sensors work similarly to the same goal, these results were expected. We emphasize here a difference related to the behavior of the algorithms with respect to the target attribute wheat aboveground biomass production. It is noticed that the Backpropagation algorithm presented better performance for sensor GreenSeeker with a correlation coefficient of 0.73 against 0.67 for the Resilient Propagation. Already, at the sensor Crop Circle, the behavior was reversed, with the Backpropagation algorithm having a correlation coefficient of 0.67 and the Propagation Resilient 0.74.

The highest correlation coefficients were obtained for both aboveground biomass production and N extraction by wheat plants (Table 5), demonstrating that Crop Circle and GreenSeeker sensors have sensitivity to accuse variation in aboveground biomass production and, indirectly, on other attributes of wheat crop.

With respect to errors, it was obtained on average, about 4.00% of the target attribute N-Leaf for the Backpropagation algorithm and 4.50% for the Resilient Propagation algorithm, 11.00% of the target attribute aboveground biomass production for Backpropagation algorithm and 10.50% for the Resilient Propagation algorithm, and 15.50% of the target attribute N extraction by wheat plants for the Backpropagation algorithm and 14.00% for the Resilient Propagation algorithm.

The performance of the algorithms on the error showed a small difference in percentage. It is related to the linear behavior of the input attributes towards target attributes. Although the Perceptron model is suitable for problems where the data input and target attributes are linearly separable, in our study by using this model the neural network learning has not submitted satisfactory. Now, when it was used the Multi Layer Perceptron model, by adding a hidden layer, the network was able to abstract learning of the data presented.

CONCLUSION

For this data set analyzed, regardless of the attribute that is aimed at estimating, using the average values of each portion, thereby reducing the number of input variables, we obtained higher correlation coefficient.

The sensor performance was similar for all attributes predicted, and demonstrated closer correlations to the aboveground biomass production and N extraction by wheat plants than with the N content in leaves, although in all statistical tests correlations were significant.

The best models established by the algorithms showed similar, but with respect to the attribute target aboveground biomass production the Backpropagation and Resilient Propagation algorithms showed better results respectively for the GreenSeeker and Crop Circle sensors.

REFERENCES

Anastasiadis, A. D., Magoulas, G. D., & Vrahatis, M. N. (2005). New globally convergent training scheme based on the resiliente propagation Algorithm. *Neurocomputing*, *64*, 253–270. doi:10.1016/j.neucom.2004.11.016

Białobrzewski, I. (2008). Neural modeling of relative air humidity. *Computers and Electronics in Agriculture*, *60*(1), 1–7. doi:10.1016/j.compag.2007.02.009

Braga, A. P., Carvalho, A. P. L., & Ludemir, T. B. (2005). Redes Neurais Artificiais. In S. O. Rezende (Ed.), *Sistemas Inteligentes: Fundamentos e aplicações* (pp. 141–168). Barueri: Manole.

Bredemeir, C., Variani, C., Almeida, D., & Rosa, A. T. (2013). Estimativa do potencial produtivo em trigo utilizando sensor óptico ativo para adubação nitrogenada em taxa variável. *Ciência Rural*, *43*(7), 1147–1154. doi:10.1590/S0103-84782013005000080

Chen, X., Xun, Y., Li, W., & Zhang, J. (2010). Combining discriminant analysis and neural networks for corn variety identification. *Computers and Electronics in Agriculture*, *71*(1), 48–53. doi:10.1016/j.compag.2009.09.003

Erdle, K., Mistele, B., & Schimidhalter, U. (2011). Comparison of active and passive spectral sensors in discriminating biomass parameters and nitrogen status in wheat cultivars. *Field Crops Research*, *124*(1), 74–84. doi:10.1016/j.fcr.2011.06.007

Feng, W., Yao, X., Zhu, Y., Tian, Y. C., & Cao, W. X. (2008a). Monitoring leaf nitrogen status with hyperspectral reflectance in wheat. *European Journal of Agronomy*, *28*(3), 394–404. doi:10.1016/j.eja.2007.11.005

Feng, W., Zhu, Y., Tian, Y., Cao, W., Yao, X., & Li, Y. (2008b). Monitoring leaf nitrogen accumulation in wheat with hyper spectral remote sensing. *Acta Ecologica Sinica*, *28*(1), 23–32. doi:10.1016/S1872-2032(08)60018-9

Freitas, A. A. C., Silva, I. N., & Souza, A. N. (2002). Aplicação de redes neurais na estimação da temperatura interna de transformadores de distribuição imersos em óleo. *Controle & Automação*, *13*(3), 266–274. doi:10.1590/S0103-17592002000300005

Gitelson, A. A. (2004). Wide Dynamic Range Vegetation Index for Remote Quantification of Biophysical Characteristics of Vegetation. *Journal of Plant Physiology*, *161*(2), 165–173. doi:10.1078/0176-1617-01176 PMID:15022830

Grohs, D. S., Bredemeier, C., Mundstock, C. M., & Polleto, N. (2009). Modelo para estimativa do potencial produtivo em trigo e cevada por meio do sensor GreenSeeker. *Engenharia Agrícola*, *29*(1), 101–112. doi:10.1590/S0100-69162009000100011

Grohs, D. S., Bredemeier, C., Polleto, N., & Mundstock, C. M. (2011). Validação de *modelo* para predição do potencial produtivo de trigo com sensor óptico ativo. *Pesquisa Agropecuária Brasileira*, *46*(4), 446–449. doi:10.1590/S0100-204X2011000400015

Hansen, P. M., & Schjoerring, J. K. (2003). Reflectance measurement of canopy biomass and nitrogen status in wheat crops using normalized difference vegetation indices and partial least squares regression. *Remote Sensing of Environment*, *86*(4), 542–553. doi:10.1016/S0034-4257(03)00131-7

Haykin, S. (2001). *Redes Neurais: princípios e práticas* (2nd ed.). Porto Alegre: Bookman.

Huete, A. R. (1988). A Soil-Adjusted Vegetation Index (SAVI). *Remote Sensing of Environment*, *25*(3), 295–309. doi:10.1016/0034-4257(88)90106-X

Inman, D., Khosla, R., Westfall, D. G., & Reich, R. (2005). Nitrogen uptake across site specific management zones in irrigated maize production systems. *Agronomy Journal*, *97*(1), 169–176. doi:10.2134/agronj2005.0169

Jacobs, R. (1988). Increased rates of convergence through learning rate adaptation. *Neural Networks*, *1*(4), 295–307. doi:10.1016/0893-6080(88)90003-2

Jiang, J., Li, A., Deng, W., & Bian, J. (2010). Construction of a New Classifier Integrated Multiple Sources and Multi-temporal Remote Sensing Data for wetlands. *Procedia Environmental Sciences*, *2*, 302–314. doi:10.1016/j.proenv.2010.10.036

Jordan, C. F. (1969). Derivation of leaf-area index from quality of light on the forest floor. *Ecology*, *50*(4), 633–666. doi:10.2307/1936256

Kitchen, N. R. (2008). Emerging technologies for real-time and integrated agriculture decisions. *Computers and Electronics in Agriculture, 61*(1), 1–3. doi:10.1016/j.compag.2007.06.007

Koch, B., Khosla, R., Frasier, M., Westfall, D. G., & Inman, D. (2004). Economic feasibility of variable rate nitrogen application utilizing site-specific management zones. *Agronomy Journal, 96*(6), 1572–1580. doi:10.2134/agronj2004.1572

Liu, M., Liu, X., Li, M., Fang, M., & Chi, W. (2010). Neural-network model for estimating leaf chlorophyll concentration in rice under stress from heavy metals using four spectral indices. *Biosystems Engineering, 106*(3), 223–233. doi:10.1016/j.biosystemseng.2009.12.008

Lofton, J., Tubana, B. S., Kanke, Y., Teboh, J., Viator, W., & Dalen, M. (2012). Estimating Sugarcane Yield Potential Using an In-Season Determination of Normalized Difference Vegetative Index. *Sensors (Basel), 12*(6), 7529–7547. doi:10.3390120607529 PMID:22969359

Malavolta, E., Vitti, G. C., & Oliveira, S. A. (1997). *Avaliação do estado nutricional das plantas: princípios e aplicações* (2nd ed.). Piracicaba: Potafos.

Mishra, A., Karimi, D., Ehsani, R., & Albrigo, L. G. (2011). Evaluation of an active optical sensor for detection of Huanglongbing (HLB) disease. *Biosystems Engineering, 110*(3), 302–309. doi:10.1016/j.biosystemseng.2011.09.003

Mullen, R. W., Freman, K. W., Raun, W. R., Johnson, G. V., Stone, M. L., & Solie, J. B. (2003). Identifying an In-Season Response Index and the Potencial to Increase wheat Yield with nitrogen. *Agronomy Journal, 95*(2), 347–351. doi:10.2134/agronj2003.0347

Neyamadpour, A., Taib, S., & Abdullah, W. (2009). Using artificial neural networks to invert 2D DC resistivity imaging data for high resistivity contrast regions: A MATLAB application. *Computers & Geosciences, 35*(11), 2268–2274. doi:10.1016/j.cageo.2009.04.004

Noh, H., & Zang, Q. (2012). Shadow effect on multi-spectral image for detection of nitrogen deficiency in corn. *Computers and Electronics in Agriculture, 83*, 52–57. doi:10.1016/j.compag.2012.01.014

Oliveira, A. C. S., Souza, A. A., Lacerda, W. S., & Gonçalves, L. R. (2010). Aplicação de redes neurais artificiais na previsão da produção de álcool. *Ciência e Agrotecnologia, 34*(2), 279–284. doi:10.1590/S1413-70542010000200002

Panda, S. S., Ames, D. P., & Panigrahi, S. (2010). Application of Vegetation Indices for Agricultural Crop Yield Prediction Using Neural Network Techniques. *Remote Sensing*, *2*(3), 673–696. doi:10.3390/rs2030673

Pathak, R. R., Lochab, S., & Raghuram, N. (2011). Improving Plant Nitrogen-Use Efficiency. *Agricultural and Related Biotechnologies*, *4*, 209–218.

Perry, E. M., & Davenport, J. R. (2007). Spectral and spatial differences in response of vegetation indices to nitrogen treatments on apple. *Computers and Electronics in Agriculture*, *59*(1-2), 56–65. doi:10.1016/j.compag.2007.05.002

Portz, G., Molin, J. P., & Jasper, J. (2012). Active crop sensor to detect variability of nitrogen supply and biomass on sugarcane fields. *Precision Agriculture*, 11–12.

Raun, W. R., Solie, J. B., Johnson, G. B., Stone, M. L., Mullen, R. W., Freman, K. W., ... Lukina, E. V. (2002). Improving nitrogen use efficiency in cereal grain production with optical sensing and variable rate application. *Agronomy Journal*, *94*(4), 815–820. doi:10.2134/agronj2002.8150

Riedmiller, M., & Braun, H. (1993). A direct adaptive meted for faster backpropagation learning: the RPROP algorithm. *Internacional Conference on Neural Network (ICNN), San Francisco: Proceedings*, 586-591.

Rossato, O. B., Aandrade-Sanchez, P., Guerra, S. P. S., & Crusciol, C. A. C. (2012). Sensores de refletância e fluorescência na avaliação de teores de nitrogênio, produção de biomassa e produtividade do algodoeiro. *Pesquisa Agropecuária Brasileira*, *47*(8), 1133–1141. doi:10.1590/S0100-204X2012000800014

Rouse, J. W. J., Haas, H. R., Schell, A. J., & Deering, W. D. (1974). Monitoring vegetation systems in the great plains with ERTS. *NASA Special Publication*, *1*, 309–317.

Shanahan, J. F., Kitchen, N. R., Raun, W. R., & Schepers, J. S. (2008). Responsive in-season nitrogen management for cereals. *Computers and Electronics in Agriculture*, *61*(1), 51–62. doi:10.1016/j.compag.2007.06.006

Sheiver, T. M., Khosla, R., & Westfall, D. G. (2011). Evaluation of two crop canopy sensors for nitrogen variability determination in irrigated maize. *Precision Agriculture*, *12*(6), 892–904. doi:10.100711119-011-9229-2

Singh, T. N., & Verma, A. K. (2012). Comparative analysis of intelligent algorithms to correlate strength and petrographic properties of some schistose rocks. *Engineering with Computers*, *28*(1), 1–12. doi:10.100700366-011-0210-5

Solari, F., Shanahan, J., Ferguson, R., Schepers, J., & Gitelson, A. (2008). Active sensor reflectance measurements of corn nitrogen status and yield potential. *Agronomy Journal, 100*(3), 571–579. doi:10.2134/agronj2007.0244

Tollenaere, T. (1990). Supersab: Fast adaptative backpropagation with good scaling properties. *Neural Networks, 3*(5), 561–573. doi:10.1016/0893-6080(90)90006-7

Trotter, T. F., Frazier, P., Trotter, M. G., & Lamb, D. W. (2008). Objective biomass assessment using an active plant sensor (CROP CIRCLE) preliminary experiences on a variety of agricultural landscapes. In R. Koshla (Ed.), *9th Conference on Precision Agriculture*. Colorado State University.

Yang, Y., Zhu, J., Zhao, C., Liu, S., & Tong, X. (2011). The spatial continuity study of NDVI based on Kriging and BPNN algorithm. *Mathematical and Computer Modelling in Agriculture., 54*(3-4), 1138–1144. doi:10.1016/j.mcm.2010.11.046

Yao, X., Zhu, Y., Tian, Y., Feng, W., & Cao, W. (2010). Exploring hyperspectral bands and estimation indices for leaf nitrogen accumulation in wheat. *International Journal of Applied Earth Observation and Geoinformation, 12*(2), 89–100. doi:10.1016/j.jag.2009.11.008

Chapter 8
Discovering Regularity Patterns of Mobility Practices Through Mobile Phone Data

Paolo Tagliolato
Politecnico di Milano, Italy

Fabio Manfredini
Politecnico di Milano, Italy

ABSTRACT

The chapter addresses the issue of analyzing and mapping mobility practices by using different kinds of mobile phone network data that provide geo-located information on mobile phone activity at a high spatial and temporal resolution. The authors present and discuss major findings and drawbacks based on an application carried out on the Milan urban region (Lombardy, Northern Italy) and suggest possible implications for policies.

INTRODUCTION

Interpretative tools for the identification of mobility practices in the contemporary metropolis are needed, not only for the some known limitations of traditional data sources but also because new forms of mobility are emerging, describing new city dynamics and time-variations in the use of urban spaces by temporary populations. In Italy, the traditional data sources for urban and mobility investigations (i.e. surveys, census) have known limitations, including the high cost of surveys, the difficulty of data updating, the difficulty of describing city dynamics and time dependent variations in intensity of urban spaces usages by temporary populations at different scales.

DOI: 10.4018/978-1-5225-5978-8.ch008

New forms of mobility, close to daily mobility, are changing the way in which urban spaces are used. They are characterized both by being based on the use of transportation system, and by the efficient appropriation of information technologies (internet, mobile phone). They intensified the density of the moves with which we can read diversified uses of the city, that traditional sources of analysis are unable to return with continuity. In Lombardy region, the systematic mobility represents only the 29% of daily travels, which are attested on 2,65 travel/day in average, with a propensity to mobility that changes in relationship to the professional condition.

As underlined by some authors (Ascher, 2004; Bourdin, 2005; Ehrenberg, 1995; Kaufmann, 2000; Scheller & Urry, 2006; Urry, 2002), the changes in the management of mobility in contemporary cities are a useful key for understanding the transformations of times, places and modes of social life and work programs, structuring the metropolitan areas.

In this perspective, mobility may represent both a tool of knowledge and a project for urban planners, provided that a better understanding of different patterns of mobility in the form of "active biographies", which increase the range of "post-fordist living and labor styles" (Nuvolati, 2003), is available.

Considering the role of mobility practices in social and spatial differentiation, it becomes important to formulate pertinent analytical approaches, aimed at describing the different densities of use of the city as a new challenge and a prerequisite for understanding the city and its dynamics.

Hence, from an analytical point of view, it becomes important to accompany the traditional quantitative approaches referred to a geographic displacement that tends to focus on movement in space and time, in an aggregate way and for limited periods, with data sources able to describe fine grain over-time variation in urban movements.

In this direction, an interesting contribution may come from mobile phone network data as a potential tool for the development of real-time monitoring, useful to describe urban dynamics, as it has been tested in several experimental studies (Ahas & Mark, 2005; Gonzalez, Hidalgo, & Barabasi, 2008; Ratti, Pulselli, Williams, & Frenchman, 2006).

The application researches were focused on two different products. Some studies dealt with aspects of representation of the data, emphasizing the most directly evocative aspects, to highlight how these data may represent the "Mobile landscapes" (Ratti et al., 2006). Other studies focused both on data-mining analysis and on the construction of instruments capable of deriving summary information and relevant data about the urban dynamics from cell-phone (Ahas & Mark, 2005).

As opposed to the more traditional methods of urban surveys, the use of aggregated and anonymous mobile phone network data has shown promise for large-scale surveys with notably smaller efforts and costs (Reades, Calabrese, Sevtsuk, & Ratti, 2007).

If we consider the observed and aggregated telephone traffic as the result of individual behaviors and habits, we can treat mobile phone data as a functional source on the *real* use of the cities, capturing, for example, traces of temporary populations, which are difficult to intercept by traditional data source, but which, at the same time, increasingly affect urban practices both quantitatively and qualitatively.

In this direction an increasing number of studies concerns the exploitation of mobile phone data in urban analysis and planning (Becker et al., 2011). In particular an interesting issue regards the classification of urban spaces according to their users' practices and behaviors (Reades et al., 2007; V Soto & Frías-Martínez, 2011). In (Víctor Soto & Frías-Martínez, 2011) the authors outline the fact that city areas are generally not characterized by just one specific use, and for this reason they introduce the use of c-means, a fuzzy unsupervised clustering technique for land use classification, which returns for each area a certain grade of membership to each class.

In this general context, we used mobile phone data provided by Telecom Italia, the main Italian operator, in order to test the potentialities of this information in describing the density of use of urban spaces at different temporal and spatial scales as a precondition:

- To identify temporary populations and different forms of mobility that structure the relationships in the contemporary city;
- To propose diversified management policies and mobility services that city users require, increasing the efficiency of the supply of public services.

MOBILE PHONE DATA

Milan is placed in an urban region which goes far beyond its administrative boundaries (see Figure 1). The core city and the whole urban area have been affected in the last 20 years by changes in their spatial structures and have generated new relationships between the center and suburbs. At the moment, the urban region of Milan is a densely populated, integrated area where 4.000.000 inhabitants live, where there are 370.000 firms and large flows of people moving daily in this wide area (Balducci, Fedeli, & Pasqui, 2010).

In order to analyze the complex temporal and spatial patterns emerging from mobile phone data, we used two different types of data provided by Telecom Italia.

Telephone Traffic Intensity: Erlang

The first data type concerns the mobile phone traffic registered by the network over the whole Milan urban region (Northern Italy).

Figure 1. Map of built-up areas in the Milan urban region (2007)
Source: Dip elaboration of DUSAF 2.1 data.

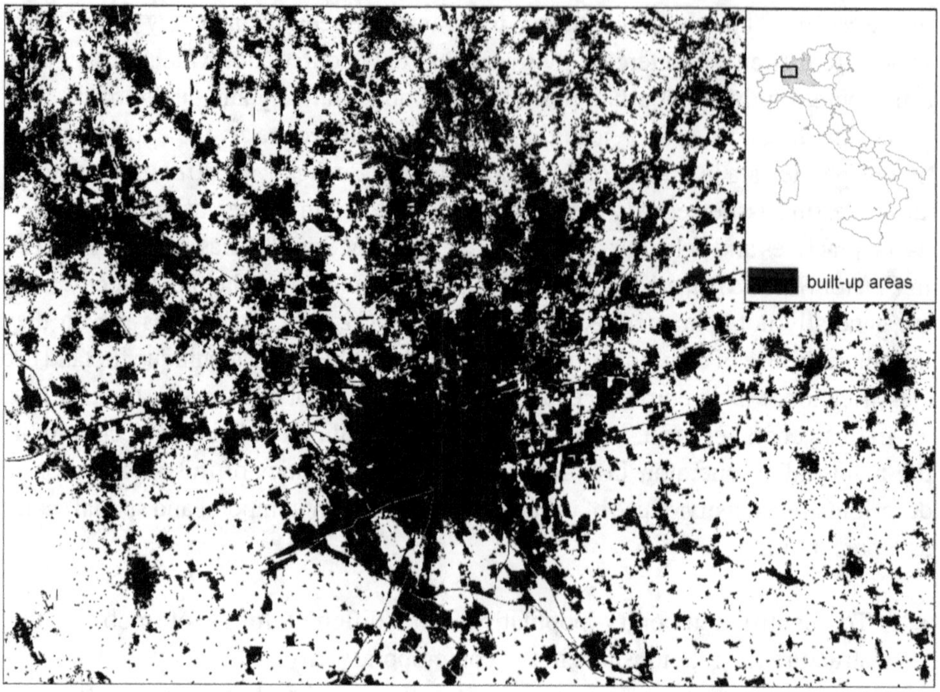

Data are expressed in Erlang, namely the average number of concurrent contacts in a time unit. In the present case, the data represent the telephonic traffic intensity every 15 minutes and was supplied by Telecom Italia in a spatialized form. From the telephone traffic recorded by each cell of the network, the telephone provider distributed the measurements, by means of weighted interpolations, throughout a tessellation of the territory in 250 meters x 250 meters squared areas (pixels) [1].

We performed time series analysis on this data along a period of 14 days in March 2009 (March 18th till March 31st), in order to evaluate specific characteristics of population behaviors at an hourly and daily base. We then applied a novel geo-statistical unsupervised learning technique aimed at identifying useful information on hidden patterns of mobile phone use. We will show that these hidden patterns regard different usages of the city in time and in space and that they are related to the mobility of individuals. The results return new maps of the region, each describing the intensity of one of the identified mobility pattern on the territory. This highlights, in our opinion, the potentials of this data for urban planning and transport research studies.

Aggregated Tracks of Users: Hourly Origin Destination Matrices

The second typology of data consists in localized and aggregated tracks of anonymized mobile phone users. It is an origin-destination datum derived from the Call Detail Record database. Italian privacy policies severely constrain the use of these data, even for research purposes. In the framework of a collaboration with Telecom Italia (T-Lab), we arrived at the definition of a datum which was free from privacy constraints, consisting in an aggregation of users' displacements based on CDR records. Telecom engineers set up a system for the automatic and blind extraction of data of this kind.

The system is fed with the CDR and a tessellation of a certain geographical region (in the present case the Lombardy Region). The output consists in time series of Origin-Destination matrices (where origin and destination zones are the tessellation's tiles), equivalent to a function $F(o,d,t) \rightarrow n$ which, at time t (t varying in within the 24 hours of a given day), assigns to origin o and destination d the number n of distinct users that performed some mobile phone activity[2] within o at time $t-1$ and a subsequent activity within d at time t.

The CDR's raw informations are available at the level of the antennas which handled the activity. The distribution of antennas in space depends on the amount of mobile phone traffic that needs to be managed. In dense urban areas we therefore observe a high density of antennas while in the suburbs the density of antennas may be very low. For the positioning of a user within a certain tile of a tessellation to be reliable, a technical constraint imposed that the tiles contained at least 13 antennas.

We defined three distinct tessellations which could gave us the possibility to map and to interpret main spatial patterns of mobile phone users' mobility: the first is more fine-grained, and it was obtained by a data-driven process taking into account the spatial distribution of antennas; the second and third, even if more coarse, were more directly related to administrative boundaries, consisting in the aggregation of adjacent municipalities' polygons.

We obtained the tessellations as follows:

- Automatic clustering of antennas (526 polygons). Each zone is an aggregation of Voronoi cells obtained from the points of location of the antennas. More precisely we proceeded by clustering the positions of the antennas by means of an agglomerative hierarchical clustering algorithm (complete linkage, Euclidean distance): we firstly cut the hierarchical tree in order to obtain 100 clusters; we then selected the groups with more than 500 antennas and we cut the corresponding sub-trees according to an *inconsistency coefficient* less than a given threshold (see e.g. (Jain & Dubes, 1998)), obtaining a final sufficiently balanced partition (i.e. with an homogeneous number of antennas per cluster). Finally, for each cluster, we calculated the polygon corresponding to the union of the Voronoi cells of its antennas. Hence we obtained the tessellation having these polygons as tiles;

- Automatic aggregation of municipalities (313 polygons): each zone is an aggregation of municipalities. Each zone contains not less than 13 antennas. An automatic procedure has been created in order to build new zones in an iterative manner.
- Manual aggregation of municipalities (202 polygons).

On these tessellations it was possible to map the direction and the intensity of mobile phone users' movements at an hourly basis. The data set was collected in different working days: five Wednesday respectively in July, August, September, October and November 2011. Using this data, we performed some analyses aimed at evaluating the overall mobility of cell phone users in the Lombardy region.

ANALYSIS

Treelet Decomposition of Erlang

Erlang measures can give insights on different aspects of the urban area to which they refer, and their analysis can be developed with various scopes: the segmentation of the area into districts characterized by homogeneous telephonic patterns; the identification of a set of reference signals able to describe the different patterns of utilization of the mobile phone network in time.

Treelet decomposition is an effective dimension reduction technique for Erlang profiles and, more generally, for data with peculiar functional features, like spikes, periodicity, outliers.

The methodology of Treelet decomposition (Manfredini et al., 2012; Vantini, Vitelli, & Zanini, 2012) allowed us to obtain: a reference basis reporting the specific effect of some activities on Erlang data; a set of maps showing the contribution of each activity to the local Erlang signal. The idea behind our approach is that different basic profiles (each being one element of the treelet decomposition basis) of city usages can concur in the same place and that the overall observed usage of a certain place is the superimposition of layers of these profiles.

We selected some results as significant for explaining specific patterns both of mobility and of city usages (commuting, nightly activities, distribution of residences, non systematic mobility). We tested their significance and their interpretation from an urban analysis and planning perspective at the Milan urban region scale.

Each of the following figures represents one of the extracted city usage profile and is organized as follows:

- Top panel depicts the considered basic profile: x-axis represents time, spanning 7 days from Wednesday to Tuesday at an hourly rate. The dotted lines correspond to 2 hours while the continuous lines separate the different days of the week; y-axis: Erlang values;
- Bottom panel depicts a map of intensity values: colors show how much the upper profile concur to explain one place's overall (telephone) use pattern.

Figure 2 is about the density of mobile phone activity late at night (in particular from midnight until 8 am). We can observe here some interesting hot spots where values are very high. For example, the exhibition district in the Northern Western side of the map. In the considered period an important Fair (the 2009 International

Figure 2. Nightly activity. Hot spots highlight the presence of night work

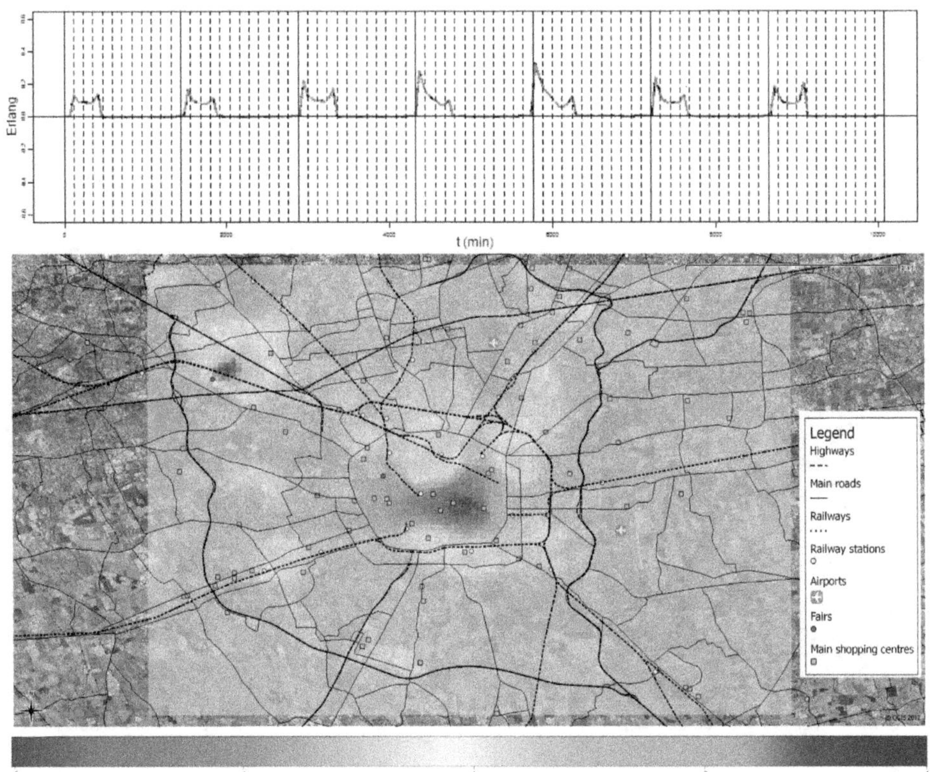

**For a more accurate representation see the electronic version.*

Milan Design Week) was held and the peak fits well with the nightly activities necessary for the mounting and the organization of the site. Another point of interest is the Fruit and Vegetable Wholesale market in the South Eastern part of the region where consistent night work happens for delivering and distributing products that come from whole Italy and abroad. The city center is characterized by a relative low value, according to the absence of relevant nightly activity inside it.

Figure 3 puts in evidence some locations with high concentration of mobile phone activity during the evening of the working days and during daytime (from 8 am until 8 pm) of the week end. It shows a significant correspondence with main residential districts of the Milan urban region. It highlights a relevant concentration of homes along the second circular ring of the city, where the density of resident population reaches the highest value of Milan, but also in some municipalities with a residential profile and social housing in the south, south-west and in the north

Figure 3. Concentration of activities during evenings of working days and during daytime (from 8 am until 8 pm) of the week end: residential districts of the Milan urban region

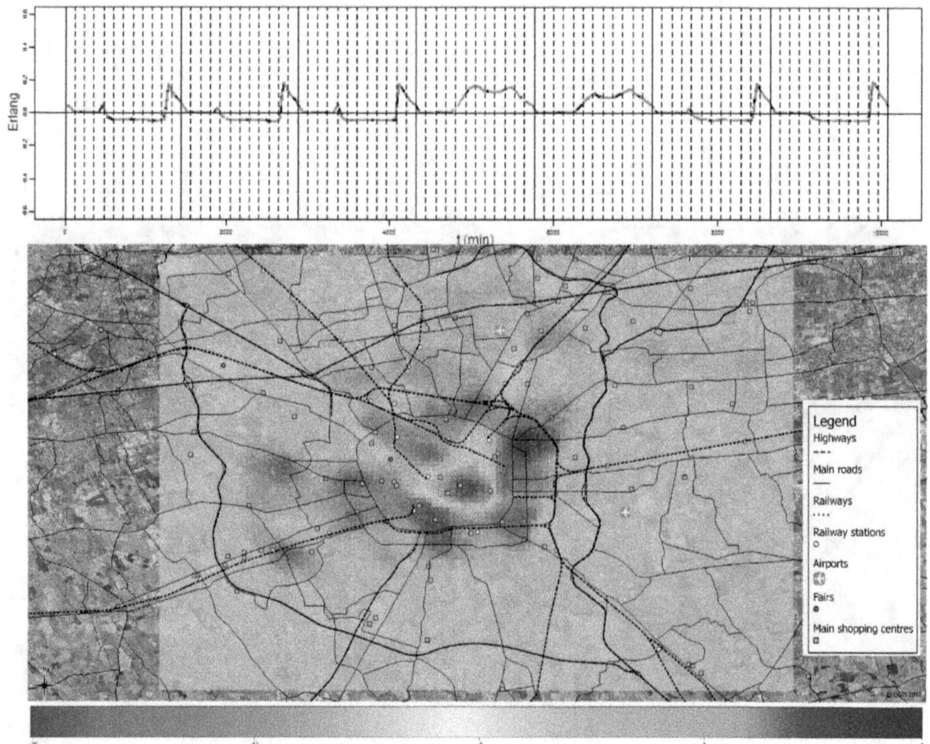

*For a more accurate representation see the electronic version.

of the metropolitan area (Corsico, Rozzano, Sesto S.G). The city center of Milan appears as a void, a fact that is consistent with the changes that occurred in the last decades, namely a gradual replacement of the residents with activities mainly related to the services sector and the commercial sector.

Figure 4 shows places with high density of activity during Saturday evening, from 8 pm until midnight. Focusing on the core city area, we notice several interesting patterns: a high activity in some places where there are many pubs and restaurants near the Milan Central Station, in the Navigli District, in the Isola Quarter and in other ambits characterized by the presence of leisure spaces (Filaforum Assago in the south of Milan), but also of activities in a continuous cycle as the hospitals. This treelet has proven to be effective in describing the temporal profile of the city lived by night populations during Saturday.

Figure 4. Density of activity during Saturday evening (8pm-midnight). Saturday night population: leisure and hospitals

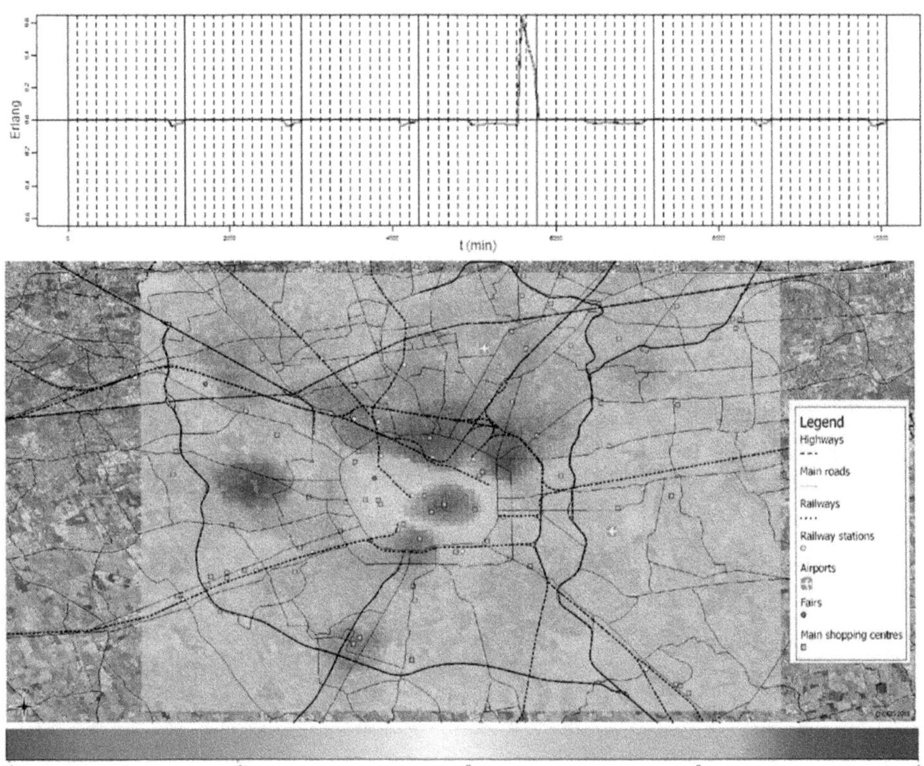

**For a more accurate representation see the electronic version.*

Figure 5 highlights another relevant mobility pattern, which is difficult to intercept through database traditionally used in urban studies: the shopping activity and, in general, the leisure activity. The map represents the density of mobile phone use during Saturday, from 10 am to 8 pm. Shopping and leisure are two of the main reasons of mobility in contemporary cities: they belong to the category of unsystematic mobility, and they significantly contribute to the even more complex mobility patterns that can be observed in the Milan urban region due to the distribution of commercial centers, commercial streets and, in general, of activities (museums, touristic sites, cinemas, just to cite some) inside and outside the city. These places attract, especially in certain days of the week, a huge amount of population coming from a vast territory that goes far beyond the administrative boundaries of the city. The map is the result of this spatial pattern and shows an important concentration of mobile phone traffic in the city center and in other several places outside the city

Figure 5. Mobility practices. Saturday (10am- 8pm), shopping and leisure activity

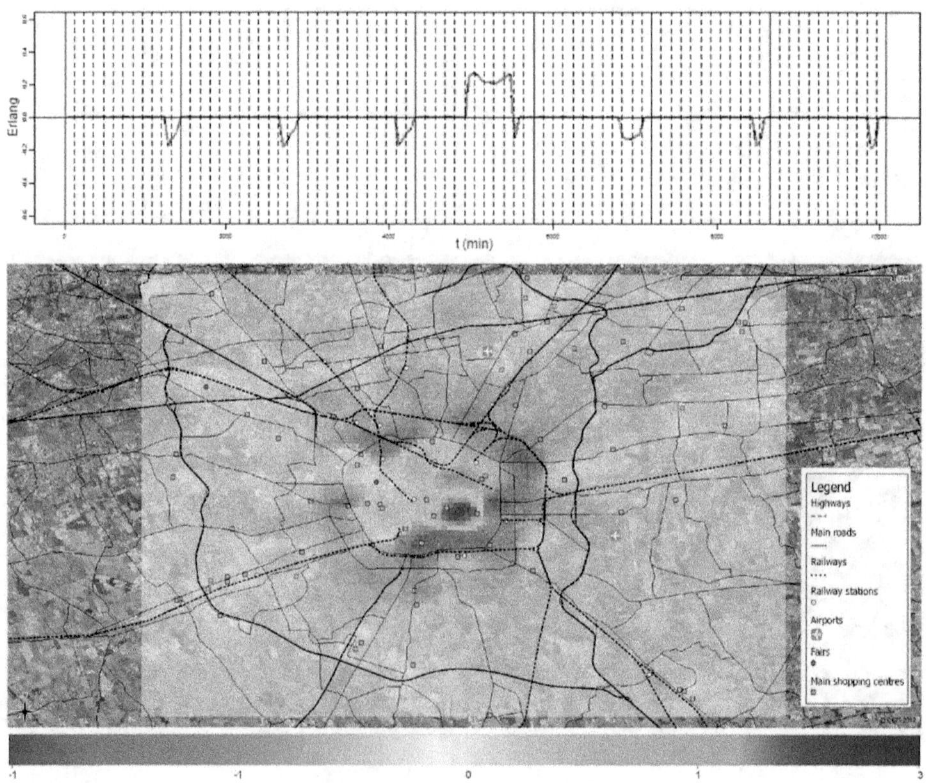

For a more accurate representation see the electronic version.

(most of them corresponding to the presence of commercial centers). The mainly residential areas, recognized in the previous Figure 3, are consequently characterized by the lowest value.

Aggregated Tracks of Mobile Phone Users

Origin-destination matrices are the traditional information for the analysis of mobility and one of the main sources for the definition of public transport policies. These matrices are generally obtained through expensive surveys. In Lombardy region the two main traditional sources on mobility are quite old: the Survey on Mobility (2002) (Regione Lombardia, Direzione Generale Infrastrutture e mobilità, 2002) and the Census data (2002). The advent of mobile phone data opened up a new perspective and in recent years several studies considered the possibility to exploit these data for the purpose of analysis of vehicular traffic (Bolla & Davoli, 2000; N Caceres, Wideberg, & Benitez, 2008; Noelia Caceres, Romero, Benitez, & del Castillo, 2012; Cayford & Johnson, 2003) and for the estimation of OD matrices (Bekhor, Cohen, & Solomon, 2011; F Calabrese, Lorenzo, Liu, & Ratti, 2011; Francesco Calabrese, Liu, Lorenzo, & Ratti, 2011). This perspective is appealing. Consider for example in table 1 the comparison of the available sources on mobility in Lombardy region, from which we can evince some pros and cons of mobile phone data:

- Mobile phone data have, at least in urban areas, superior spatial resolution than conventional surveys, permitting to obtain finer visualization of mobility practices and to generate customized regions of analysis;
- The temporal resolution of mobile phone data is very high; it allows to monitor in time different practices at an hourly, daily or seasonal basis;
- Mobile phone data lack of information regarding the means of transport used. It is therefore possible to derive only indirect indications about the traffic on main roads, by means of interpretation of derived maps;
- Conventional surveys are expensive and it is possible to guess that, when informations derived by mobile phone data will be available on the market, their cost will be relatively low, being them already collected by providers both for accounting and for network monitoring.

The aforementioned constrains imposed by the Italian law on the use of CDR data make unfeasible to reproduce the tentative evaluation of OD matrices suggested for example by (Bekhor et al., 2011): while in that case the produced matrices were directly derived by tracks of a selected sample of cell phone users, the data at our disposal were aggregated OD counts, with no reference to the users generating

Table 1. Comparison of the available sources on mobility in Lombardy

	Survey on Mobility (OD) Lombardy Region 2002	Census (on Commuting) ISTAT 2001	Mobile Phone Data
Sample	750K interviews	All residents	Mobile phone users ~1.5M per day
Type of movement	All	Study and work	All
Reference Period	"Typical" working day of 2002 (one Wednesday)	One working day of October 2001	Every day
Updates	No	Census 2011 (results not yet available)	Continuous
Information on vehicle	Yes	Yes	No
Spatial resolution	Municipalities, aggregation of minor municipal districts, subdivision of major municipalities	Municipalities	Variable aggregation of cells
Temporal resolution	24 hours	7am-10 am	Hourly or sub-hourly
Cost	Expensive	Very expensive	Not known

them. To better explain the difference: the first data type let for example to infer the "home" and the "work" position of the users (resp. the most recurring position at night and weekday): on this basis one can aggregate the number of users living in city O and commuting to city D; the data at our disposal do not permit these kind of evaluation: we just know that a certain number of users in the course of a hour of the day moved from city O' to city D', but we don't know if O' is the residence of the counted users and D' their final destination. We just know the *flow*.

Despite this apparent limitation even these "legal constrain free" OD data are very appealing for urban analysis and investigations on mobility.

In the following of this section we propose some applications of the data.

The analysis of the activity of mobile phone users permitted to put in evidence the main hourly distribution of origin destination movements of a huge sample of people (more than one million per day).

We started from the hourly origin destination matrices (October, 19th 2011) of mobile phone users among the 526 zones (tiles) of the more fine grained tessellation. For each zone it was available a set of directed connections towards the other and for each connection it was available the number of traced users.

Our goal was to find a synthetic visualization. Our proposal is to consider "prevalent fluxes" of mobility at different hours of a typical working day defined as the sum vector of all the fluxes moving from each zone.

More specifically we associated to each origin destination flux, the vector applied to the centroid of the origin tile, directed to the centroid of the destination tile and having the magnitude given by the value in the origin destination matrix corresponding to the selected origin and destination. We then calculated, for each tile, the sum vector of all the vectors applied to their centroid. It is characterized by two dimensions: the magnitude, which is function of the magnitudes of the original vectors, and the angle which expresses the prevalent direction of the flux.

A set of maps of the sum vector moving from each zone at different hours has been produced in order to highlight the main patterns of mobility during a typical working day (October 19th, 2011). The interested reader can visit our interactive web version of the map showing prevalent fluxes of mobility in a working day at the following URL: http://www.ladec.polimi.it/maps/od/fluxes.html .

The maps cover a wide area that goes from Milan in the West to Brescia in the East and comprises many populated city regions (Pavia in the South, Monza in the North and Bergamo between Milan and Brescia).

The length and the thickness of the arrow are proportional to the magnitude of the sum vector and they are related to the prevalent fluxes of mobile phone users at specific hours. The convergence of travels toward the main centers during the morning, the more complex direction of movements during the afternoon, are some interesting phenomena emerging from our analysis. The maps represent also the main infrastructures (railways and highways) and the zones of the tessellation, colored according to the intensity of the outgoing fluxes (light red color corresponds to lower intensity while dark red corresponds to higher intensity).

The maps show *the variability* of phenomena that conventional data sources, such as census data, cannot give for a typical day. A broad use of the territory and an articulation of daily moves are visible every hour. The maps can be used as meaningful tools for monitoring the use of the infrastructural networks and of the urban spaces. On the one hand, the morning map (9pm; Figure 6) confirms a polarization of movements towards the main centers offering job opportunities and highlights also the most commonly used infrastructures. On the other hand, the aggregated flows of mobile phone users in the afternoon (at 5 pm; Figure 7) allow to recognize significant places for shopping and leisure, that are attended after work. This type of information is difficult, if not impossible, to monitor through conventional data at a comparable spatial and temporal resolution.

Figure 6. Aggregated flows of mobile phone users: 9 am – 2011-10-19

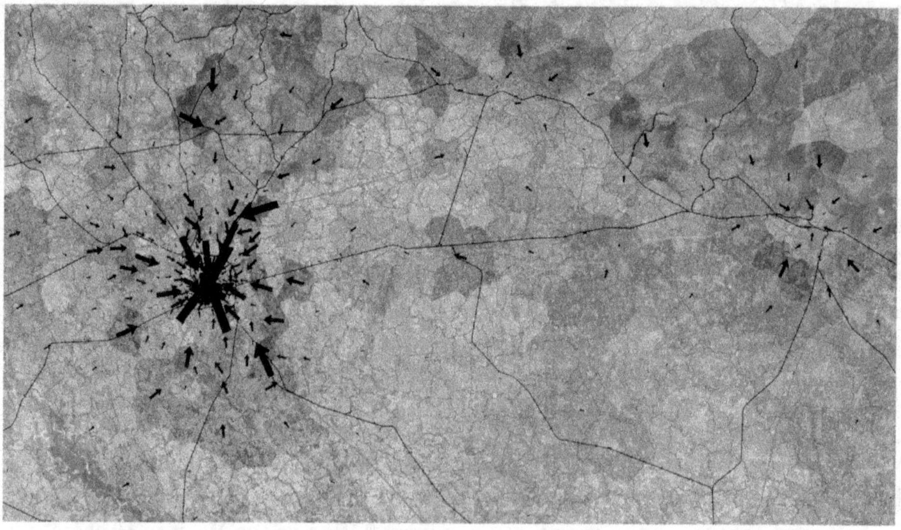

*For a more accurate representation see the electronic version.

Figure 7. Aggregated flows of mobile phone users: 5 pm – 2011-10-19

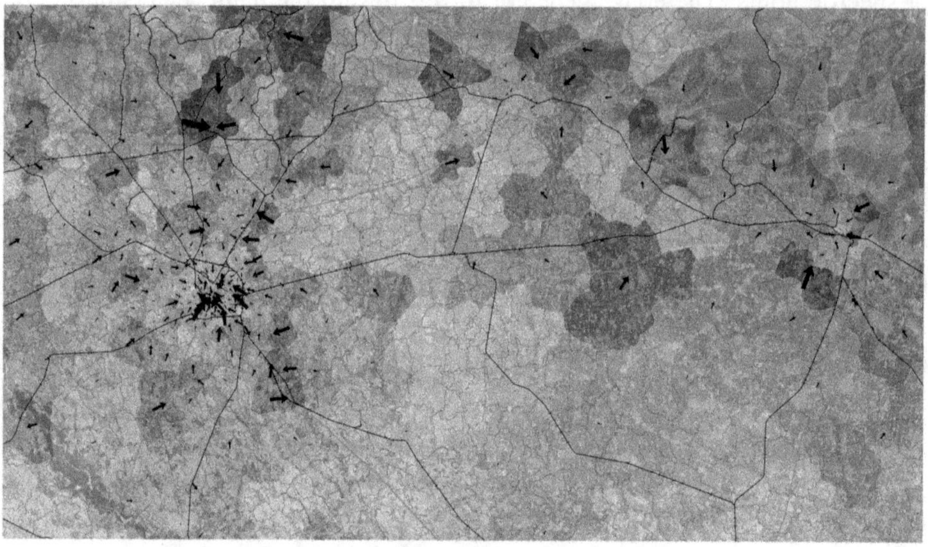

*For a more accurate representation see the electronic version.

The automatic aggregation of municipalities (313 polygons) has been used for producing maps of mobile phone users' fluxes directed from each zone towards the others. We used this tessellation in order to have a better visualization of the complex and more relevant directed connections between the zones.

Only flows of more than 100 users are shown by oriented lines that connect the centroid of each origin zone with the centroid of the destination zones. In order to better visualize the overall patterns of mobility in the region, in and out flows from/ to Milan city were excluded.

The map (Figure 8) evidences relevant relations and fluxes towards the main cities of the Lombardy Region, but also some interesting patterns in the Northern Milan area where a high density of huge connections emerges.

It is also evident a linear element of interconnected centers along some important infrastructural corridors (i.e. the Sempione road in the northern western side of the region and along the highway in the western side of the map between Bergamo and Brescia). These maps show the complexity of daily mobility patterns that modify the hierarchical structure of the cities where traditionally the physical relationship between jobs and homes was the main reason of mobility. The density of fluxes at 5 pm (Figure 9) describes not only the return home, but also the unsystematic mobility related to individual habits, as an effect of the diversified uses of the Milan urban region.

Figure 8. Origin destination fluxes of mobile phone users: 8 am, 2011–10-19

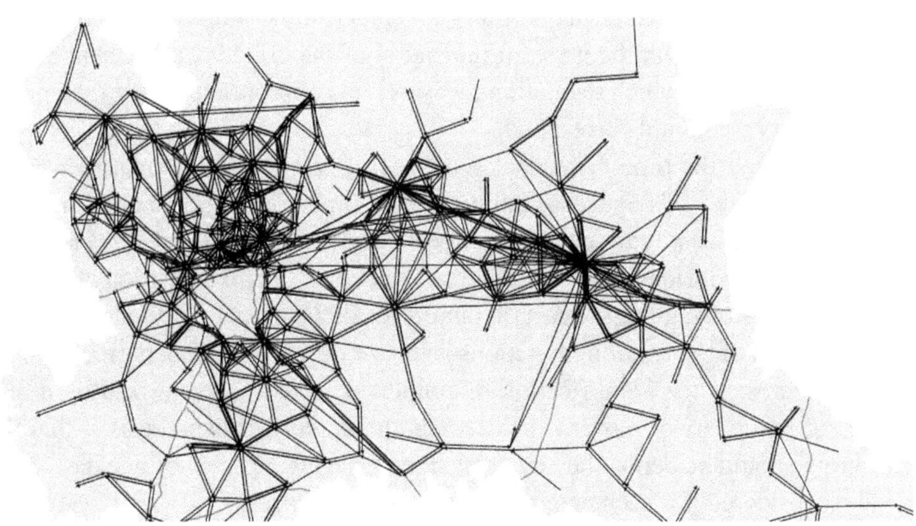

**For a more accurate representation see the electronic version.*

Figure 9. Origin destination fluxes of mobile phone users: 5 pm, 2011-10-19

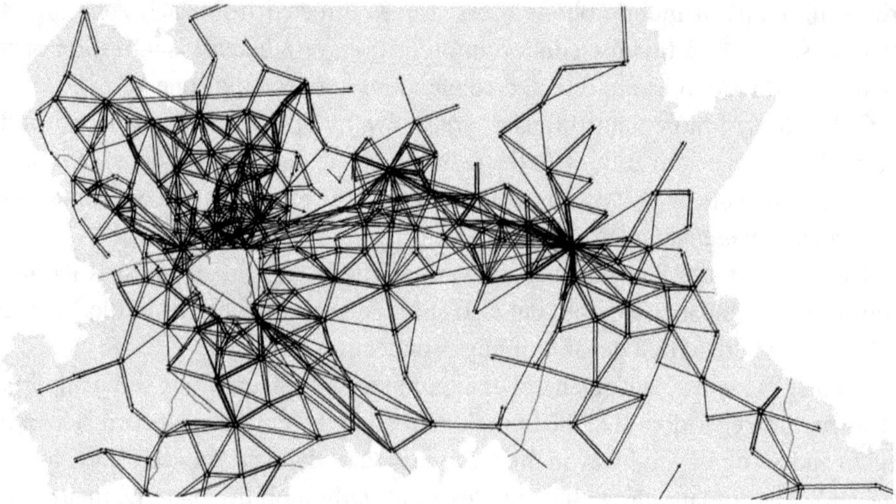

**For a more accurate representation see the electronic version.*

An interesting question concerns the *regularity* of people movements. It has been recently proven, by the examination of the tracked positions of individuals, that "human trajectories show a high degree of temporal and spatial regularity" (Gonzalez et al., 2008). Conversely, instead of focusing on individual people and observing how they move through places, we can study recurrent mobility practices by focusing on places and observing how people pass through them.

From the perspective of public transport policies, the individuation of regular mobility practices could be a useful indication of the effective catchment area of an urban context, to which regulation measures and appropriate rates of the public transport service should correspond.

We analyzed the time dependent sequence of cell phone OD matrices at our disposal, related to different tessellations of the Lombardy Region. For each couple (origin, destination) we considered the associated time-varying signal $F_{od}(time)$. More specifically we considered the time series obtained by the juxtaposition of four out of the five days (excluding August) at our disposal: 24 hours for 4 Wednesday. The (time) domain of the resulting functions is the set of integer numbers from 1 to 96.

Our interest is the identification of couples of zones characterized by a high degree of *regularity* of movements between them, and our hypothesis is that this regularity should be denoted by cyclical signals having a period of one day.

We considered, for each od couple, the seasonal process

$$F_{od}(t)=\Sigma\beta_i D_i(t)+\varepsilon(t)$$

where the sums are taken on i varying in {1,…,24} and the D_i are seasonal (i.e. hourly) indicator variables.

We computed for each origin-destination couple the coefficient of determination (or multiple R^2) resulting from fitting the seasonal means model to the corresponding OD-fluxes data. The R^2 is a measure of the quality of the model fit. We took this as a synthetic indication of the regularity of the signal.

Figure 10 depicts some examples of the original cell phone OD data and the fit of the seasonal means model. The curves are relative to four OD couples having Milan, the Lombardy region capital, as the destination: they are the first four best fitted curves, the first (bottom left) representing Milan internal fluxes.

The plot in Figure 11 depicts in colors the value of R^2 for all the origin-destination couples in all three tessellations (resp. 202, 313 and 526 tiles). It is clear the emergence of a symmetric pattern of the R^2: if two zones are linked by regular movements in one direction, it is likely to observe a regularity also in the opposite direction.

Finally, in figure 12 we map the areas of the Lombardy Region (considering here the 202 OD zones tessellations) for which the flows towards Milan are more

Figure 10. Examples of OD fluxes directed towards Milan (solid lines) and fit of the seasonal means model (dashed lines). The four best fitted curves (sorted from bottom left to top right according to R^2). Sampling rate is 1 hour. Each colored region in tones of gray represents one day of data, each day being one Wednesday in a different month of the year 2011.

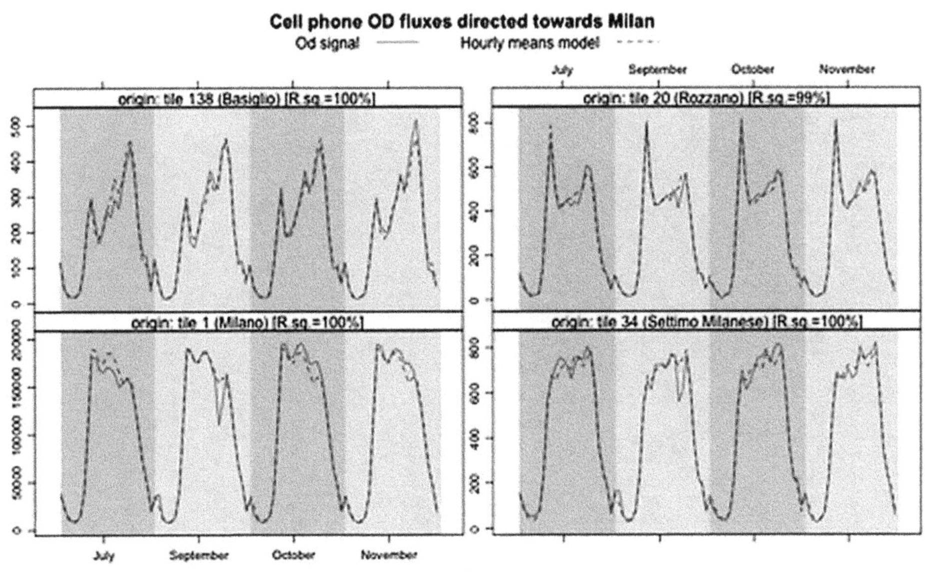

Figure 11. R squared values of the fit of each origin-destination signal to the seasonal means model.

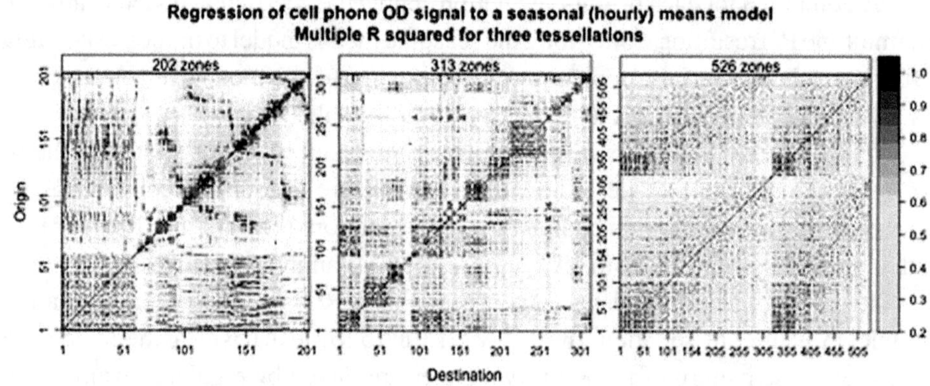

Figure 12. Regularity of fluxes directed towards Milan: pap of the R squared values for the 202 zones tessellation of the Lombardi Region. The perimeter of the institutional boundaries of the management of the local public transport is superimposed.

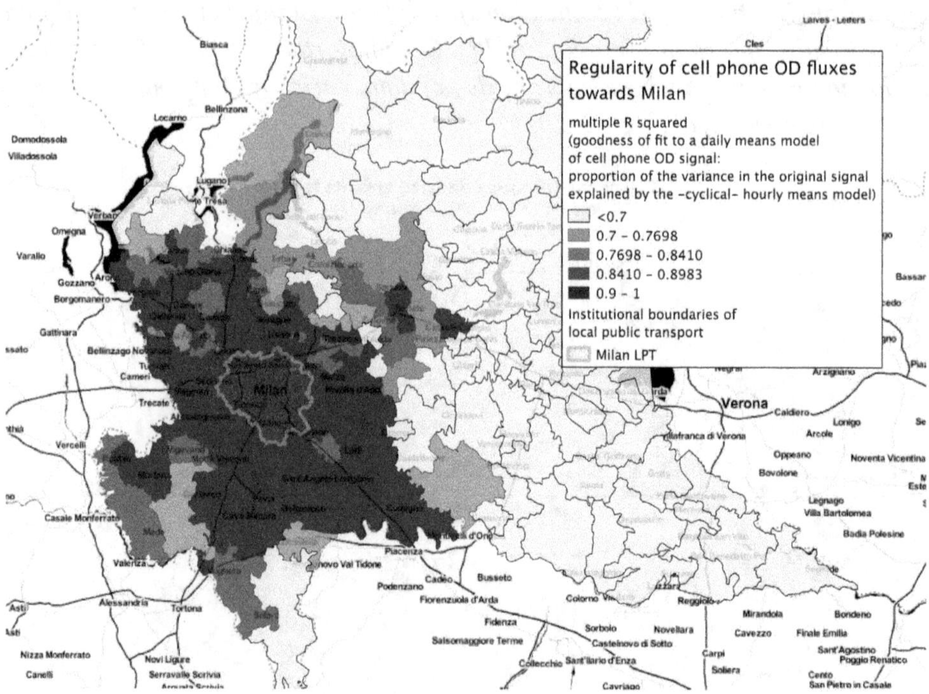

*For a more accurate representation see the electronic version.

regular, varying the days. The institutional boundaries for the management of the Local Public Transport are superimposed to the map: if we compare this perimeter and the zones with high regularities, we can easily notice the inconsistence in the boundaries.

The inadequacy of the institutional boundaries in the Milan Urban Region (as in other urban regions in the whole Italy) has been one of the subjects of the more broad cultural and political debate on how policy demands generated from *temporary populations* should be dealt by public institutions. It is a problem of equity and efficiency: the impact on the budget of the Municipality due to groups of people, which are part of its economic jurisdiction but that at the same time are not voting nor tax-payers is not negligible. The net cost of these population incurred by the Milan municipality is estimated in 295,52 M euros (Bernareggi, 2013). Reciprocally temporary populations use urban services governed by local administrations they did not vote (Martinotti, 1993).

The individuation of the *actual* boundaries of the provenience of these populations could contribute to the definition of more fair and efficient policies.

IMPLICATIONS FOR POLICIES

The research allowed us to test the potential of mobile phone data in explaining relevant urban usage and mobility patterns at the Milan urban region scale and in understanding the dynamic of temporary populations, two important topics that can be hardly intercepted through traditional data sources. This opens new implications for the urban research community which needs to elaborate new strategies to integrate traditional data with user generated data, such as mobile phone activity, in order to achieve a better comprehension of urban usages, in time and in space.

The presented data and methodology let the recognition of effective mobile populations in the urban environment. This knowledge can be exploited by decision makers for the definition of specific policies directed to temporary populations, which are more and more important in contemporary cities, otherwise ignored.

Describing the trends of use of urban spaces, the maps of mobile phone data give important information for mobility policies: the lack of coincidence between the mobility practices in the peak hours in the morning and in the afternoon when the chains of displacements are very articulate and complex, allows to recognize not only the variability in mobility practices, but also the places where these practices are occurring.

The commuters between 8 am and 9 am, become city users between 5 pm and 7 pm. This phenomenon strictly affects land use and can pose new questions and indications for transport policy.

Indeed, if we overlay the boundary of the institutional management of local public transport in the Milan area with the areas of mobility practices, taken from the mobile phone data (Figure 12), we can observe the "deep structural effects of the mobility of people on urban policies" and the obvious disconnection between fixed jurisdictions and "mobile factors" (Estebe, 2008).

The variability in the space-time of use of urban spaces resulting by mobile phone data is also revealed by:

- the spaces of night leisure that define a geography of places densely crowded at Saturday night, that is quite different from the territories of night work during the week (Monday to Friday night);
- the shopping and leisure spaces during the weekend (between 10 am and 8 pm) show the inner city center of Milan, but also some commercial malls along the ring roads;
- the space of the residence, where most significant call-densities are concentrated in the evenings and on weekends and represent the "negative copy" of the work places;
- the spaces of temporary events (International Design Week) that attract a significant portion of tourists and city users who visit several places in the city that are not identified by traditional sources and are not limited to traditional exhibition spaces.

The same data helps us to question some interpretations in the literature on the erratic behaviors of metropolitan populations, on the nomadism that characterizes the contemporary practices, that surveys on mobile phone data have already undertaken (Gonzalez et al., 2008). Some research about a significant sample of mobile phone data have, in fact, contested interpretations of nomadism of contemporary populations.

If they confirm the high density of commuting, they also show the strong recursion of the paths. In other words we move more during the day, but according to the known and usual paths.

REFERENCES

Ahas, R., & Mark, U. (2005). Location based services - new challenges for planning and public administrations? *Futures*, *37*(6), 547–561. doi:10.1016/j.futures.2004.10.012

Ascher, F. (2004). Les sens du mouvement. Modernité et mobilités. In S. Allemand, F. Ascher, & J. Lévy (Eds.), Les sens du mouvement. Modernité et mobilités dans les sociétés urbaines contemporaines (pp. 20–34). Paris: Belin.

Balducci, A., Fedeli, V., & Pasqui, G. (2010). *Strategic planning for contemporary urban regions: city of cities : a project for Milan*. Burlington, VT: Ashgate.

Becker, R. A., Caceres, R., Hanson, K., Loh, J. M., Urbanek, S., Varshavsky, A., & Volinsky, C. (2011). A Tale of One City: Using Cellular Network Data for Urban Planning. *IEEE Pervasive Computing*, *10*(4), 18–26. doi:10.1109/MPRV.2011.44

Bekhor, S., Cohen, Y., & Solomon, C. (2011). Evaluating long-distance travel patterns in Israel by tracking cellular phone positions. *Journal of Advanced Transportation*. doi:10.1002/atr.170

Bernareggi, G. M. (2013). L'istituzionalizzazione della città metropolitana di Milano – Aspetti economici. In A. Ceriani (Ed.), *Aggiornamento della ricerca "Gli enti locali nella transizione verso il federalismo-Effetti ordinamentali della Spending review in Lombardia"* (pp. 75–84). Éupolis Lombardia.

Bolla, R., & Davoli, F. (2000). Road traffic estimation from location tracking data in the mobile cellular network. *2000 IEEE Wireless Communications and Networking Conference. Conference Record (Cat. No.00TH8540), 3*. doi:10.1109/WCNC.2000.904783

Bourdin, A. (2005). Les mobilités et le programme de la sociologie. *Cahiers Internationaux de Sociologie*, *118*(1), 5–21. doi:10.3917/cis.118.0005

Caceres, N., Romero, L. M., Benitez, F. G., & del Castillo, J. M. (2012). Traffic Flow Estimation Models Using Cellular Phone Data. *IEEE Transactions on Intelligent Transportation Systems*, *13*(3), 1430–1441. doi:10.1109/TITS.2012.2189006

Caceres, N., Wideberg, J. P., & Benitez, F. G. (2008). Review of traffic data estimations extracted from cellular networks. *Intelligent Transport Systems, IET*, *2*(3), 179–192. doi:10.1049/iet-its:20080003

Calabrese, F., Liu, L., Lorenzo, G. D., & Ratti, C. (2011). Estimating Origin-Destination flows using opportunistically collected mobile phone location data from one million users in Boston Metropolitan Area. *IEEE Pervasive Computing*. doi:10.1109/MPRV.2011.41

Calabrese, F., Di Lorenzo, G., Liu, L., & Ratti, C. (2011). Estimating Origin-Destination Flows Using Mobile Phone Location Data. *IEEE Pervasive Computing*, *10*(4), 36–44. doi:10.1109/MPRV.2011.41

Cayford, R., & Johnson, T. (2003). Operational parameters affecting use of anonymous cell phone tracking for generating traffic information. In *Institute of transportation studies for the 82th TRB Annual Meeting* (Vol. 1). Academic Press.

Ehrenberg, A. (1995). *L'individu incertain*. Paris: Calmann-Lévy.

Estebe, P. (2008). *Gouverner la ville mobile*. Paris: PUF. doi:10.3917/puf.esteb.2008.01

Gonzalez, M. C., Hidalgo, C. A., & Barabasi, A.-L. (2008). Understanding individual human mobility patterns. *Nature*, *453*(7196), 779–782. doi:10.1038/nature06958 PMID:18528393

Jain, A., & Dubes, R. (1998). *Algorithms for Clustering Data*. Upper Saddle River, NJ: Prentice-Hall.

Kaufmann, V. (2000). *Re-thinking mobility*. Aldershot, UK: Ashgate.

Regione Lombardia, Direzione Generale Infrastrutture e mobilità. (2002). *Indagine Origine/Destinazione regionale 2002 - sintesi*. Author.

Manfredini, F., Pucci, P., Secchi, P., Tagliolato, P., Vantini, S., & Vitelli, V. (2012). *Treelet decomposition of mobile phone data for deriving city usage and mobility pattern in the Milan urban region*. Academic Press.

Martinotti, G. (1993). *Metropoli: la nuova morfologia sociale della città*. Bologna: Il Mulino.

Nuvolati, G. (2003). Resident and Non-resident Populations: Quality of Life, Mobility and Time Policy. *The Journal of Regional Analysis & Policy*, *33*(2), 67–83.

Ratti, C., Pulselli, R. M., Williams, S., & Frenchman, D. (2006). Mobile Landscapes: Using location data from cell phones for urban analysis. *Environment and Planning. B, Planning & Design*, *33*(5), 727–748. doi:10.1068/b32047

Reades, J., Calabrese, F., Sevtsuk, A., & Ratti, C. (2007). Cellular Census: Explorations in Urban Data Collection. *IEEE Pervasive Computing*, *6*(3), 30–38. doi:10.1109/MPRV.2007.53

Scheller, M., & Urry, J. (2006). The new mobilities paradigm. *Environment & Planning A*, *38*(2), 207–226. doi:10.1068/a37268

Soto, V., & Frías-Martínez, E. (2011). Robust Land Use Characterization of Urban Landscapes using Cell Phone Data. *The first workshop on pervasive Urban Applications (PURBA)*.

Soto, V., & Frías-Martínez, E. (2011). Automated land use identification using cell-phone records. In *Proceedings of the 3rd ACM international workshop on MobiArch* (pp. 17–22). Academic Press. 10.1145/2000172.2000179

Urry, J. (2002). *Sociology beyond societies: Mobilities for the twenty-first century*. Routledge.

Vantini, S., Vitelli, V., & Zanini, P. (2012). Treelet Analysis and Independent Component Analysis of Milan Mobile-Network Data: Investigating Population Mobility and Behavior. *Analysis and Modeling of Complex Data in Behavioural and Social Sciences - Joint Meeting of of the Italian and the Japanese Statistical Societies.*

ENDNOTES

1 More formally we can define the Erlang E_{xj} relevant to the pixel x and to the j-th quarter of an hour as: $E_{xj} = \frac{1}{15}\int_{15(j-1)}^{15j} N_x(t)\,dt$ where $N_x(x)$ is the number of mobile phone using the network within pixel x at time t, hence E_{xj} is the temporal mean over the j-th quarter of an hour of the number of mobile phones using the network within pixel x.

2 With mobile phone activity we intend each interaction of the device with the mobile phone network (i.e., calls received or made, SMSs sent or received, internet connections, etc.).

Chapter 9
Environmental Monitoring Based on the Wireless Sensor Networking Technology:
A Survey of Real-World Applications

Eirini Karapistoli
University of Macedonia, Greece

Ioanna Mampentzidou
University of Macedonia, Greece

Anastasios A. Economides
University of Macedonia, Greece

ABSTRACT

This chapter investigates real-life environmental monitoring applications based on wireless sensor networks (WSNs). Wireless sensor networking is an emerging technology, which has been adopted by many scientific fields in order to accurately and effectively monitor climate phenomena such as air pollution, destruction phenomena, etc. It has also been widely used in agriculture as well as in horticulture for field monitoring. In this chapter, the authors provide a critical overview of the basic components existing WSN deployments use. They also categorize these deployments, 111 in total, into five different field categories in order to provide a general view of the technologies used, the conditions under which the deployments were conducted, and much more. Then, five easy-to-use guides are provided discussing basic considerations for deploying WSNs in each of these fields. In order to showcase the usefulness of consulting the resulted guides, the authors consider representative application scenarios for each of these field deployments.

DOI: 10.4018/978-1-5225-5978-8.ch009

INTRODUCTION

The curiosity of humankind for the natural environment was the driving force that led him to search and learn issues related to complex environmental phenomena. Thanks to that and through detailed monitoring, nowadays we have the knowledge to predict events or prevent them from happening. At the very beginning, every physical condition and parameter was measured by some analog devices, which, at that time, were very innovative, but too costly and not very accurate. Following on, the use of digital data loggers replaced these analog devices, and despite being less expensive and more easy to use, this data logging technology was still inefficient. Recent technological advances, namely continuing miniaturization of electronics, the availability of large data storage and computational capacity, and the pervasive connectivity of the Internet, led to the development of tiny sensor devices with sensing, processing, and communicational capabilities that were able to provide with accurate local measurements of the monitored parameters. These devices, which are called wireless sensor nodes, when deployed in an area, form a Wireless Sensor Network. WSNs constitute a powerful and promising tool for monitoring events (Akyildiz, Su, Sankarasubramaniam, & Cayirci, 2002). This new approach of gathering information from the environment could provide with the much-needed feedback between the monitored field, the local climate conditions, and the human's decisions of treating the field.

The initial development of WSN was motivated by military applications such as enemy monitoring and tracking, force protection, battlefield surveillance, etc. Nowadays, WSNs are used in many other application fields such as agriculture, environmental monitoring (e.g., air-water pollution, greenhouse phenomena, monitoring of oceans, volcanoes, forests, etc.), health monitoring, home automation, and more. In this article, we consider the case of WSNs being deployed for environmental monitoring purposes. Such WSNs are also referred to as Environmental Sensor Networks (ESNs) (Martinez, Hart, & Ong, 2004), (Corke, Wark, Jurdak, Hu, Valencia, & Moore, 2010). Depending on the application, ESNs can be employed to perform habitat monitoring (Polastre, 2003), flooding-landslide-earthquake detection (Tan, Xing, Chen, Song, & Huang, 2013) monitoring of volcanic eruptions (Song, Huang, Xu, Ma, Shirazi, & Lahusen, 2009), (Huang, Song, Xu, Peterson, Shirazi, & LaHusen, 2012), microclimate monitoring for farms and rain forests (Wark, et al., 2008), cattle monitoring and control (Kwong, et al., 2009), and much other.

While there is an endless list of scientific papers discussing WSN-based environmental monitoring applications, these works do not provide deployment guidelines or design considerations with regard to the deployed sensor network. Hence, someone willing to deploy a WSN for these applications is left alone to choose the technologies to be used. In order to fill the gap, this paper provides generic guidelines on how to deploy WSNs in five different real-life environmental monitoring applications. In achieving this objective, the paper discusses the different field deployments summarizing at the same time the software and hardware components used in each of them. *Using this information, it then proposes five easy-to-use guides for deploying WSNs in these deployment fields.* Finally and more importantly, it showcases the usefulness of consulting the resulted guides, by considering representative real-life application scenarios for each of the five different field deployments. It is expected that a number of stakeholders including researchers, scientists, *environmentalists,* agronomists, farmers, etc., will be benefited from consulting the produced guides.

To the best of our knowledge, limited work has been conducted in this largely extensive investigated research area. In (Giannopoulos, Goumopoulos, & Kameas, 2009), the authors present their experience in deploying a WSN for environmental monitoring applications, and based on the gained experience, they introduce guidelines for these types of deployments. *There do not provide similar guidelines for alternate field deployments such as for the agricultural field. This is also the case for the works conducted in* (Oliveira & Rodrigues, 2011), (Barrenetxea, Ingelrest, Schaefer, & Vetterli, 2008), (Bri, Garcia, Lloret, & Dini, 2009) where the authors present design guidelines for building a WSN for environmental monitoring purposes. In these papers, the *application alone is an indirect guide. On the other hand, our proposed guide refers to the deployment itself, providing practical, and at the same time generic guidelines. Based on these characteristics, the major contribution of this paper lies on the fact of becoming a useful and practical deployment guide for environmentalists, scientists, and general users who are* willing to deploy a WSN for such applications and *are not necessarily experts at this wireless networking technology. In addition to the above, the goal of using these guides is to contribute in making the WSN technology part of the user's everyday working life, making this technology even more attractive and easy to use.*

The remainder of the paper is organized as follows. Section II frames basic concepts relevant to the wireless sensor networking technology. It also summarizes the hardware and software components employed by the sensor nodes. Based on the collected information of Section II, Section III pinpoints existing WSN deployments in five different field applications. Section IV describes five easy-to-use guides for deploying WSNs in the aforementioned field deployments. Finally, Section V summarizes the major findings of this work.

WIRELESS SENSOR NETWORKS

Network Components

As shown in Figure 1, a typical WSN consists of a large set of inexpensive wireless sensor nodes and a base station (BS) (Akyildiz, Su, Sankarasubramaniam, & Cayirci, 2002). A base station, also referred to as the sink, is the data collection point of a WSN. It functions as a network-layer gateway (GW) to another network, a powerful data processing or storage center, or an access point for human interfaces. Moreover, it can be used as a connection to disseminate control information into the network or extract data from it. In most applications, this special type of node is more powerful than sensor nodes and it is assumed to have unlimited available energy. Typically, a sensor network will contain many sensor nodes for each sink. In fact, there may only be a single sink for the entire sensor network. Upon receiving the collected sensory data from the wireless nodes, the sink is then tasked with forwarding the data to the task manager node via the Internet or via satellite communications.

The wireless sensor nodes are tiny devices capable of a) interacting with their environment by sensing and/or controlling some physical parameters, b) performing some data processing, and c) communicating with other connected nodes in the network for accomplishing a common task. Each sensor node also has some storage capacity to store the collected data. These small, battery-powered motes, however, have limited energy capacity and computational processing capability requiring mechanisms to minimize their energy consumption and ensure a long-lasting and unattended operation in remote geographic areas without the need for replacement/recharging the battery. Regarding their placement, sensor nodes are densely deployed inside the observation phenomenon either in an ad hoc or in a pre-planned manner. For deterministic deployments, the sensors are manually placed and data is routed

Figure 1. A typical WSN

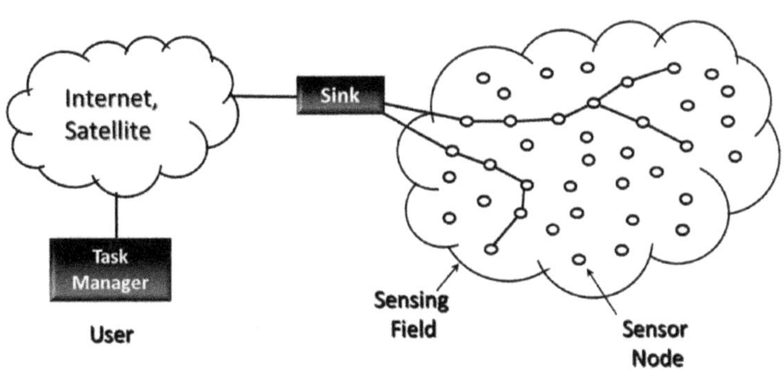

through pre-determined paths. In contrast, random node deployment involves the sensor nodes being scattered randomly creating an ad hoc infrastructure.

The typical architecture of a sensor node is shown in Figure 2. Each sensor node consists of five main components, namely a microcontroller unit, a transceiver unit, a memory unit, a power unit, and a sensing unit (Hill, 2003). The sensor board might consist of different types of sensors ranging from low sampling rate magnetic sensors to thermal, visual, infrared, acoustic, or radar sensors. As it will be seen, each of these components becomes a determinant design choice while engineering a WSN for a specific field deployment. Among the issues affecting their design is the way power is consumed by these nodes. Figure 2 shows that the node spends most of its energy in communicating with other nodes. Hence, in order to minimize power consumption it is desirable to have the node's transceiver in the sleep mode as much as possible.

Figure 2. (left) Basic sensor node architecture, (right) Typical power consumption of the node's hardware components
TI CC2420 Datasheet, 2010.

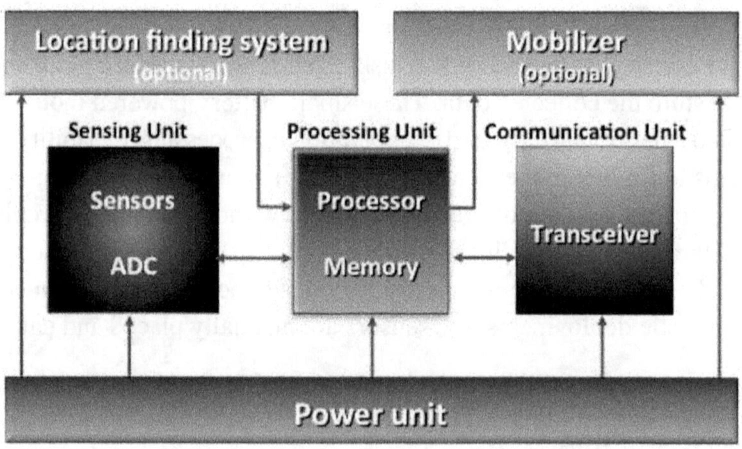

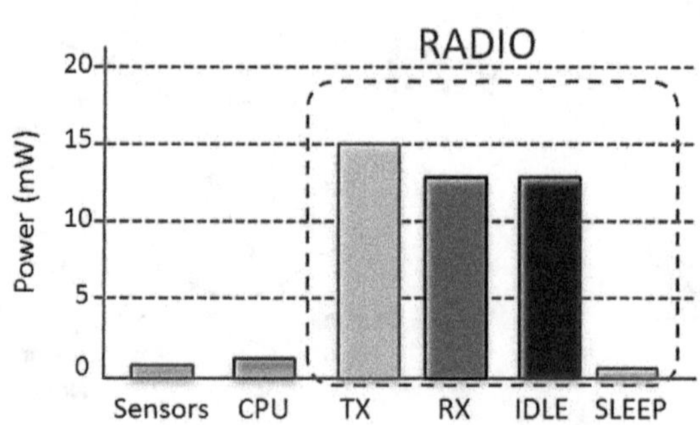

Communication Patterns

Regarding communication, WSNs exhibit a wide variety of communication patterns. The major communication patterns include the following (Demirkol, Ersoy, & Alagoz, 2006):

- Node to BS communication (*many-to-one*), when for instance, sensor nodes transmit sensor readings and alerts to the BS.
- BS to node communication (*one-to-many*), when the base station multicasts requests for sensor readings.
- BS to all nodes (*one-to-all*), e.g., when the BS broadcasts routing beacons or queries to the entire network.
- Communication amongst a defined set of nodes organized in a *cluster* (*one-to-one*).

WSNs are also engineered to support three network topologies, namely *star*, *tree*, and *mesh*, in which the sensor nodes are engaged in single hop or multi hop communications (Hou, Li, & Stojmenović, 2005). Multi-hop message rely is necessary when the sink is out of the transmission range of the sources. In the star topology, every node in the WSN is connected directly to its sink node only. While being simple in its implementation, it is not recommended in deployments with many sensor nodes and large distances between the nodes and the sink. This is because in case of link failure between one node and the sink there is no alternative route of communication. The tree-based topology on the other hand, is an ideal option for WSNs with large distances between the sink node and the rest of the nodes. Of course, it has its own disadvantages, for example, the nodes that are close to sink are being overcharged, hence collisions and delays are unavoidable. In addition, if a node fails to operate for an unknown reason, then the communication with its children will be lost. Finally, in the mesh topology, every node is connected with a maximum number of peer nodes. It is the specification of a fully mesh topology and with the appropriate routing algorithms that ensures the recovery of the network from breakdowns (Al-Karaki & Kamal, 2004).

Data gathering is another crucial communication ingredient of a WSN (Demirkol, Ersoy, & Alagoz, 2006). In WSNs, data gathering is triggered by queries or events. Accordingly, it could take place either in a *push mode* or in a *pull mode*. In the push mode, the information holders, namely the sensor nodes, actively send data to the sink node in response to significant changes in environmental phenomena they detected. In the pull mode instead, the communication is initiated by the users. The sensor nodes respond to the queries imposed by the users by reporting the collected data to the sink node. Based on the above categorization, a WSN can be classified as *event-driven*, *query-driven*, *time-driven* (i.e., by sending reports in a periodic manner), or as a *hybrid* system.

Challenges in Deploying WSNs

Designing a WSN for environmental monitoring applications is influenced by many challenging factors (Puccinelli & Haenggi, 2005). Next, we summarize some of the challenges that affect their design.

- **Power Consumption:** Energy conservation is a central design consideration in WSNs. This is because sensor nodes are autonomous devices that usually derive their power from a battery mounted on each node. It becomes necessary to have an energy-saving mechanism in every component of the WSN system to prolong the lifetime of each node and accordingly, the lifetime of the deployed network as a whole.
- **Node Deployment:** *Deployment* is an important *issue* in *WNSs*, and affects the performance of the network directly. The large number of nodes expected in sensor network deployments and the unpredictable nature of deployment conditions, introduce important scalability and reliability concerns.
- **Data Collection and Management:** Sensor deployments can generate far more data than can be managed by the traditional methods used in field research, placing data quality and control a challenging issue for individuals to effectively monitor.
- **Fault Tolerance:** Some sensor nodes may fail or be blocked due to lack of power, physical damage, or environmental interference. The failure of sensor nodes should not affect the overall task of the sensor network. If many nodes fail, various communication protocols must accommodate formation of new links and routes to the data collection point, namely the BS. Otherwise, multiple levels of redundancy may be needed in a fault-tolerant sensor network.
- **Wireless Connectivity:** Wireless communication in WSNs is quite unpredictable because of the low-power RF transceivers the wireless sensor nodes are equipped with. Moreover, while sensor nodes are expected to be highly connected (recall that node density is expected to be high), this, however, may not prevent the network topology from being variable and the network size from being shrinking due to sensor node failures, leading to connectivity issues.
- **Coverage:** In WSNs, each sensor node obtains a certain view of the environment. A sensor's given view of the environment is limited both in range and in accuracy, meaning that it can only cover a limited physical

area of the environment. Hence, area coverage is also an important design parameter in WSNs.

- **Self-Organization:** A WSN is typically an ad hoc network, which requires every sensor node to be independent and flexible enough as well as to be self-organizing and self-healing according to different situations. While this may be seen as an asset, on the other hand it means that there is no fixed infrastructure available for the purpose of network management.
- **Maintenance:** Depending on the function and deployment area of the particular sensor network, the sensor nodes may be left unattended for long periods. This fact necessitates the periodic replacement of batteries/nodes in the sensor field. Given the associated high maintenance costs, an appropriate design is needed to take deployment and maintenance needs into consideration.
- **Security:** WSNs are exposed to security challenges as well. Depending on the application, the sensor nodes may be deployed in an environment open to adversaries, bad weather, and so on. The likelihood that an unattended sensor suffers a physical attack in such an environment is therefore high. From the system point of view, it is critical that the information provided by the nodes be authenticated and the integrity verified.

With all these challenges in mind, in the following section, we analyze the available network design options both in terms of software and hardware. The analysis that follows takes into account every single WSN deployment surveyed in this work, 111 in total.

WSN COMPONENTS IN FIVE ENVIROMENTAL MONITORING APPLICATIONS

Major Design Choices: Node Platforms and Operating System (OS)

We begin our analysis with the first design option one needs to consider when deploying a WSN for *real-life environmental monitoring applications*, which is the sensor node platform. The Mica2 (Mica2 Datasheet, 2005), MicaZ (MICAz Datasheet, 2005) and Imote2 (Imote2 Datasheet, 2008) platforms were very popular in the early deployments. Their hardware platform consists of Processor/Radio boards (MPR) commonly referred to as Motes. These battery-powered devices run the Crossbow's XMesh self-forming, low-power networking stack. In addition to running the XMesh stack, each Mote runs the open-source TinyOS operating system, which provides low-level event and task management (Levis, et al., 2005). Despite the great success they enjoyed, these motes, which were developed by the Crossbow technology, are no longer commercially available, meaning that one needs to adapt his/her decisions by choosing other Mica-like platforms for his/her deployment. In recent years, the TelosB-compatible TMote Sky platform (Moteiv Tmote Sky, 2004) has been one of the most demanding platforms due to its ultra-low power consumption, versatility, and reasonable price range. Tmote Sky is an MSP430-based board with an 802.15.4-compatible CC2420 radio chip, a 1 MB external serial flash memory, and two light sensors. Another node platform, the TinyNode trademarked by Shockfish SA (Dubois-Ferriere, Meier, Fabre, & Metrailler, 2006), is currently being used in application requiring long communication range. TinyNode features the Semtech XE1205 radio transceiver, which operates in the 868-870 MHz band and provides a 76Kb/s channel and a range of 500 m using the on-board antenna (ranges of up to 1km are feasible using an external quarter-wavelength omnidirectional antenna). The TinyNode platform comes with full TinyOS support including a complete radio stack, support for over-the-air network reprogramming with Deluge (Hui & Culler, 2004), and bridging software for GPRS/GSM data transfer. A second platform that is increasingly being used in applications dealing with water quality monitoring or animal tracking is the CSIRO's Fleck node (Flecks Datasheet, 2011). Similar to, and inspired by, the Mica-2 mote, the Flecks were designed to overcome limitations in those devices. The Fleck3 nodes use the Atmega 128 processor with 128 KB of program flash memory, 4 KB of RAM, and a stream-based Nordic nRF905 radio transceiver at 915 MHz, which provides a 72Kb/s channel and a range of 500 m.

Recently, two low power embedded platforms, namely the BeagleBoard (The BeagleBoard technology, 2008) and its successor the Pandaboard (The Pandaboard technology, 2010), found their way in being used as gateway nodes in environmental monitoring applications using the WSN technology. By leveraging the many advantages of the open-source Linux ecosystem, and their ultra low cost ($45 only), these platforms have started to become attractive and have been implemented in several long-term monitoring deployments (Mingli & Yihai, 2013), (Alkandari, Alnasheet, Alabduljader, & Moein, 2012), (Rao, Marshall, Gubbi, Palaniswami, Sinnott, & Pettigrovet, 2013). These fully featured, high-performance Single Board Computers (SBCs) are composed of an OMAP 3530 processor from Texas Instruments that can run up to 720MHz. The survey revealed that a number of deployments, especially those developing sensor nodes for monitoring destruction phenomena, design the sensor node platform from scratch. The reason behind building a customized platform is to serve application-specific requirements, such as specific sensing and packaging constraints (Jelicic, Razov, Oletic, Kuri, & Bilas, 2011). In the latter category one may find that the Arduino platform (The Arduino platforms, 2009), an open-source electronics prototyping platform, is intended for such designers and hobbyists. In majority, the aforementioned node platforms use the Atmel's ATMega AVR architecture (Atmel AVR Microcontrollers, 1984) or the Texas Instruments MSP430 16-bit flash microcontroller (MUC) (Texas Instruments Microcontrollers, 1980). The former is used in the Mica-family motes, while the latter is the core of the TelosB platform.

Several approaches exist aimed at managing the node's battery storage. In the default mode of operation, all the energy for the device will come from non-rechargeable, alkaline batteries. Another option is to use rechargeable 1.2-V 2700-mAh NiMH batteries working in combination with monocrystalline solar panels capable of supplying them with charging current. Finally, several platforms combine both rechargeable and non-rechargeable batteries in each device. Although energy harvesting is envisioned as the only way for sustainable sensing systems, it was interesting to witness that power sources other than batteries or static power networks are rarely used. As expected, more than 80% of deployments have powered motes present in the network (BSs and gateways) or at least nodes with increased energy budget. Usually, these motes are capable of running at 100% duty cycle without sleep mode activation. The target lifetime is very dynamic among the surveyed deployments, and it ranges from few hours to several years. Apparently, low duty cycle applications with sampling rate below 1 Hz are the most popular. Long-living deployments instead, use a duty-cycle below 1%, meaning that sleep mode is used 99% of the time. As a consequence, OSs should provide effective routines for duty cycling incurring low computational overhead at the same time.

The most popular on-board sensors are the temperature, light and accelerometer sensors. Depending on the monitored phenomena, several other physical, chemical, and biological sensor modalities are also chosen, namely soil moisture sensors, wind speed and direction sensors, accelerometers, etc. With regard to the radio transceivers, the TI CC1000 radio was in demand in the early deployments. However, in recent years, the TI CC2420 radio is by far the most popular one (TI CC2420 Datasheet, 2010). The IEEE 802.15.4/ZigBee standard for low rate wireless personal area networks (WPANs) (IEEE-802.15.4, 2011) (http://www.eol.ucar.edu/isf/facilities/isa/internal/CrossBow/DataSheets/mica2.pdf, 2005)(used in CC2420 and other radio chips) is currently dominating the communications between the sensor motes. Several platforms support other integrated Radio Frequency (RF) interfaces as well, such as Bluetooth or Wi-Fi radio interfaces. Most of these transceivers provide received signal strength indication (RSSI) or link quality indication (LQI) information, which is necessary in many applications supported by WSNs.

Table 1 outlines the node platform distribution usage in each identified deployment field. The percentages shown below were calculated taking into account the respective WSN deployments surveyed in each of the five deployment fields (34, 18, 21, 16, and

Table 1. Node platform usage

Node Platforms	WSN Deployment Fields				
	Agriculture	**Natural Environment**	**Air-Water Pollution**	**Destruction Phenomena**	**Livestock and Wild Animals**
Crossbow MICA2, MICAZ, Imote2 Motes	19%	49%	17%	44%	34%
Moteiv Tmote Sky	11%	15%	2%	12%	16%
Shockfish TinyNode	4%	21%	-	-	-
CSIRO Fleck nodes	2%	-	18%	-	22%
Custom	39%	12%	34%	30%	18%
Other	25%	3%	29%	14%	10%

22, for the agricultural, environmental, air-water pollution, destruction phenomena, and animal monitoring deployments respectively).

Despite theoretical assumptions about huge networks consisting of thousands of nodes, only few deployments contained more than 200 nodes. 80% percent of the surveyed deployments contain 50 or fewer nodes, while the 34% consist of less than 10 nodes. Almost all networks have a sink node or a base station to collect the sensory data. Moreover, a significant part of the surveyed deployments used multiple sinks in their architecture. Nearly half of the deployments use a regular mote connected to a PC (usually a laptop) as a base station hardware solution. Finally, nearly 47% of the deployments operate in a meshed network, setting up ad hoc networks to wireless transfer environmental data they have collected. The second most popular network topology that is implemented is the star (one-hop) network, especially when the observation phenomenon is spatially restricted.

Since 2002, the TinyOS (Levis, et al., 2005), an open source, flexible, and component-based operating system designed for sensor networks, has been used in nearly 42% of all the deployments. There are several reasons behind that (Mallikarjuna & Kunz, 2011). First, TinyOS has a large community supporting it; therefore, device drivers and protocols are well tested. Second, it has a very low memory footprint that fits in 400 bytes. TinyOS is written using the nesC language and it component library includes network protocols, distributed services, sensor drivers, and data acquisition tools. Another state-of-the-art operating system is the Contiki OS (Dunkels, Gronvall, & Voigt, 2004). Contiki is a lightweight open source OS written in C. Contiki is a highly portable OS and it is build around an event-driven kernel. A typical configuration consumes 2 kilobytes of RAM and 40 kilobytes of ROM.

A full Contiki installation includes features like: multitasking kernel, preemptive multithreading, proto-threads, TCP/IP networking, IPv6, a Graphical User Interface (GUI), a web browser, etc.

The MANTIS Operating System (MantisOS) instead, provides a new multithreaded operating system for WSNs (Bhatti, et al., 2005). MantisOS has a footprint of 500 bytes, which includes kernel, scheduler, and network stack. Its key feature is that it is portable across multiple platforms, i.e., one can test MantisOS applications on a PDA or a PC, and afterwards, the application can be ported to the sensor node. Other OSs that are being use include the LiteOS (LiteOS operating System, 2011), a Unix-like operating system designed for WSNs at the University of Illinois at Urbana-Champaign, and the Fleck operating system (FOS), a cooperative thread-based operating system for the Fleck wireless sensor module. An interesting piece of evidence is that in 11% of the total deployments, a self-made or customized OS is used, emphasizing the still actual WSN user need for a flexible and easily usable OS. Table 2 summarizes the OS distribution usage in each WSN deployment field.

Most of the nodes programmed under the TinyOS use the inbuilt proprietary media access control (MAC) and routing protocols of its networking stack, including the powerful BMAC protocol, and the MintRoute protocol. BMAC is a low-power MAC protocol based on low-power listening (LPL), In LPL, nodes wake up periodically (every 57 msec) for a short period (3 msec) to check for communication activities and to receive messages. No surprisingly, a large number of deployments (nearly half of them) use custom-developed MAC protocols (either CSMA-based or TDMA-based) proving that data link layer problems are very application-specific.

Table 2. Commonly used OSs

Operating Systems (OSs)	WSN Deployment Fields				
	Agriculture	**Natural Environment**	**Air-Water Pollution**	**Destruction Phenomena**	**Livestock and Wild Animals**
TinyOS (Levis, et al., 2005)	41%	71%	7%	40%	49%
FleckOS (FOS)	-	-	13%	-	-
MantisOS (Bhatti, et al., 2005)	-	-	-	16%	-
Contiki OS (Dunkels, Gronvall, & Voigt, 2004),	-	-	-	-	6%
MansOS (Strazdins, Elsts, & Selavo, 2010)	-	-	-	-	11%
Custom	14%	3%	20%	16%	1%
Other	45%	26%	60%	28%	33%

Although Internet Protocol version 6 (IPv6) is a widely discussed protocol for the Internet of Things (IoT) (Vasseur & Dunkels, 2010) and modifications such as the 6loWPAN (Z. & Bormann, 2010) for resource-constrained devices have been developed, the protocol is very novel and, it was not widely used yet; only 5% of the surveyed deployments currently implement the IPv6 networking stack in their operating system (Su, Shao, Vause, & Tang, 2013), (Christin, Reinhardt, Mogre, & Steinmetz, 2009), (Ramos, Foster, FeliciCastell, Fos, & Solano, 2013). However, it can be expected that this number will increase in the coming years since by transforming the sensor devices into smart objects, we realize the vision of ambient networks where many different devices will collect and process information from many different sources to both control physical processes and interact with human users. TinyOS and Contiki OS already provide IP communication, both for the IPv4 and IPv6 stack.

Another particularly promising tendency that has been exploited by several recent deployments is the transformation of smart phones from simple personal communication and computing platforms, to sensing platforms (Mittal, Chetan, Jayaraman, Jagyasi, Pande, & Balamuralidhar, 2012). According to (Miluzzo, Lu, Peebles, Choudhury, & Campbell, 2010), sensor-equipped mobile phones will revolutionize many sectors including that of environmental monitoring. Mobile phone sensing is a complementary and radically different sensing approach in which smartphones fulfill the sensing role by acting as super sensor nodes and gateways (Zoller, Reinhardt, Wachtel, & Steinmetz, 2013). The PEIR project (Mun, et al., 2009) is one such example of participatory, ubiquitous sensing that uses sensors in phones to build a system that enables personalized environmental impact reports by tracking how the actions of individuals affect both their exposure and their contribution to problems such as carbon emissions. Table 3 provides an overview of the previously discussed design choices. More specifically, it lists the possible types of sensors, power sources, optional hardware, communication modalities, networking protocols as well as the different network topologies WSN deployments may use.

Five Environmental Monitoring Applications and Their Components

Agricultural Monitoring

By the time the agriculture domain began to incorporate the wireless sensor networking technology to support its operations, Precision Agriculture (PA) started to flourish (Baggio, 2009), (Blackmore, 1994). Precision Agriculture is the science of precise understanding, estimating and evaluating crops condition with the aim of determining the proper use of fertilizer, and the real needs of irrigation both

Table 3. Deployments: used sensors, power sources, optional hardware, communication modalities, and network protocols

Other Design Choices	**Sensors**	Temperature (air/surface) l Relative humidity l Soil moisture l Salinity l pH l Wind speed and direction l Solar radiation l Light and ultraviolet intensity l Carbon monoxide (CO) l Carbon Dioxide (CO_2) l Methane (CH_4) lAtmospheric pressure l Two-axis accelerometers l Three-axis magnetometer
	Power Sources	Non-rechargeable batteries (Alkaline) l Rechargeable batteries (Lithium Ion, NiMH) l Solar panels
	Optional Hardware	GPS l Gyroscope l Digital Compass l lActuators l Motors l Geophones l Camera, audio and DSP
	Communication Modalities (Radio Transceivers)	RF @ 433 or 868 MHz (Semtech XE1205, TI CC1000, Nordic nRF905) l 802.15.4 @ 2.4 GHz (TI CC2420) WiFi l Bluetooth l GPRS/3G l Satellite links
	Networking Protocols	CSMA l TDMA l MintRoute l CTP l Diffusion l LPL l IPv4 l IPv6 (6LowPAN)
	Deployment Types	Randomly deployed l Static deployment l Air-drop deployment l Attached to animals
	Network Topologies	One-hop l Multi-hop l Mesh l Multi-mesh l Predefined tree

during the sowing and harvesting periods (Shaikh, Shaikh, & Islam, 2010), (Hu, Dinh, Corke, & Jha, 2012). All these functions can effectively be realized using the WSNs technology, since the low power needs and low data rate capabilities of this technology perfectly suite this field of research. Horticulture can also benefit from the use of this technology as indicated in (Riquelme, Soto, Suardíaz, Sánchez, Iborra, & Vera, 2009).

The best way to gather the requirements for deploying a WSN system for PA is to realize what actions the user would like the system to perform. Every single deployment has its own needs, which are imposed by the type of the monitored crop or plant or by other special application-related design requirements, and definitely, by the budget one can afford. In this direction, choosing the parameters that need to be measured in order to have precision agriculture comes first (Lee, Alchanatis, Yang, Hirafuji, Moshou, & Li, 2010). Depending on the crop type, micrometeorological parameters like air temperature, air humidity, wind speed and direction, precipitation, as well as other weather-related data both around as well as inside the deployment field (either it is an open field or a greenhouse) need to be collected (Chaudhary, Nayse, & Waghmare, 2011), (Pahuja, Verma, & Uddin, 2013), (Bencini, 2010). This is because the forecasts about a region, where for example the vineyard is located, do not relate with the climate in the field because fields with crops always have different climate, known as the *microclimate*. The micrometeorological parameters outside the field are typically monitored with the installation of weather stations (Kotamäki, et al., 2009), (Langendoen, Baggio, & Visser, 2006), (Garcia-Sanchez, Garcia-Sanchez, & Garcia-Haro, 2011). Except from these parameters, there are in-field factors that have to be measured as well, and usually these include the air temperature (T), relative humidity (RH), soil T/moisture, salinity, PH, etc. (Zhang, et al., 2010), (Tik, Khuan, & Palaniappan, 2009)

- **Node Platforms:** The choice of hardware and software always depends on the available technology at the time of purchase. Starting with the node platform, many applications made use of the Crossbow MicaZ, Mica2 motes (Cao, Chena, Zhangb, & Suna, 2008), (Vellidis, Tucker, Perry, Kvien, & Bednarz, 2008), (Terzis, et al.), (Kumar, et al., 2009), which are no longer available. Other commercially available platforms that deployments used were the Tmote Sky motes (Xia, Tang, Shi, Fan, & Li, 2011), the Fleck3 platform (Hu, Dinh, Corke, & Jha, 2012), the Waspmote (Libelium Waspmote, 2013), the Sensinode (Ahonen, Virrankoski, & Elmusrati, 2008), the TNOde (Langendoen, Baggio, & Visser, 2006), (Mittal, Chetan, Jayaraman, Jagyasi, Pande, & Balamuralidhar, 2012) motes and the LiveNode (Hou, et al., 2007). Most nodes were built around the MSP430 and ATMega microcontrollers. In most cases, the node platforms had on-board radio transceiver, memory, and an antenna, so upon purchasing them, these components were ready for use (Kim, Yang, Kang, & Kim, 2014). As analyzed before, the TinyOS (Levis, et al., 2005) was the most coveted OS choice for these types of applications.
- **Power Sources:** Regarding the power unit, the sensor nodes were powered using non-rechargeable batteries or energy harvesters (Mafuta, Zennaro, Bagula, Ault, Gombachika, & Chadza, 2012). In most agricultural deployments, batteries were rechargeable using renewable energy in the form of solar panels. However, in long-term deployments, the aforementioned batteries while being rechargeable for efficient use of power and unattended operation, they were also combined with protocols and algorithms in order to better regulate the use of power in the system. In these latter protocols, one may find power management and power saving techniques like the duty-cycling technique, which necessitates that nodes sleep and wake up periodically. In some cases, specialized algorithms, such as the Delta compression algorithm, were also used to achieve data packet size reduction (Langendoen, Baggio, & Visser, 2006).
- **Communication Modalities:** As far as the communication is concerned, the Radio Frequency (RF) transmission appeared to be the most suitable form of wireless communication (Pierce & Elliott, 2008) with the ZigBee protocol

based on the IEEE 802.15.4 protocol (Mancuso & Bustaffa, 2006), (Hiu, Meng, & Wang, 2009), (Morais, Fernandes, Matos, Serôdio, Ferreira, & Reis, 2008) being the most commonly adopted standard. RF is mainly used for the node-to-node and node-to-base station (BS) short-range communications. Typically, radio frequencies have longer wavelengths and lower frequencies compared to the Bluetooth technology making them ideal for computationally lightweight communications. Moreover, Bluetooth has lower range and allows a different network topology than the one required by the environmental WSNs. Wireless Fidelity (Wi-Fi) is another way of wireless, long-range communication employed usually between the BS and the remote PC server (Martin, Moisan, Paris, & Nicolas, 2008), (Alippi, Boracchi, Camplani, & Roveri, 2012). Cellular communication is quite popular in agricultural WSNs as well, since most of the deployment areas exploit the existing GSM/GPRS infrastructure (Kotamäki, et al., 2009), (Martin, Moisan, Paris, & Nicolas, 2008), (Tseng, et al., 2008), (Matese, et al., 2013), (Xia, Tang, Shi, Fan, & Li, 2011). Finally, Ethernet and RS232 links (Mancuso & Bustaffa, 2006), (Hu, Dinh, Corke, & Jha, 2012), (Xia, Tang, Shi, Fan, & Li, 2011) were also used to connect the gateways nodes to the GPRS modules. Regarding the topology and architecture that is being used in WSN-based agricultural applications, the mesh topology is the most commonly used with the nodes being organized in clusters to decrease the power consumption (Langendoen, Baggio, & Visser, 2006). Another topology that is being used is the tree-based and the grid one, both requiring multi-hop communications. In other applications, with combined (mesh) topologies were also adopted (Hu, Shen, Yang, & Lv, 2010), (Wang, Wang, Qi, Xu, Chen, & Wang, 2010).

- **Data Collection and Management:** In the deployments of this field, special attention was given to the data collection and transmission modules. In almost every agricultural deployment, both the data collection and reporting was set to be time-driven, such as in (Cao, Chena, Zhangb, & Suna, 2008), (Dursun & Ozden, 2011), (Ayday & Safak, 2009). Time-driven data collection allows a user to periodically acquire the complete picture of the crop status, and act accordingly. The sensing intervals varied from one minute to one hour (Aquino-Santos, Apolinar, Edwards-Block, & Virgen-Ortiz, 2011), although a number of agriculturists (Goumopoulos, Kameas, & O'Flynn, 2007) suggest that sensing measurements should be reported every five (5) minutes. On demand sensing task is not an option for these types of application requirements. This is also the case for event-driven strategies, which are mostly used when monitoring other environmental phenomena such as volcanoes, earthquakes, forest fires, etc., because in these cases the monitored subject is the event itself. However, in one single case, the event-driven sensing strategy was used in an agricultural deployment as well (Cao, Chena, Zhangb, & Suna, 2008).

- **Security:** When speaking about security in precision agriculture applications, we are mostly interested about the network reliability that can be threatened due to node failures or because of the harsh weather conditions. To avoid malfunctions and errors, a common practice is to use redundant nodes. However, several researchers do not take security for granted. As such, towards investigating existing agricultural applications based on the WSN technology, we observed that in some cases the authors used either specific protocols such as the MintRoute routing protocol[1] (Woo, Tong, & Culler, 2003), or route maintenance strategies (Cao, Chena, Zhangb, & Suna, 2008), or even data encryption to prevent possible competitors from network invasion. Several other projects also focused on the detection of ''intruders'' (i.e., insects, rabbits, birds, etc.), which may destroy the crops. In these deployments, the WSN detected these insects using special traps and video cameras (Langendoen, Baggio, & Visser, 2006), (Liu, Zhang, & Richards, 2009), (Tseng, et al., 2008), (Martin, Moisan, Paris, & Nicolas, 2008).
- **Maintenance:** Finally, yet importantly, it is the cost and maintenance issue associated with the WSN deployment. Generally speaking, in most deployments, the cost of the tiny sensor nodes is not publicly available. However, research studies that present deployment economics refer to costs that range from 20$ to 150$ per node. The variation in cost is due to the different platforms, and the numerous companies manufacturing sensor nodes. The maintenance needs of these deployments were also not released, because the sensor nodes in all deployments were considered to be installed in special enclosures for protection from the natural elements, and the conditions under which they were deployed, as such requiring no special maintenance mechanisms.

Some additional issues, but equally important with the aforementioned ones, are the following. Prior to any deployment, test field deployments either in labs or under real outdoor conditions were always conducted for evaluating the overall system performance. In addition, simulations were used for the same purpose. The evaluation of the WSN, which is another crucial issue, was performed using metrics such as the RSSI (Received Signal Strength Indication), LQI (Link Quality Indicator) indicators as well as the PRR (Packet Reception Rate) and MDR (Message Delivery Rate) values. Generally, in most deployments, the deployed WSN managed to face the different challenges and problems, and at the same time operated until the scheduled time, serving the purposes of the particular deployment. However, in one case the whole WSN malfunctioned, and did not manage to recover giving the opportunity to the team that worked over the project to give instructions for a successful deployment according to their experience (Langendoen, Baggio, & Visser, 2006).

Monitoring of the Natural Environment

WSN deployments aimed at monitoring the natural environment have been set to mostly monitor forested environments including canopy closures, permafrost areas, as well as glaciers. Because of the monitored environment, the areas under research are usually large, measured in hectares (ha). Moreover, these areas are unreachable most time of the year, so the WSN must be robust, and power efficient to withstand the harsh environmental conditions. In these types of deployment, apart from the soil moisture other factors that are typically being measured are the air temperature and humidity, surface temperature, solar radiation, wind speed and direction, etc., (Zheng, Liu, Chen, Chuang, Chen, & Jiang, 2013), (Ramos, Foster, FeliciCastell, Fos, & Solano, 2013), (Cardell-Oliver, Smettem, Kranz, & Mayer, 2005).

In monitoring rock glaciers (Ingelrest, Barrenetxea, Schaefer, Vetterli, Couach, & Parlange., 2010), (Barrenetxea, Ingelrest, Schaefer, & Vetterli, 2008) or glaciers (Padhy, Martinez, Riddoch, Ong, & Hart, 2005), (Martinez, Hart, & Ong, Deploying a Wireless Sensor Network in Iceland, 2009) sensor nodes are placed inside the rocks or are buried under the ice. As apparent, any physical deformation or surface movement may seriously damage the nodes. This is why the nodes used in these deployments have a high rate of failure, and as such, it is important to constantly check the health of the WSN using status messages, battery voltage updates, and link quality indications (Wark, et al., 2008), (Jelicic, Razov, Oletic, Kuri, & Bilas, 2011).

- **Node Platforms:** The Mica mote family together with the TinyNode platforms by Shockfish are the preponderant hardware choice, providing the lowest power state-of-the-art module, and full support of the TinyOS operating system, which also dominates these deployments over other operating systems. Several deployments instead use custom-developed platforms (Chen & Lu, 2013), (Elsts, et al., 2012), (Jelicic, Razov, Oletic, Kuri, & Bilas, 2011). In general, these deployments require that nodes be employed with a Global Positioning System (GPS). A GPS is used for localization of the nodes, and for obtaining local and global timestamps of the collected data packets. As already revealed, the controller, the radio, the antenna, and the memory are all integrated in the node platform, so their characteristics depend on the platform choice. Finally, these environments necessitate that a weatherproof enclosure is available, while in many cases the boxes that contain the sensor nodes need to, at least, have IP67 rating or to be made of polyester (Martinez, Basford, De Jager, & Hart, 2012).

- **Power Sources:** Different types and sizes of batteries are used for power supply, but in almost all cases, these batteries have high-energy budget to extend the lifetime of the network as much as possible (Cardell-Oliver R. S., 2005). In order to satisfy the requirement for long-term deployment (recall that in some deployments replacing the batteries may be infeasible), the use of solar cells and panels combined with rechargeable batteries is also frequent (Ingelrest, Barrenetxea, Schaefer, Vetterli, Couach, & Parlange., 2010). Of course, the implementation of different power management techniques, including duty cycling mechanisms or the use of aggregation/compression algorithms, remains an essential power-saving strategy. By using these techniques, the deployed WSN could manage to function for months or even years.
- **Communication Modalities:** From a communication viewpoint, RF with the use of the IEEE 802.15.4/ZigBee standard (Talzi, Hasler, Gruber, & Tschudin, 2007) is again considered the most effective way of communication between the sensor nodes and the BS. GSM/GPRS connections (Ingelrest, Barrenetxea, Schaefer, Vetterli, Couach, & Parlange., 2010) are also used for backup purposes, i.e., either in case other connections fail, or when the remote server needs to reconfigure the BS. Finally, the communication between the gateway and the BS is implemented using serial connections including RS232 ports (Yang, Zhang, Li, Huang, Fu, & Acevedo, 2010), Ethernet or ISDN dial up connections (Rice & Bales, 2010), (Padhy, Martinez, Riddoch, Ong, & Hart, 2005).

The topologies used in these types of deployment are mostly mesh, hierarchical, tiered and tree-based topologies. Tiered topologies are often self-supported in order not to affect the other tiers in cases of failure (recall that monitoring natural environments usually involves harsh conditions). In most deployments, the installation of the sensor nodes is pre-planned and is made by the scientists themselves in an organized and strategic way. In one specific, and at the same time, demanding application, as the one envisioned in the Permafrost project (Talzi, Hasler, Gruber, & Tschudin, 2007) the team members of the project needed to even undertake regular alpine safety training courses in order to avoid risking their lives when working in the Swiss Alps.

- **Data Collection and Management:** As far as data collection module is concerned, the measured parameters are sensed from the sensor nodes in pre-determined, fixed intervals that vary from few minutes to several hours, leading to time-based reporting systems like those met in agriculture. There are some deployments where the sensing is reactive triggered for instance by rain events, i.e. when a node senses soil moisture over a pre-determined threshold, then it will start to sense more frequently reducing the data collection time interval (Tolle, et al., 2005). In these cases, the network is hybrid, combining the characteristics of both the time-based and the event-driven systems. Among the necessary protocols and algorithms used here, numerous deployments use the MintRoute (Tolle, et al., 2005), Collect Tree Protocol (CTP) (Jiang, Zhou, G.. Liu, & Wang, 2010), the Dozer multi-hop protocol or the Spanning Tree Protocol (STP) (Talzi, Hasler, Gruber, & Tschudin, 2007), combined with filtering and compression algorithms.

- **Maintenance:** As apparent, in these types of deployment, the sensor nodes remain unattended for long periods under harsh conditions, therefore, maintenance tasks need to be undertaken several times during a year. Sending a maintenance team for these tasks is suggested as an option by many of the aforementioned deployments, however, such an action may incur an additional cost to the already high cost of purchasing the WSN components. The cost may also rise in case safety-training lessons need to be undertaken first.

Air-Water Pollution Monitoring

The goal of monitoring the quality of the air or the water (usually the drinking water) is to evaluate its pollution level by measuring the concentration of dangerous gases or water pollutant (Zia, Harris, Merrett, Rivers, & Coles, 2013). The WSN technology promises to provide reliable and accurate air/water assessments. Existing deployments have been mostly installed in borehole wells at the perimeter of landfill sites (Fay, et al., 2011), (Kiernan, Beirne, Fay, & Diamond, 2008) or in rivers, lakes and dams (Chang & Bonnet, 2010), (O'Flyrm, et al., 2007).

In some cases, these deployments were even tested in labs with the use of special chambers. The duration of these deployments varied from few days to several months (Collins, Orpen, Maher, Cleary, Fay, & Diamond, 2011). In most deployments, the area that is covered by the WSN is not mentioned. However, from the number of nodes being used, one can estimate that the area size, apart one case (Le Dinh, Hu, Sikka, Corke, Overs, & Brosnan, 2007), is relatively small. Moreover, the monitored subject is either one particular chemical element (Shepherd, Beirne, Lau, Corcoran, & Diamond, 2007), or several well-known pollutants (Choi, Kim, Cha, & Ha, 2009), (Capella, Bonastre, Ors, & Peris, 2010). In particular, in air pollution monitoring the factors that are measured typically are the methane (CH_4) and the carbon dioxide (CO_2), while in water quality assessments, the typical measurements include; dissolved oxygen, water pH, salinity, phosphate, and chlorophyll concentrations (Yaacoub, Kadri, Mushtaha, & Abu-Dayya, 2013), (Wen, Jiang, Sun, Juang, & Lin, 2013), (Antoin, Javier, Brendan, John, & Fiona, 2012).

- **Node Platforms:** In the deployments of this category, the sensor node platforms being used are again the obsolete Mica2 and Mica2Dots motes (Shepherd, Beirne, Lau, Corcoran, & Diamond, 2007). Nodes of the Fleck family are also being used by deployments located in Australia (Dunbabin & Corke, A framework for marine sensor network & autonomous vehicle interaction, 2010). In many cases, nodes were equipped with GPS modules as well (Capella, Bonastre, Ors, & Peris, 2010), (Le Dinh, Hu, Sikka, Corke, Overs, & Brosnan, 2007), (Alkandari, Alnasheet, Alabduljader, & Moein, 2012). Protective cases were used in almost all types of deployment (Alippi, Camplani, Galperti, & Roveri, 2011). In addition, in two projects (Le Dinh, Hu, Sikka, Corke, Overs, & Brosnan, 2007), (Ong, Yang, Mukherjee, Wang, Surender, & Grimes, 2004), robotic vehicles have also been used (Dunbabin, Alistair, & James, 2009), (Wang, Ren, Shen, & Liu, 2010). These vehicles included, amongst others, board platforms, GPS modules, compasses, and several power supply modules such as batteries and solar panels. In almost half of the deployments, the OS running on top of the platforms is not mentioned, while in others, the FleckOS (FOS) (Dunbabin, Alistair, & James, 2009) and the Linux OS were among the choices due to existence of Linux-based components (Mingli & Yihai, 2013), (Rao, Marshall, Gubbi, Palaniswami, Sinnott, & Pettigrovet, 2013). Finally, few deployments relied on the TinyOS (Le Dinh, Hu, Sikka, Corke, Overs, & Brosnan, 2007), (Wang, Ren, Shen, & Liu, 2010).
- **Power Sources:** As for the power supplies, different types and sizes of batteries as well as large solar panels were used (Alippi, Camplani, Galperti, & Roveri, 2011). In one particular case, a car battery was used to operate the sensing nodes. Besides using solar panels, various other energy harvesters were employed including regulators, adapters, power outlets, and UPSs (Capella, Bonastre, Ors, & Peris, 2010), (Choi, Kim, Cha, & Ha, 2009). To further reduce the power consumption, electrical switches for manually turning off the gas sensors, or power management techniques, such as duty cycling, were implemented as well (Capella, Bonastre, Ors, & Peris, 2010).
- **Communication Modalities:** Regarding communication, the RF and Bluetooth technologies were most employed to support the communications between the nodes and the BS or the gateways (Le Dinh, Hu, Sikka, Corke,

Overs, & Brosnan, 2007). In several deployments, the latter nodes also had access to the Ethernet and GSM networks in order to send messages through the Internet (Collins, Orpen, Maher, Cleary, Fay, & Diamond, 2011), (Yaacoub, Kadri, Mushtaha, & Abu-Dayya, 2013), (Somov, Baranov, Spirjakin, Spirjakin, Sleptsov, & Passerone, 2013). A time-based data-gathering model was popular in these types of deployment, with the sensing intervals to be varying from several minutes to few hours.

The network topology used in water quality assessments is mostly tiered-based consisting of both static and mobile nodes. The static nodes are placed in fixed positions (usually they are scattered on the surface of the water or are anchored in the sea floor) (Kadri, Yaacoub, Mushtaha, & Abu-Dayya, 2013). Mobility is enabled when static nodes are mounted on robotic boats or Autonomous Surface Vehicles (ASVs) (Le Dinh, Hu, Sikka, Corke, Overs, & Brosnan, 2007). Regarding air pollution assessments, the WSN deployments were simpler in their installation and consisted only of a small number of nodes that relied single hop communications.

- **Maintenance:** Maintenance is a necessary task in these deployments as well, especially for the nodes that are in direct contact with the water. This is because the water includes substances that may severely affect the node's functionality. In one particular deployment, the team needed to install an anti-bio fouling mechanism, i.e., a mechanism that provides an automatic cleaning process to the node, in order to avoid the need of performing any kind of maintenance (Capella, Bonastre, Ors, & Peris, 2010). Finally, the cost of the sensor nodes is almost identical with the one identified in the previous deployments. However, the use of the robotic vehicles slightly differentiated, in fact increased, the cost of the entire network.

Monitoring of Destruction Phenomena

WSNs have also been used to monitor destruction phenomena, such fires, earthquakes, volcanic activities, etc. In these types of deployment node placement is crucial. Accordingly, the installation of the wireless sensor nodes is carefully pre-engineered taking into account the peculiarities of the deployment region. In monitoring destructive natural phenomena, the commonly used sensors measure air temperature (T), relative humidity (RH), wind speed and direction (in case fire behavior prediction is needed). In volcano monitoring, the sensors that are being used are seismic, infrasonic, and geophones measuring volcanic activity, ground deformation, etc. (The GlacsWeb project, 2012), (Huang, Song, Xu, Peterson, Shirazi, & LaHusen, Real-world sensor network for long-term volcano monitoring: Design and findings, 2012), (Girard, Beutel, Gruber, Hunziker, Lim, & Weber, 2012). Web cameras are also used in some occasions to capture images from the monitored areas. In most cases, the data-reporting scheme was time-driven, with the sensing intervals taking values between few minutes to 15 minutes. Uninterrupted reporting must be ensured in these deployments since the firefighters must have up-to-date information with regard to the prevailing conditions, given also the unstable nature of fire (Hartung, Han, Seielstad, & Holbrook, 2006), (Yoon, Noh, Lee, Teguh, Honma, & Shin, 2012), (Zhu, Dong, & Yuan, 2012).

- **Node Platforms:** On the hardware side, several custom-developed platforms made their appearance, namely the Linksys WRT54GL and Soekris net4801 motes (Lloret, Garcia, Bri, & Sendra, 2009), (Hartung, Han, Seielstad, & Holbrook, 2006). In volcanic deployments, the Crossbow motes (Micaz, Mica2, iMote2) and the Tmote Sky motes were the preferred platforms (Huang, Song, Xu, Peterson, Shirazi, & LaHusen, 2012), (Werner-Allen, Lorincz, Johnson, Lees, & Welsh, 2006), (Werner-Allen, et al., 2006), (Tan, Xing, Chen, Song, & Huang, 2013). Sensor nodes implemented the MantisOS, Linux and TinyOS (Lloret, Garcia, Bri, & Sendra, 2009). Many protocols and algorithms are used for proper and accurate functioning of their system including the Z-SYNC hybrid time synchronization protocol that combines GPS & FSTP merits, the MultihopOasis data collection routing protocol, Cascades data dissemination protocol, STA/LTA (short term average over long term average) algorithm, etc. (Doolin & Sitar, 2005). The MPEG4 compression algorithm was used for compressing the captured images and videos.
- **Power Sources:** Regarding the node's power supply, large rechargeable batteries and solar panels were used to cover the energy requirements. In one case, a diesel generator was used as a secondary power supply of the BS (Werner-Allen, Lorincz, Johnson, Lees, & Welsh, 2006). Duty cycling was also used to manage power consumption, together with regulators and negative voltage converters.
- **Communication Modalities:** Node communication is mostly based on the ZigBee/IEEE 802.15.4, and the Wi-Fi/IEEE 802.11g standards (Alippi C., Camplani, Marullo, & Roveri, 2012). In several cases, GSM modules (The GlacsWeb project, 2012), (Wang, Zhang, Li, Cui, & Jing, 2010) or a satellite dish was used for covering communication in remote regions (Hartung, Han,

Seielstad, & Holbrook, 2006). In monitoring volcanic activities, the sensor nodes and any other network component are placed on tripods, called station nodes, and are either statically positioned around the volcano or are air-dropped (targeting the crater or the area around the volcano flank) (Huang, Song, Xu, Peterson, Shirazi, & LaHusen, 2012), (Song, Huang, Xu, Ma, Shirazi, & Lahusen, 2009).

- **Maintenance:** In most deployments, the maintenance costs were low due to the network's self-organizing and self-healing nature (Song, Huang, Xu, Ma, Shirazi, & Lahusen, 2009). However, the reported costs for purchasing the WSN components were higher because of the need to use protective cases for the sensor nodes. Recall that the possibility of node damage is really high, and as such protective cases need to be included in almost all deployments in order to protect the sensor nodes from the destructive natural phenomena.

Livestock and Wild Animal Monitoring

Deployments of this type monitor frog species and birds to cows and carnivores (Kwong K. H., et al., 2012), (Anthony, et al., 2011). Although some of them were designed to monitor wild animals, they were also tested with domestic ones (Zviedris, Elsts, Strazdins, Mednis, & Selavo, 2010), (Dyo, et al., Evolution and Sustainability of a Wildlife Monitoring Sensor Network, 2010). The area being monitored is typically large in size, as animals usually travel, except from the case of cattle monitoring. One such deployment showing Fleck nodes built into collars worn by cows can be found in (Butler, Corke, Peterson, & Rus, 2004). The factors that are usually measured, except for the environmental ones such as temperature, humidity, light, etc., are the presence of species, vocalizations, and the location of the animal. These parameters are measured using GPS sensors, motion detection and identification sensors, three-axis magnetometers (electronic compass) (Hu, et al., 2009), three-axis accelerometers, and microphones for vocalization detection, (Stephen, Michie, & Andonovic, 2013), (Larios, et al., 2013), (Ehsan, et al., 2012).

- **Node Platforms:** Regarding the H/W, all known platforms, namely the Crossbow platforms (Mica2, Imote), the Tmote Sky mote and the Fleck platforms, were used (Guo, Corke, Poulton, Wark, Bishop-Hurley, & Swain, 2006). Few deployments also used custom-based platforms such as the Carnivore platform (Rutishauser, et al., 2010) or the platform produced by the CiNet Company (Hakala, Tikkakoski, & Kivelä, 2008), and the works cited in (Jurdak, et al., 2013), (Larios, et al., 2013). Moreover, two hybrid

RFID/WSN-based systems were also implemented for animal monitoring purposes (Markham, Trigoni, Macdonald, & Ellwood, 2012), (Dyo, et al., 2010). As analyzed before, the MCU, radio transceiver and the memory by being embedded in the platforms, are chosen indirectly on the basis of the platform in use. GPS is used in many deployments, since the location of the animals is amongst the factors that is measured the most.

The sensors are being chosen very carefully in order to take accurate measurements. In all cases, the nodes are mobile since they are embedded onto the collars, which are worn by the monitored animals. Collar design issues, such as the size and weight of the collar and its installation on the animals, follow specific requirements and must be proportional to the physiology and weight of the animal (Wark, et al., 2007). In some deployments, stimuli coming from vibrations, noise and light electroshock were also used, following strict requirements related to animal ethics and welfare (Wark, et al., 2007). Similar to the previous deployments, another important issue to be considered is the protection of the sensor node from physical damage. Due to the fact that the sensor nodes/collars are installed on the animals or the habitats, depending on the animal species and its size, it is very likely that these nodes will be damaged. Once again, the TinyOS operating system is the dominant OS being used by the sensor nodes. However, there are some others OSs that are being used including the ContikiOS, MansOS, and ImpalaOS.

- **Power Supplies:** Similar to the previous deployments, various types and sizes of batteries were used to power the sensor nodes. Many protocols and algorithms were also implemented to enable the proper functioning and effective power management of the nodes.
- **Communication Modalities:** The communication between the sensor nodes/collar, and the sink node is typically implemented using RF and Wi-Fi links (Lenka & Mohapatra, 2012), (Nadimi, Jørgensen, Blanes-Vidal, & Christensen, 2012). In these types of deployment, 3G/GPRS and satellite links are less frequently used (Demirbas, Chow, & Wan, 2006), (Wijesinghe, Siriwardena, & Dias, 2013). The data monitoring and reporting follows a time-driven approach where the sensing interval ranged from few seconds to several hours. In case of large habitat monitoring (Polastre, 2003), (Mainwaring, 2002), (Szewczyk, Mainwaring, Polastre, Anderson, & Culler, 2004), the network is deployed using a cluster tree-based topology, while the star topology was preferred in case of wild animal monitoring, (Nadimi, Jørgensen, Blanes-Vidal, & Christensen, 2012).

- **Maintenance:** The cost of purchasing sensor nodes for animal monitoring as well as all related maintenance issues are not mentioned in most surveyed deployments. In few deployments the authors simply stated that the necessary equipment is expensive enough because of the components being used (Kwong K. H., et al., 2009), (Rutishauser, et al., 2010).

GUIDELINES FOR DEPLOYING WSNS IN REAL-LIFE ENVIRONMENTAL MONITORING APPLICATIONS

The following sections present basic guidelines for deploying wireless sensor networks in the different field deployments reviewed earlier. These guidelines resulted after surveying an exhaustive number of existing WSN-based applications. In discussing basic considerations relevant to the deployment, the guide covers issues such as type of sensors used, node platforms, OSs, transceivers, network topologies, installation and maintenance issues, etc. In developing the resulted guides we considered representative application scenarios for each of the five field deployments. Our findings are summarized in a Table following the presentation of each guide.

Basic Guidelines for Deploying WSNs in Agricultural Monitoring Applications

- **General Deployment Issues:** Towards deploying a WSN for agricultural monitoring purposes someone has to firstly consider the budget that is available in order to cover the purchase of the necessary equipment. Moreover, he or she must decide whether the deployment will be in an open field or in a greenhouse. The interested person must also find out the requirements of the crop that will be monitored or the peculiarities of the monitored area in order for instance to be aware of the potential intruders, i.e. rabbits. Decisions need to be made as to whether the WSN will be proactive, meaning that the network will require human intervention when some critical tasks like irrigation and fertilization need to take place. Finally, the monitored parameters need to be selected from the beginning in order for the appropriate sensors to be purchase.
- **Hardware and Software:** In choosing the hardware/software and the types of sensors that will be used, as already revealed, there are many commercial platforms available. The Crossbow Mica motes seem to be suitable for this type of deployments. However, since these motes are no longer available,

one should choose alternate Mica-like platforms, such as the Fleck nodes. In addition, in order to attach more sensors, a sensor board must be adopted. Another crucial option one needs to consider relates to the power consumption of the node. For improving the radio coverage, external antennas can be used providing additional coverage for each sensor. Appropriate protocols and algorithms must also be implemented, including communication, routing, synchronization protocols and maybe compression algorithms, for the efficient operation of the WSN. Regarding the operating system that the sensor nodes will run, this could be the TinyOS, which is commonly used and is compatible with many commercial platforms as well as with the Mica family motes.

- **Communication Issues – Topology:** In general, the placement of the sensor nodes in agricultural WSNs must be dense. Dense deployments allow for all the necessary measurements to be captured correctly in order to have complete, accurate and reliable knowledge of the monitored area. Otherwise, there is no need to deploy such an expensive system. For a small deployment (e.g. 5 to 10 sensor nodes), a star topology with single hop communication can be implemented. A BS and a PC-based server are also needed to further analyze or even display the collected data. The necessity of buying such equipment directly depends on the size of the deployment area. For large deployments with over 20 sensor nodes, the single hop star topology is not recommended because the increased radio ranges will consume higher power. Hence, a cluster or tree-based multi hop approach seems to be more appropriate. The communication of the nodes with the BS will be realized over radio frequency (RF), while the BS will communicate with the PC using either Wi-Fi connection or through radio modems for long distances. It is important that the BS will be placed close to the field deployment. There may be a need to strengthen the signal, so, some repeaters may need to be deployed. This depends on the distance between the sensor network and the BS as well as the BS and the server. In addition, end users may communicate directly with the server through the Internet using web browsers as well as GUI tools for the visualization of the sensor readings. The connection between the end users and the server is usually established through 3G/GPRS or standard Ethernet connections depending on the communication infrastructure around the deployment area.
- **Sensing Issues:** Sensor types and measured factors: Some of the common and most critical measured factors in WSN-based agricultural monitoring applications are the soil moisture, temperature, relative humidity, ambient light, wind speed and direction. In addition to these, there are several other

factors to measure such as leaf T and atmospheric pressure, which depends on the crop being monitored. The option of the type of sensor to be used (i.e., Sensirion SHT75 or SHT71 for RH and T) depends on various factors such as sensor accuracy, resolution, range, power consumption, precision, cost, etc. Another issue in choosing the necessary sensors is that some platforms internally provide with some of them, while others do not. For example, the TmoteSky node, and the Fleck3 nodes provide with onboard humidity, temperature, and light sensors, while TNode does not. This means that in case the later mote is chosen, separate sensors must be purchased and embedded through the sensor boards. Regarding data collection and transmission, this is usually performed in a time-based manner. As expected, the sensing interval depends on the crop type. However, according to existing deployments, sensing intervals of 5 or 10 minutes are sufficient for the system to be operable for the whole crop season.

- **Power Supply Issues:** Regarding power supply, it is evident that someone who is interested in setting a WSN for agricultural monitoring applications, would like the network to maintain its operation for the whole crop season, which depends on the nature of the crop itself. For example, potato crop needs about 3 to 4 months from sowing and cultivation to harvesting. This means that the network must remain functional for at least four months without the need for replacing or recharging the node batteries. In case battery replacement or the use of renewable energy sources such as solar panels is an option, then the network may be engineered with less stringent criteria. Once again, the battery size depends on the node platform used. Regarding the battery chemistry, namely Li-ion or lead-acid, the choice must be done based on how the battery behaves in specific environmental conditions under which it operates. The most important issue however, is the implementation of power saving or power management techniques through the use of appropriate energy-efficient protocols and algorithms.
- **Maintenance - Safety Issues:** Because of the long-term nature of the deployment, maintenance issues regarding the WSN system need to be considered. Accordingly, the sensor nodes must be put into protective cases preventing them from being exposed to moisture, mud, etc. These cases have ratings in the form of IP00 with 00 meaning no protection. In general, the first digit means protection against solid objects, while the second means protection against liquids and every level has its own definition. For example, in IP67, the 6 digit means total protection against dust and the 7 digit means protection against the effects of temporal immersion until 1m underwater (Goumopoulos, Kameas, & O'Flynn, 2007).

Having discussed the general issues and concerns surrounding the deployment of WSNs in agricultural monitoring applications, we are now in position to summarize our findings in a simple, generic guide that is depicted in Table 4. Without loss of generality, we consider a representative PA scenario, in which the farmer needs to monitor his potato field using the WSN technology in order to perform irrigation. In this specific example, suppose that the size of the monitored field is about 100m2 and that the farmer needs to deploy the WSN for the whole crop season, about 4 months. Moreover, the farmer requires that the system not only transmits the measured parameters to his PC, but also performs some irrigation tasks, a fact that necessitates the existence of actuator nodes as well.

Basic Guidelines for Deploying WSNs in Environmental Monitoring Applications

- **General Issues:** This section focuses on providing deployment guidelines for WSN-based environmental monitoring applications, such as applications relevant to canopy closure, permafrost and glacier study. Depending on the deployment site, variations exist with regard to the area size. In ESNs, generally, the monitored areas are large in space. Proper evaluation of the monitored area must be done, that is the place where the nodes will be installed, the distance between them and between the nodes and the sink node or the gateway. Moreover, all the possible weather-related data must be known for the whole duration for which the WSN will be deployed. In addition, in some environments like glaciers, special surveys must be done to determine sub-glacial circumstances, which include geophysical anomalies such as the existence of a river. Following these assessments, decisions must be made regarding the parameters that will be measured. This decision is based on the environment where the WSN system will be deployed. For example, if the monitored environment is a headwater catchment, the probable measurement parameters, among others, will be soil moisture, the temperature, and the relative humidity (RH). After selecting the parameters, the network topology must be selected. The decision regarding the topology will be based on the number of sensor nodes that will be used, on their energy constraints, and the overall network cost.
- **Hardware and Software:** The hardware choice is again based on the aforementioned constraints. However, according to the existing deployments the commonly used platforms are the Crossbow Berkeley motes and the TinyNode developed by Shockfish, which are proved to withstand the harsh environments. The coverage of sensor nodes in the area depends on

Table 4. Generic guidelines for deploying WSNs in agricultural monitoring applications

Monitored Object	Crop
Deployment Duration	One crop season (3-4 months)
Deployment Area	100 m^2 (approximately)
Average Node Count	Considering 1 sensor node per 1m^2, then 100 nodes or at least 80-90 nodes are needed. Ten of these nodes should act as actuators connected with sprinklers for irrigation. Two nodes should act as sink nodes and the rest as sensing nodes.
Monitored Parameters	-T -RH -Soil moisture (Note that not every single node will carry all the sensors)
Sensors	Either internal or external. Representative examples include: • Hydra-Probe II for soil moisture. • MTS-420 for ambient light. • SensirionSHT11 for T & Humidity.
Node Platform	Moteiv 's Tmote Sky
Microcontroller	Texas Instruments MSP430 featuring 10kB of RAM, 48kB of flash, and 128B of information storage. For the wireless communication, use the TI CC2420 radio.
Radio Transceiver	
Memory Size	
Power Supply	Two AA batteries. For long-term deployments, rechargeable batteries are also needed in combination with solar panels for recharging.
Operating System	Contiki, TinyOS, and MantisOS are all supported by the aforementioned node platform
Communication Modalities	RF links for node-to-node and node-to-sink communication. Wi-Fi links between the GW and the BS. *Note that the BS is in close distance with the deployed WSN
Data Reporting Scheme	Time-driven data gathering with a time interval equal to 5min.
Network Topology	Tree-based topology with multi-hop communications
Architecture	The network consists of sensor nodes and actuator nodes, several sink nodes, a powered PC-based BS, and a server for data storage
Installation	The sensor nodes can be placed in a grid in order to cover the entire monitoring area. The soil moisture sensor shall be buried 20 – 40cm underground. The actuators shall be placed somewhere in between the sensor nodes, while the sink nodes shall be within radio range
Protocols/Algorithms	Customized MAC and proactive routing protocols need to be appropriately chosen. Reduced duty cycling and sleep/wake up mode is also advised (around 10%).
Waterproof Case	Yes
Maintenance Tasks	Maintenance of field crop is needed to increase the quality of the collected sensory data especially in large-scale applications.

the monitored environment, which means that when monitoring glaciers or permafrost it doesn't have to be necessary a large or dense deployment, because the variations in these kind of environments occur in large distances. Of course the higher the number of sensor nodes, the larger the area coverage, but this is not something that will affect the obtained results. On the other hand, if someone is interested in setting up a WSN to monitor the canopy closure estimation or the microclimate of a forest or more specific some type of a tree, as is the case in (Tolle, et al., 2005), then the nodes must be densely placed and must be higher in number. Lastly, the nodes must be enclosed in protective cases. In addition, a GPS receiver is needed in order to perform node localization and data time stamping. The software to be used also depends on the deployment environment, although there are standard protocols and algorithms that are essential for the proper functioning of the WSN. Regarding the OS, the TinyOS can be used since the largest number of sensor platforms supports it.

- **Sensing Issues:** The sensing interval is strongly dependent on the constraints mentioned above and the environment under observation. In addition, the sensing interval can be adaptive. For instance, if someone wants to measure the soil properties before and after natural phenomena, such as before and after raining, the sensing interval could be smaller and then be adjusted to its initial value. Also, the data in an ESN does not necessarily need to be gathered in real-time. Hence, the measured data can be stored temporarily in the motes' local memory and then be relayed to the BS once every 10 to 15min or even once per day. By reducing the time a sensor node needs to communicate with other nodes, we minimize the energy consumption and extend the network lifetime.
- **Communication Issues:** As far as the communication is concerned, RF is the standard way for the wireless motes to communicate with. Cellular network technologies, such as 3G/WiMaX, or the Ethernet technology can be used for the remote connection of the user with the server and the WSN itself. Finally, the sink node can be attached to the BS, which can be a laptop, or a PC, through a serial port.
- **Power Supply Issues:** The power supply is the most important issue for the survival of the WSN and its long-term deployment. One solution regarding the battery choice is the non-rechargeable multi-cell Li-SOC12 batteries. With the proper modification in the source code for automatic power management and sleep wake cycle, these batteries can achieve a network lifetime equal to 4-5 years. Lithium Thionyl Chloride cells can also be used, since these cells due to their high energy density and good low temperature characteristics can theoretically last for at least 10 years. Finally, energy consumption

Table 5. Guidelines for deploying WSNs in environmental monitoring applications

Monitored Object	Canopy closure, permafrost areas, glaciers
Deployment Duration	Typically, the network is deployed for long time, i.e. over a month
Deployment Area	Depends on the deployment site. Approximately an area of > 1 km^2
Average Node Count	10 to 20 sensor nodes on average
Monitored Parameters	• Air T • RH • Soil moisture & T • Solar radiation • Air pressure *Note that not every node will have to carry all the sensors.
Sensors	Either internal or external like: • Hydra-Probe II for EC-TM or soil moisture. • SensirionSHT11 for T & humidity. • Intersema MS5540B for Air pressure
Node Platform	Shockfish's TinyNode
Microcontroller	TI MSP430 microcontroller with 10kB of RAM, 48kB of flash, and 128B of memory. Semtech XE1205 RF chip operating at 868 – 915 MHz
Radio Transceiver	
Memory Size	
Power Supply	Li-SOC12 with modifications in the source code or 3.6V Lithium Thionyl Chloride cells
Operating System	TinyOS
Communication Modalities	RF or Wi-Fi from node-to-node and node-to-GW Wi-Fi, between GW and BS For remote connection to the server use 3G/GPRS or Ethernet links
Data Reporting Scheme	Hybrid (depending on the monitored object). When in time-driven mode, sensing occurs every 5-10min and data transmission every 15 min.
Network Topology	Mostly tiered-based (hierarchical) multi hop topology
Architecture	15-20 powered motes and one gateway node with GPRS extension.
Installation	The sensor nodes should be placed in strategic places. The BS should be in close proximity,
Protocols/Algorithms	TDMA-like node synchronization. Simple spanning tree protocol for routing. 10% duty cycles with sleep periods of 5 min.
Waterproof Case	IP67
Maintenance Tasks	Due to the envisaged long-term operation and the harsh conditions, revisiting the monitored site several times is recommended.

minimization can be achieved if data compression is used, if the sensing interval length is large or if duty cycling is used.

- **Maintenance – Safety Issues:** Considering the cost issue, the cost of deploying an ESN is maybe higher than in other applications mostly due to the fact that the sensor nodes either have to be mounted on rocks or trees, requiring extra equipment. Also the WSN must be even more robust, self recovering, able to cover large distances and able to keep data temporally in node's memory due to the fact that the nearest BS server will be kilometers away from the deployment site. All these requirements necessitate strong that a stronger MCU and radio transceiver are in place and the memory size is big enough. Maintenance represents also part of the total deployment cost.

Basic Guidelines for Deploying WSNs Towards Monitoring Air-Water Pollution

- **General Issues:** As already revealed, initial assessment is essential and unavoidable. Beginning with the monitoring subject, namely air or water pollution, the deployment area includes landfill gas detection, lake and river deployment. In landfill monitoring the deployment is simple, because the sensor nodes are placed inside boreholes and then with a star topology e.g. with a sink node, the information is gathered and the monitoring process is done. In water quality monitoring, if the deployment is large, i.e. it includes 50 sensor nodes or more, there must be also mobile nodes for easy and smooth gathering of the sensed parameters.
- **Hardware and Software:** Regarding the hardware, the Crossbow Mica platforms can be used or the Fleck platform, which is proved to be efficient when deployed in water. GPS modules can also be used, however one must take into consideration its power consumption, which is high enough. The sensor nodes, which are put at the water surface, have their sensors underwater, since they are deployed for water pollution monitoring purposes. Accordingly, they must be placed properly and with the appropriate weight underneath in order to be as stable as possible and not carried away from their initial placement. The waterproof case is always a must, especially when using the nodes in the water, where they so-called buoys float. If one could afford to use ASVs as mobile nodes, then the aforementioned components can also be included. The node OS can either be the FOS, if Fleck platforms are used, or the TinyOS if Mica platforms are used. Finally, in case of PC-based sink/gateway nodes, the Linux OS is an effective solution.
- **Power Supply Issues:** Regarding the power supply, many types of batteries can be used, i.e. rechargeable for sufficient power combined with two solar panels, depending on the deployment components used in the WSN and the duration of the deployment as well as on the number of sensor nodes and

whether the nodes are mobile or immobile. Management techniques are necessary once again for stable, balanced and long term functioning of the WSN.

- **Sensor Types and Measured Factors – Sensing Issues:** In this deployment case, the real-time data collection is not the most critical component as is the fact that the data must be accurate. Therefore, the sampling/sensing time can be set to every hour or even four times a day. If there are mobile nodes, then these nodes may download the data at regular intervals.
- **Communication Issues – Topology:** Since short-range communication is required, RF and Bluetooth can be used. For nodes that are underwater, the communication is implemented through acoustic waves, due to their better propagation in the water, so the uplink communication includes both RF and acoustic modules, the underwater nodes include only the acoustic module while the BS includes only an RF module (Ong, Yang, Mukherjee, Wang, Surender, & Grimes, 2004). The preferred topology in these types of deployment is the tiered, multi hop.
- **Maintenance – Safety Issues:** Maintenance issues definitely exist, unless internal cleaning mechanisms are used. Moreover, the deployment cost is almost the same with previous mentioned cases, however it may dramatically rise with the use of the ASVs.

Basic Guidelines for Deploying WSNs Towards Monitoring Destruction Phenomena

- **General Issues:** When deploying a WSN for monitoring a hazardous or landslide event, the phenomenon to be monitored, extends to a different coverage area. So depending on the deployment area, there will be different landscape and terrain features, a fact that affects the design of the WSN.
- **Hardware and Software:** The platform choice can be one of the Mica families. The Mica family platforms are being used in many deployments proving their reliability and robustness. The use of protective boxes is a critical issue and must be implemented because of the ambient hazardous conditions. The OS choice is accordingly based on the platform in use. Hence, the use of TinyOS, which is compatible with many existing platforms, is an efficient option. The use of proper protocols and algorithms must be considered because the WSN must be robust, self-healing and self-maintaining due to the fact that the frequent visits in the deployment area will difficult.
- **Communication and Sensing Issues:** The sensing task is mostly time-based in landslide and wild fire monitoring, while in volcanoes, it is event-driven based on predetermined thresholds. By relying on existing projects,

Table 6. Guidelines for monitoring air-water pollution using WSNs

Monitored Object	Lakes, rivers, dams, Gas emissions in borehole wells, Atmospheric pollution in cities
Deployment Duration	Months are required in order to obtain sufficient information about the pollution of air or water
Deployment Area	Depends on the deployment site. Approximately an area of >10 square km
Average Node Count	Depends on the deployment site. Maybe 5-10 nodes if the deployment area is small, and around 30-40 is deployment area is large.
Monitored Parameters	• For air pollution the most common factors measured are methane (CH_4) and carbon dioxide (CO_2). • For water pollution, dissolved oxygen, water pH, salinity, phosphate and chlorophyll concentrations, barometers + other depending on the monitored elements. *Note that not every node should carry all the sensors.
Sensors	Either internal or external like: • Hydra-Probe II for soil moisture. • SensirionSHT11 for T & humidity. • OGS-401F2 NDIR CO2 Sensor for CO_2 detection.
Node Platform	Fleck series of nodes
Microcontroller	Atmel-1281 MCU with 28KB ISP flash memory, 8KB SRAM, 4KB EEPROM nRF905 radio chip
Radio Transceiver	
Memory Size	
Power Supply	Li-ion rechargeable or lead acid. Two solar panels and sleep wake cycle and one of the power techniques used in existing deployments.
Operating System	FleckOS (FOS)
Communication Modalities	RF for the nodes on the surface of the water and acoustic waves for the communication between the underwater sensor nodes and the nodes on the surface. Wi-Fi links, between GW and BS GSM/GPRS for remote connection
Data Reporting Scheme	Time-driven reporting; every one hour or every four times per day. If mobile nodes are used, they will have to gather the data in regular intervals.
Network Topology	Multi-tiered, hierarchical topology
Architecture	In water pollution the nodes are scattered at water surface, not in long distances from one each other, as well as below the water surface, while for air pollution monitoring the nodes must be placed all around the site under research in strategic places.
Installation	A set of fixed sensors nodes and multiple sinks shall be installed at fix positions (mobile nodes should also be considered in some cases)
Protocols/Algorithms	TDMA-like node synchronization. Multi-hop routing
Waterproof Case	YES
Maintenance Tasks	In case sensor nodes are placed underwater, biofouling is recommended.

hierarchical, multi hop topologies are better suited for these deployments. Once again, the sensors in use depend on the monitoring subject. In any case, the chosen ones must as accurate as possible. Finally, RF and Wi-Fi are the most common ways of communication in these deployment fields.

- **Power Supply Issues:** As for the power supply, depending on deployment duration and the components used in the system, there are many solutions of what batteries to use. Solar panels are indicated for use if there are appropriate weather conditions. Still, power management techniques shall be considered an integral part of the WSN as well.
- **Maintenance – Safety Issues:** Due to difficult circumstances, maintenance tasks are almost infeasible. Hence, the system must be robust, self-organizing, and self-healing with the possibility of quick recovery from bugs.

Basic Guidelines for Deploying WSNs Towards Monitoring Livestock and Wild Animals

- **General Issues:** One of the primary issues one needs to consider when deploying a WSN system in a farm to monitor his/her livestock is the deployment period and the budget. In general, one of the attracted characteristics of WSNs is the fact that this technology is easy to deploy even for those who are not familiar with this technology.
- **Hardware and Software:** After the budget considerations, H/W decisions must be made. Based on the existing deployments, Fleck1 and Fleck2 platforms with their embedded components are a good choice especially for animal tracking and control due to their design. The number of collars depends, firstly, on the number of animals to be monitored and on the radio coverage, which in this case can accommodate large number of collars increasing the cost in terms of money and power consumption. The installation part in this paradigm is easy enough as long as the animals are cattle. The monitored parameters depend on the nature of monitoring. Both environmental metrics such as humidity, light intensity, etc. can be used or metrics related to the health of the cattle and animals. Other factors that can be measured among others are presence of the animal, vocalization and location. In case someone wants to control the behavior of bulls, he or she must also include actuator nodes, which will apply some kind of light stimuli to prevent for instance the bulls from fighting in mating season, i.e. where they are extremely aggressive (Wark, et al., 2007). Regarding the SW, and since the operating system is the Flecktm platform, the TinyOS operating system is better suited for these deployments.

Table 7. Guidelines for monitoring destruction phenomena using WSNs

Monitored Object	Forests, Volcanoes, Mountains
Deployment Duration	Long-term deployment. The system can be set all year around
Deployment Area	Large deployments measured in hectares (ha)
Average Node Count	20-25 stations
Monitored Parameters	Commonly monitored parameters include air temperature (T), relative humidity (RH), wind speed and direction, seismic waves, infrasound, velocity of motion
Sensors	Maybe internal of the platform or external with the use of sensor board • Seismic sensors • Infrasonic sensors • Accelerometer and geophone sensors and others
Node Platform	Tmote Sky
Microcontroller	Texas Instruments MSP430 featuring 10kB of RAM, 48kB of flash, and 128B of information storage. For the wireless communication, use the TI CC2420 radio.
Radio Transceiver	
Memory Size	
Power Supply	Due to long-term deployment, rechargeable batteries are recommended. Solar panels for recharging *Still, note that the event-driven mode is energy saving
Operating System	TinyOS
Communication Modalities	RF between nodes No cellular infrastructure is available near volcanoes Long distance radio modems for remote communication to BS
Data Reporting Scheme	Event-triggered data reporting
Network Topology	Tree-based multi hop topology
Architecture	One base station and a numbers of sensors spatially distributed in the monitored area
Installation	The sensor nodes have to be placed strategically (in some occasions, a helicopter may be needed to drop the sensor nodes – puts an extra cost)
Protocols/Algorithms	A multi-hop time-synchronization protocol to establish a network-wide time base. 100% duty-cycle.
Waterproof Case	YES
Maintenance Tasks	Due to difficult circumstances, maintenance tasks are almost infeasible. Hence, the system must be robust, self-organizing, and self-healing with the possibility of quick recovery from bugs.

- **Communication and Sensing Issues:** Both the RF and the IEEE 802.15.4 technologies are suited for the communication between the nodes and the BS. For smaller WSNs, the communication can be single hop while for larger ones it has to be multi hop, because the distance between some nodes and the sink node will be long enough. Indeed, when monitoring animals like cows, carnivores or European badgers (Meles meles) with collars, the animals will be spread out in the woods, thus the transmission of the data have to be implemented when two or more collared animals will be in range of communication between them or with the BS thus forming a star topology (Rutishauser, et al., 2010). Finally, the parameters should be sensed mostly using a time-driven paradigm since in these types of deployments we would like to know the temporal changes of the measured factors throughout the deployment, while the sampling rate (Hz) varies depending on the nature of the deployment.

- **Power Supply Issues:** For the power supply, batteries shall be used like the NiMH ones with the combination of algorithmic approaches such as duty cycle or sleep/wake up modes. If the deployment is for long period, then the use of natural sources for battery charging must be used, with the most common to be solar panels, if the climate allows it. Lastly, risk assessment has to be considered, since there will be cases of node damages as explained in (Wark, et al., 2007).
- **Maintenance – Safety Issues:** Someone who wants to detect and identify animal species must include detector nodes with PIR sensors and camera nodes for identification (Demirbas, Chow, & Wan, 2006). Thus, additional equipment, i.e. cameras, must be purchased. On the other hand, for simply monitoring stationary nodes and mobile nodes, collars must be used.

Next, we provide a Table that summarizes the guidelines proposed in each WSN deployment field (Table 9). By doing this, we attempt to highlight the similarities as well as differences between the provided guides.

CONCLUSION

This paper *explored real-life environmental monitoring applications based on WSNs. The conducted analysis* highlighted that this emerging technology is an effective way for improving the way we monitor the environment in terms of precision and quality. After categorizing the *environmental monitoring* applications *into five different field categories*, we then discussed the *basic hardware components that existing WSN deployments use*. F*ive easy-to-use guides were introduced discussing basic considerations for deploying WSNs in each of these deployment fields. The generated guides covered various issues, such as sensor node platforms, operating*

Table 8. Guidelines for monitoring livestock and wild animals using WSNs

Monitored Object	Domestic animals, wild animals, birds
Deployment Duration	Non very lengthy deployments (it ranges from few hours to several days)
Deployment Area	In wild animal monitoring, the area is usually large and it is measured in hectares. The farm size instead is relatively smaller.
Average Node Count	No more that 20 sensor nodes
Monitored Parameters	• Temperature, humidity, light, • Presence of species • Vocalizations • Animal location
Sensors	• Temperature, humidity, light sensor • Ultrasonic sensors • Motion detectors
Node Platform	Fleck series of nodes
Microcontroller	Atmel-1281 MCU with 28KB ISP flash memory, 8KB SRAM, 4KB EEPROM nRF905 radio chip
Radio Transceiver	
Memory Size	
Power Supply	A sensor node attached to livestock must not have lengthy battery life to minimize costly maintenance. Thus, non-rechargeable batteries are recommended.
Operating System	Fleck OS
Communication Modalities	RF between nodes and BS In monitoring wild animals, satellite connections maybe needed for remote communication
Data Reporting Scheme	Time-based reporting. Sensing time depends on the power supply and the application, i.e. using actuators the sensing must be frequent
Network Topology	Star when monitoring collared animals Tree-based topology for habitat monitoring
Architecture	Tiered architecture with a number of sensor nodes connected with a BS that provides WSN connectivity and data logging
Installation	Typically, animals wearing collar mounted sensors
Protocols/Algorithms	MAC, routing, and synchronization protocols are needed to achieve low duty cycle (less than 5%)
Waterproof Case	YES
Maintenance Tasks	Light maintenance is needed like changing the batteries and replacing any damaged collars.

systems (OSs), topologies, installation and maintenance issues, and much more. Based on our understanding, we also highlighted *issues and concerns surrounding the application of WSNs in each deployment category.* In the future, we intent to extend this guide with even more application-specific instructions that will provide better guidance to *environmentalists, scientists, and general users who are* willing to

Table 9. A summary of the guidelines proposed in each WSN deployment field

	WSN Deployment Fields				
	Agriculture	**Natural Environment**	**Air-Water Pollution**	**Destruction Phenomena**	**Livestock and Wild Animals**
Monitored Object	Crops	Canopy closures, Permafrost areas, Glaciers	Lakes, Rivers, Dams,	Forests, Volcanoes, Mountains	Domestic animals, Wild Animals, Birds
Deployment Duration	3-4 months	> 1 month	~1 year	~ 1 year	Few days
Deployment Area	$100m^2$	> 1 Km^2	> 10 Km^2	Several hectares	Few hectares
Average Node Count	80-90	10-20	< 40	< 25	<20
Special Sensors	Soil moisture sensors	Solar radiation sensors	CO_2, CH_4 detectors, Salinity sensors	Seismic sensors, Accelerometers, geophones	Ultrasonic sensors, Motion detectors
Node Platforms	TmoteSky	TinyNode	Fleck	TmoteSky	Fleck
Power Supply	Combination	Rechargeable batteries	Rechargeable batteries	Combination	Non-rechargeable batteries
Operating System	TinyOS, Contiki	TinyOS	FleckOS	TinyOS, Contiki,	FleckOS
Communication Modalities	RF	Wi-Fi, 3G/GPRS	RF/Wi-Fi, 3G/GPRS	Wi-Fi, 3G/GPRS / Satellite	RF/Wi-Fi
Data Reporting Scheme	Time-driven	Hybrid	Time-driven	Event-triggered	Time-driven
Network Topology	Tree-based	Tiered-based multi-hop	Multi-tiered	Tiered-based multi-hop	Star single-hop
Ease of Installation	Easy	Not always easy	Medium	Difficult	Medium
Damage probability	Medium	High	Increased	Very High	Medium to High
Maintenance Tasks	Medium	Heavy	Medium	Infeasible	Light

deploy a WSN for such applications and *are not necessarily experts at this wireless networking technology*. Overall, we believe that technological advances in sensors, sensor data logging and communication, and software management of sensor networks will continue to provide transformative potential for new and innovative avenues of WSN-based ecological research in ways previously not possible.

REFERENCES

Ahonen, T., Virrankoski, R., & Elmusrati, M. (2008). Greenhouse Monitoring with Wireless Sensor Network. In *Proceedings of the IEEE/ASME International Conference on Mechtronic and Embedded Systems and Applications* (pp. 1-6). Beijing, China: IEEE.

Akyildiz, I. F., Su, W., Sankarasubramaniam, Y., & Cayirci, E. (2002). Wireless Sensor Networks: A Survey. *Computer Networks*, *38*(4), 393–422. doi:10.1016/S1389-1286(01)00302-4

Al-Karaki, J. N., & Kamal, A. E. (2004). Routing techniques in Wireless Sensor Networks: A Survey. *Wireless Communications*, *11*(6), 6–28. doi:10.1109/MWC.2004.1368893

Alippi, C., Boracchi, G., Camplani, R., & Roveri, M. (2012). Wireless sensor networks for monitoring vineyards. In G. Anastasi, B. E., E. Di Nitto, C. Ghezzi, L. Tanca, & E. Zimeo (Eds.), Methodologies and Technologies for Networked Enterprises (pp. 295–310). Berlin: Springer-Verlag Heidelberg. doi:10.1007/978-3-642-31739-2_15

Alippi, C., Camplani, R., Galperti, C., & Roveri, M. (2011). A Robust, Adaptive, Solar-Powered WSN Framework for Aquatic Environmental Monitoring. *Sensors Journal, IEEE*, *11*(1), 45–55. doi:10.1109/JSEN.2010.2051539

Alippi, C., Camplani, R., Marullo, A., & Roveri, M. (2012). A real-time monitoring framework for landslide and rock-collapse forecasting. In S. Mukhopadhyay (Ed.), *Smart Sensing Technology for Agriculture and Environmental Monitoring* (pp. 285–302). Berlin: Springer-Verlag Berlin Heidelberg. doi:10.1007/978-3-642-27638-5_14

Alkandari, A., Alnasheet, M., Alabduljader, Y., & Moein, S. (2012). Water monitoring system using wireless sensor network (wsn): Case study of kuwait beaches. *Proceedings of the Second International Conference on Digital Information Processing and Communications (ICDIPC)*. 10.1109/ICDIPC.2012.6257270

Anthony, D., Bennett, W. P., Vuran, M. C., Dwyer, M. B., Elbaum, S., & Lacy, A. (2011). Sensing through the continent: Towards monitoring migratory birds using cellular sensor networks. *Proceedings of the 11th International Conference on Information Processing in Sensor Networks*. doi: 10.1145/2185677.2185747

Antoin, L., Javier, T., Brendan, O., John, W., & Fiona, R. (2012). Deploy: A long term deployment of a water quality sensor monitoring system. *Sensor Review*, *1*(12), 29–38.

Aquino-Santos, R., Apolinar, G.-P., Edwards-Block, A., & Virgen-Ortiz, R. A. (2011). Developing a New Wireless Sensor Network Platform and Its Application in Precision Agriculture. *Sensors (Basel)*, *11*(1), 1193–1211. doi:10.3390110101192 PMID:22346622

Atmel AVR Microcontrollers. (1984). Retrieved from http://www.atmel.com/products/microcontrollers/avr/default.aspx

Ayday, C., & Safak, S. (2009). Application of Wireless Sensor Networks with GIS on the Soil Moisture Distribution Mapping. *Proceedings of the 16th International Symposium GIS Ostrava.*

Baggio, A. (2009). Wireless Sensor Networks in Precision Agriculture. *Proceedings of the ACM Workshop on Real-World Wireless Sensor Networks.* Retrieved from www.sics.se/realwsn05/papers/baggio05wireless.pdf

Barrenetxea, G., Ingelrest, F., Schaefer, G., & Vetterli, M. (2008). The Hitchhiker's Guide to Successful Wireless Sensor Network Deployments. *Proceedings of the 6th ACM Conference on Embedded Network Sensor Systems, (SenSys).* 10.1145/1460412.1460418

Bencini, L. D. (2010). Wireless sensor networks for on-field agricultural management process. In G. V. Merrett & Y. K. Tan (Eds.), *Wireless Sensor Networks: application-centric design* (pp. 1–18). InTech - Open Access Company. doi:10.5772/13001

Bhatti, S., Carlson, J., Dai, H., Deng, J., Rose, J., Sheth, A., ... Han, R. (2005). MANTIS OS: An embedded multithreaded operating system for wireless micro sensor platforms. *Mobile Networks and Applications*, *10*(4), 563–579. doi:10.100711036-005-1567-8

Blackmore, S. (1994). *Precision Farming: An Introduction* (Vol. 23). Outlook on Agriculture Journal.

Bri, D., Garcia, M., Lloret, J., & Dini, P. (2009). Real Deployments of Wireless Sensor Networks. *Proceedings of the third International Conference on Sensor Technologies and Applications (SENSORCOMM).* 10.1109/SENSORCOMM.2009.69

Butler, Z., Corke, P., Peterson, R., & Rus, D. (2004). Virtual Fences for Controlling Cows. *Proceedings of the IEEE International Conference on Robotics and Automation, (ICRA).* doi: 10.1109/ROBOT.2004.1302415

Cao, X., Chena, J., Zhangb, Y., & Suna, Y. (2008). Development of an integrated wireless sensor network micro-environmental monitoring system. *ISA Transactions*, *47*(3), 247–255. doi:10.1016/j.isatra.2008.02.001 PMID:18355827

Capella, J., Bonastre, A., Ors, R., & Peris, M. (2010). A Wireless Sensor Network approach for distributed in-line chemical analysis of water. *Talanta*, *80*(5), 1789–1798. doi:10.1016/j.talanta.2009.10.025 PMID:20152412

Cardell-Oliver, R., Smettem, K., Kranz, M., & Mayer, K. (2005). Field Testing a Wireless Sensor Network for Reactive Environmental Monitoring. *International Journal of Distributed Sensor Networks*, *1*, 149–162. doi:10.1080/15501320590966422

Chang, M., & Bonnet, P. (2010). Monitoring in a High-Arctic Environment: Some Lessons from MANA. *Pervasive Computing, IEEE*, *9*(4), 16–23. doi:10.1109/MPRV.2010.53

Chaudhary, D., Nayse, S., & Waghmare, L. (2011). Application of Wireless Sensor Networks for Greenhouse parameter Control. *Precision Agriculture International Journal of Wireless & Mobile Networks*, *3*(1), 140–148. doi:10.5121/ijwmn.2011.3113

Chen, P., & Lu, Z. (2013). A web-based indoor environment monitoring system using wireless sensor networks. *Proceedings of the Fifth International Conference on Computational and Information Sciences (ICCIS).* 10.1109/ICCIS.2013.529

Choi, S., Kim, N., Cha, H., & Ha, R. (2009). Micro Sensor Node for Air Pollutant Monitoring: Hardware and Software Issues. *Sensors (Basel)*, *9*(10), 7970–7987. doi:10.339091007970 PMID:22408489

Christin, D., Reinhardt, A., Mogre, P., & Steinmetz, R. (2009). Wireless sensor networks and the Internet of Things: Selected Challenges. *Proceedings of the 8th GI/ITG KuVS Fachgespräch Drahtlose Sensornetze*, 31-34.

Collins, F., Orpen, D., Maher, D., Cleary, J., Fay, C., & Diamond, D. (2011). Distributed Chemical Sensor Networks for Environmental Sensing. *Proceedings of the 2nd International Conference on Sensor Device Technologies and Applications*, 58-62.

Contiki: The Open Source OS. (2006). Retrieved from http://www.contiki-os.org

Corke, P., Wark, T., Jurdak, R., Hu, W., Valencia, P., & Moore, D. (2010). Environmental Wireless Sensor Networks. *Proceedings of the IEEE*, *98*(11), 1903–1917. doi:10.1109/JPROC.2010.2068530

Demirbas, M., Chow, K. Y., & Wan, C. S. (2006). INSIGHT: Internet-Sensor Integration for Habitat Monitoring. *Proceedings of the IEEE International Symposium on a World of Wireless, Mobile and Multimedia Networks (WoWMoM)*, 553-558.

Demirkol, I., Ersoy, C., & Alagoz, F. (2006). MAC Protocols for Wireless Sensor Networks: A Survey. *Communications Magazine, IEEE*, *44*(4), 115–121. doi:10.1109/MCOM.2006.1632658

Doolin, D. M., & Sitar, N. (2005). *Wireless sensors for wildfire monitoring. In Sensors and Smart Structures Technologies for Civil, Mechanical, and Aerospace Systems*. SPIE. doi:10.1117/12.605655

Dubois-Ferriere, H., Meier, R., Fabre, L., & Metrailler, P. (2006). TinyNode: a comprehensive platform for wireless sensor network applications. *Proceedings of the Fifth International Conference on Information Processing in Sensor Networks (IPSN)*. 10.1145/1127777.1127831

Dunbabin, M., Alistair, G., & James, U. (2009). An autonomous surface vehicle for water quality monitoring. *Proceedings of the Australasian Conference on Robotics & Automation*, 1-6.

Dunbabin, M., & Corke, P. (2010). A framework for marine sensor network & autonomous vehicle interaction. *Proceedings of the OCEANS 2010 IEEE*. 10.1109/OCEANSSYD.2010.5603677

Dunkels, A., Gronvall, B., & Voigt, T. (2004). Contiki-A Lightweight and Flexible Operating System for Tiny Networked Sensors. *Proceedings of the 29th Annual IEEE Conference on Local Computer Networks*. 10.1109/LCN.2004.38

Dursun, M., & Ozden, S. (2011). A wireless application of drip irrigation automationsupported by soil moisture sensors. *Scientific Research and Essays*, *6*, 1573–1582.

Dyo, V., Ellwood, S. A., Macdonald, D. W., Markham, A., Scellato, S., & Trigoni, N. (2010). Evolution and Sustainability of a Wildlife Monitoring Sensor Network. *Proceedings of the 8th ACM Conference on Embedded Networked Sensor Systems (SenSys)*, 127-140. 10.1145/1869983.1869997

Ehsan, S., Bradford, K., Brugger, M., Hamdaoui, B., Kovchegov, Y., & Johnson, D. (2012). Design and analysis of delay-tolerant sensor networks for monitoring and tracking free-roaming animals. *Wireless Communications. IEEE Transactions on*, *11*(3), 1220–1227.

Elsts, A., Balass, R., Judvaitis, J., Zviedris, R., Strazdins, G., & Mednis, A. (2012). SADmote: a robust and cost-effective device for environmental monitoring. *Proceedings of the 25th International Conference on Architecture of Computing Systems (ARCS)*. 10.1007/978-3-642-28293-5_19

Fay, C., Doherty, A., Beirne, S., Collins, F., Foley, C., Healy, J., ... Diamond, D. (2011). Remote Real-Time Monitoring of Subsurface Landfill Gas Migration. *Sensors (Basel)*, *11*(7), 6603–6628. doi:10.3390110706603 PMID:22163975

Flecks Datasheet. (2011). Retrieved from http://www.snm.ethz.ch/snmwiki/pub/uploads/Projects/Flecks.pdf

Garcia-Sanchez, A.-J., Garcia-Sanchez, F., & Garcia-Haro, J. (2011). Wireless sensor network deployment for integrating video-surveillance and data-monitoring in precision agriculture over distributed crops. *Computers and Electronics in Agriculture*, *75*(2), 288–303. doi:10.1016/j.compag.2010.12.005

Giannopoulos, N., Goumopoulos, C., & Kameas, A. (2009). Design Guidelines for Building a Wireless Sensor Network for Environmental Monitoring. *Proceedings of the 13th Panhellenic Conference on Informatics*. 10.1109/PCI.2009.17

Girard, L., Beutel, J., Gruber, S., Hunziker, J., Lim, R., & Weber, S. (2012). A custom acoustic emission monitoring system for harsh environments: Application to freezing-induced damage in alpine rock walls. *Geoscientific Instrumentation. Methods and Data Systems*, *1*(2), 155–167.

Goumopoulos, C., Kameas, A., & O'Flynn, B. (2007). Proactive Agriculture: An Integrated Framework for Developing Distributed Hybrid Systems. *Proceedings of the 4th International Conference on Ubiquitous Intelligence and Computing*, 214-224. 10.1007/978-3-540-73549-6_22

Guo, Y., Corke, P., Poulton, G., Wark, T., Bishop-Hurley, G., & Swain, D. (2006). Animal Behaviour Understanding using Wireless Sensor Networks. *Proceedings of the 31st IEEE Conference on Local Computer Networks (LCN)*. 10.1109/LCN.2006.322023

Hakala, I., Tikkakoski, M., & Kivelä, I. (2008). Wireless Sensor Network in Environmental Monitoring - Case Foxhouse. *Proceedings of the 2nd International Conference on Sensor Technologies and Applications (SENSORCOMM)*. 10.1109/SENSORCOMM.2008.27

Hartung, C., Han, R., Seielstad, C., & Holbrook, S. (2006). FireWxNet: a multi-tiered portable wireless system for monitoring weather conditions in wildland fire environments. *Proceedings of the 4th International Conference on Mobile systems, applications and services*. 10.1145/1134680.1134685

Hill, J. (2003). *System Architecture for Wireless Sensor Networks*. Berkeley, CA: University of California.

Hiu, L., Meng, Z., & Wang, M. (2009). A Wireless Sensor Network for Cropland Environmental Monitoring. *Proceedings of the International Conference on Network Security, Wireless Communications and Trusted Computing (NSWCTC)*. doi: 10.1109/NSWCTC.2009.306

Hou, J. C., Li, N., & Stojmenović, I. (2005). *Topology Construction and Maintenance in Wireless Sensor Networks*. New York, NJ: John Wiley & Sons, Inc. doi:10.1002/047174414X.ch10

Hou, K., De Sousa, G., Zhou, H., Chanet, J., Kara, M., Amamra, A., & ... . (2007). LiveNode: LIMOS versatile embedded wireless sensor node. *Journal of Harbin Institute of Technology*, *7*, 140–144.

Hu, J., Shen, L., Yang, Y., & Lv, R. (2010). Design and Implementation of Wireless Sensor and Actor Network for Precision Agriculture. *Proceedings of the IEEE International Conference on Wireless Communications, Networking and Information Security (WCNIS)*. 10.1109/WCINS.2010.5544151

Hu, W., Bulusu, N., Chou, T. C., Jha, S., Taylor, A., & Tran, V. N. (2009). Design and evaluation of a hybrid sensor network for cane toad monitoring. *ACM Transactions on Sensor Networks*, *5*(1), 1–28. doi:10.1145/1464420.1464424

Hu, W., Dinh, L., Corke, P., & Jha, S. (2012). Outdoor sensornet design and deployment: Experiences from a sugar farm. *IEEE Pervasive Computing*, *11*(2), 82–91. doi:10.1109/MPRV.2010.70

Huang, R., Song, W.-Z., Xu, M., Peterson, N., Shirazi, B., & LaHusen, R. (2012). Real-World Sensor Network for Long-Term Volcano Monitoring: Design and Findings. *IEEE Transactions on Parallel and Distributed Systems*, *23*(1), 321–329. doi:10.1109/TPDS.2011.170

Hui, J. W., & Culler, D. (2004). The dynamic behavior of a data dissemination protocol for network programming at scale. *Proceedings of the 2nd international conference on Embedded networked sensor systems (SenSys)*. 10.1145/1031495.1031506

IEEE-802.15.4. (2011). *IEEE Standard for Local and Metropolitan Area Networks-Part 15.4: Low-Rate Wireless Personal Area Networks (LR-WPANs)*. Retrieved from http://ieeexplore.ieee.org/xpl/articleDetails.jsp?reload=true&arnumber=6185525

Imote2 Datasheet. (2008). Retrieved from http://web.univ-pau.fr/~cpham/ENSEIGNEMENT/PAU-UPPA/RESA-M2/DOC/Imote2_Datasheet.pdf

Ingelrest, F., Barrenetxea, G., Schaefer, G., Vetterli, M., Couach, O., & Parlange, M. (2010). SensorScope: Application-specific sensor network for environmental monitoring. *ACM Transactions on Sensor Networks*, *6*(2), 1–32. doi:10.1145/1689239.1689247

Jelicic, V., Razov, T., Oletic, D., Kuri, M., & Bilas, V. (2011). MasliNET: A Wireless Sensor Network based environmental monitoring system. *Proceedings of the 34th International Convention (MIPRO)*, 150-156.

Jiang, X., Zhou, G., Liu, Y., & Wang, Y. (2010). Wireless sensor networks for forest environmental monitoring. *Proceedings of the 2nd International Conference on Information Science and Engineerrng (ICISE)*. 10.1109/ICISE.2010.5691483

Jurdak, R., Sommer, P., Kusy, B., Kottege, N., Crossman, C., & Mckeown, A. (2013). Camazotz: multimodal activity-based GPS sampling. *Proceedings of the 12th ACM International conference on Information processing in sensor networks (IPSN)*. doi: 10.1145/2461381.2461393

Kadri, A., Yaacoub, E., Mushtaha, M., & Abu-Dayya, A. (2013). Wireless sensor network for real-time air pollution monitoring. *Proceedings of the 1st International Conference on Communications, Signal Processing, and their Applications (ICCSPA)*. 10.1109/ICCSPA.2013.6487323

Kiernan, B. M., Beirne, S., Fay, C., & Diamond, D. (2008). Landfill gas monitoring at borehole wells using an autonomous environmental monitoring system. *Proceedings of the 5th International Conference on Climate Change and Global Warming*, 166-171.

Kim, Y.-D., Yang, Y.-M., Kang, W.-S., & Kim, D.-K. (2014). On the design of beacon based wireless sensor network for agricultural emergency monitoring systems. *Computer Standards & Interfaces*, *36*(2), 288–299. doi:10.1016/j.csi.2011.05.004

Kotamäki, N., Thessler, S., Koskiaho, J., Hannukkala, A. O., Huitu, H., Huttula, T., ... Järvenpää, M. (2009). Wireless in-situ Sensor Network for Agriculture and Water Monitoring on a River Basin Scale in Southern Finland: Evaluation from a Data User's Perspective. *Sensors (Basel)*, *9*(4), 2863–2883. doi:10.339090402862 PMID:22574050

Kumar, S., Iyengar, S. S., Lochan, R., Wiggins, U., Sekhon, K., Chakraborty, P., & ... . (2009). Application of Sensor Networks for Monitoring of Rice Plants: A Case Study. *Proceedings of the 4th International Symposium on Innovations and Real-time Applications of Distributed Sensor Networks (IRADSN)*, 1-7.

Kwong, K. H., Wu, T., Goh, H. G., Stephen, B., Gilroy, M. P., Michie, C., & Andonovic, I. (2009). Wireless sensor networks in agriculture: Cattle monitoring for farming industries. *PIERS Online*, *5*(1), 31–35. doi:10.2529/PIERS081001110156

Kwong, K. H., Wu, T.-T., Goh, H. G., Sasloglou, K., Stephen, B., Glover, I., ... Andonovic, I. (2012). Practical considerations for wireless sensor networks in cattle monitoring applications. *Computers and Electronics in Agriculture*, *81*, 33–44. doi:10.1016/j.compag.2011.10.013

Langendoen, K., Baggio, A., & Visser, O. (2006). Murphy Loves Potatoes. Experiences from a Pilot Sensor Network Deployment in Precision Agriculture. *Proceedings of the 20th IEEE International Parallel Distributed Processing Symposium (IPDPS)*. 10.1109/IPDPS.2006.1639412

Larios, D. F., Barbancho, J., Sevillano, J. L., Rodrıguez, G., Molina, F. J., Gasull, V. G., et al. (2013). Five years of designing wireless sensor networks in the doñana biological reserve (Spain): An applications approach. *Sensors, 13*(9), 12:044–12:069.

Le Dinh, T., Hu, W., Sikka, P., Corke, P., Overs, L., & Brosnan, S. (2007). Design and Deployment of a Remote Robust Sensor Network: Experiences from an outdoor Water Quality Monitoring. *Proceedings of the IEEE Conference on Local Computer Networks*. doi: 10.1109/LCN.2007.39

Lee, W., Alchanatis, V., Yang, C., Hirafuji, M., Moshou, D., & Li, C. (2010). Sensing technologies for precision specialty crop production. *Computers and Electronics in Agriculture*, *74*(1), 2–33. doi:10.1016/j.compag.2010.08.005

Lenka, S., & Mohapatra, A. (2012). A novel approach of cattle health monitoring system based on wireless sensor network for early detection of disease. *Journal of Grid and Distributed Computing*, *2*(1), 2249–7056.

Levis, P., Madden, S., Polastre, J., Szewczyk, R., Whitehouse, K., Woo, A., ... Culler, D. (2005). Tinyos: An operating system for sensor networks. *Ambient Intelligence*, *35*, 115–148. doi:10.1007/3-540-27139-2_7

Libelium Waspmote. (2013). Retrieved from http://www.libelium.com/products/waspmote/

LiteOS operating System. (2011). Retrieved from https://www.linuxliteos.com

Liu, Y., Zhang, J., & Richards, M. (2009). Towards Continuous Surveillance of Fruit Flies Using Sensor Networks and Machine Vision. *Proceedings of the 5th International Conference on Wireless Communications, Networking and Mobile Computing*. 10.1109/WICOM.2009.5303034

Lloret, J., Garcia, M., Bri, D., & Sendra, S. (2009). A Wireless Sensor Network Deployment for Rural and Forest Fire Detection and Verification. *Sensors (Basel), 9*(11), 8722–8747. doi:10.339091108722 PMID:22291533

Mafuta, M., Zennaro, M., Bagula, A., Ault, G., Gombachika, H., & Chadza, T. (2012). Successful deployment of a wireless sensor network for precision agriculture in Malawi. *Proceedings of the 3rd IEEE International Conference on Networked Embedded Systems for Every Application (NESEA).* 10.1109/NESEA.2012.6474009

Mainwaring, A. C. (2002). Wireless Sensor Networks for Habitat Monitoring. *Proceedings of the 1st ACM International Workshop on Wireless Sensor Networks and Applications, (WSNA).* 10.1145/570738.570751

Mallikarjuna, A., & Kunz, T. (2011). Operating Systems for Wireless Sensor Networks: A Survey. *Sensors, 11(6)*, 5900-5930.

Mancuso, M., & Bustaffa, F. (2006). Wireless Sensors Network For Monitoring Environmental Variables in a Tomato Greenhouse. *Proceedings of the IEEE International Workshop on Factory Communication Systems.* 10.1109/WFCS.2006.1704135

Markham, A., Trigoni, N., Macdonald, D. W., & Ellwood, S. A. (2012). Underground localization in 3-D using magneto-inductive tracking. *Sensors Journal, IEEE, 12*(6), 1809–1816. doi:10.1109/JSEN.2011.2178064

Martin, V., Moisan, S., Paris, B., & Nicolas, O. (2008). Towards a Video Camera Network for Early Pest Detection in Greenhouses. *Proceedings of the Endure International Conference Diversifying Crop Protection*, 1-5.

Martinez, K., Basford, P., De Jager, D., & Hart, J. (2012). A wireless sensor network system deployment for detecting stick slip motion in glaciers. *Proceedings of the IET Conference on Wireless Sensor Systems (WSS).* 10.1049/cp.2012.0580

Martinez, K., Hart, J., & Ong, R. (2004). Environmental sensor networks. *Computer, 37*(8), 50–56. doi:10.1109/MC.2004.91

Martinez, K., Hart, J. K., & Ong, R. (2009). Deploying a Wireless Sensor Network in Iceland. *Proceedings of the 3rd International Conference on Geosensor Networks.* 10.1007/978-3-642-02903-5_13

Matese, A., Vaccari, F. P., Tomasi, D., Di Gennaro, S. F., Primicerio, J., Sabatini, F., & Guidoni, S. (2013). Crossvit: Enhancing canopy monitoring management practices in viticulture. *Sensors (Basel), 13*(6), 7652–7667. doi:10.3390130607652 PMID:23765273

Mica2 Datasheet. (2005). Retrieved from http://www.eol.ucar.edu/isf/facilities/isa/internal/CrossBow/DataSheets/mica2.pdf

MICAz Datasheet. (2005). Retrieved from http://www.openautomation.net/uploadsproductos/micaz_datasheet.pdf

Miluzzo, E., Lu, H., Peebles, D., Choudhury, T., & Campbell, A. (2010). A survey of mobile phone sensing. *Communications Magazine, IEEE, 48*(9), 140–150. doi:10.1109/MCOM.2010.5560598

Mingli, Y., & Yihai, S. (2013). *Marine data collection based on embedded system with wired and wireless transmission*. University of Agder.

Mittal, A., Chetan, K., Jayaraman, S., Jagyasi, B., Pande, A., & Balamuralidhar, P. (2012). mKRISHI: Wireless sensor network platform for precision agriculture. *Proceedings of the 6th International Conference on Sensing Technology (ICST)*. 10.1109/ICSensT.2012.6461755

Morais, R., Fernandes, M. A., Matos, S., Serôdio, C., Ferreira, P., & Reis, M. J. (2008). A ZigBee multi-powered wireless acquisition device for remote sensing applications in precision viticulture. *Computers and Electronics in Agriculture, 62*(2), 94–106. doi:10.1016/j.compag.2007.12.004

Moteiv Tmote Sky. (2004). Retrieved from http://www.snm.ethz.ch/snmwiki/pub/uploads/Projects/tmote_sky_datasheet.pdf

Mun, M., Reddy, S., Shilton, K., Yau, N., Burke, J., & Estrin, D. (2009). PEIR: the personal environmental impact report, as a platform for participatory sensing systems research. *Proceedings of the 7th international conference on Mobile systems, applications, and services (MobiSys)*. 10.1145/1555816.1555823

Nadimi, E. S., Jørgensen, R. N., Blanes-Vidal, V., & Christensen, S. (2012). Monitoring and classifying animal behavior using zigbee-based mobile ad hoc wireless sensor networks and artificial neural networks. *Computers and Electronics in Agriculture, 82*, 44–54. doi:10.1016/j.compag.2011.12.008

O'Flyrm, B., Martinez, R., Cleary, J., Slater, C., Regan, F., Diamond, D., & ... . (2007). SmartCoast: A Wireless Sensor Network for Water Quality Monitoring. *Proceedings of the 32nd IEEE Conference on Local Computer Networks (LCN)*. doi: 10.1109/LCN.2007.34

Oliveira, L. M., & Rodrigues, J. J. (2011). Wireless Sensor Networks: A Survey on Environmental Monitoring. *Journal of Communication, 6*, 143–151.

Ong, K. G., Yang, X., Mukherjee, N., Wang, H., Surender, S., & Grimes, C. A. (2004). A wireless sensor network for long-term monitoring of aquatic environments: Design and implementation. *Sensor Letters*, *2*(1), 48–57. doi:10.1166l.2004.023

Padhy, P., Martinez, K., Riddoch, A., Ong, H. L., & Hart, J. K. (2005). Glacial Environment Monitoring using Sensor Networks. *Proceedings of the International Conference on Real-World Wireless Sensor Networks*, 10-14.

Pahuja, R., Verma, H., & Uddin, M. (2013). A wireless sensor network for greenhouse climate control. *Pervasive Computing, IEEE*, *12*(2), 49–58. doi:10.1109/MPRV.2013.26

Pierce, F. J., & Elliott, T. V. (2008). Regional and on-farm wireless sensor networks for agricultural systems in Eastern Washington. *Computers and Electronics in Agriculture*, *61*(1), 32–43. doi:10.1016/j.compag.2007.05.007

Polastre, J. (2003). *Design and Implementation of Wireless Sensor Networks for Habitat Monitoring*. Berkeley, CA: University of California.

Puccinelli, D., & Haenggi, M. (2005). Wireless sensor networks: Applications and challenges of ubiquitous sensing. *Circuits and Systems Magazine, IEEE*, *5*(3), 19–31. doi:10.1109/MCAS.2005.1507522

Ramos, M., Foster, A., Castell, S. F., Fos, V. G., & Solano, J. J. P. (2013). Gatherer: An environmental monitoring application based on IPv6 using wireless sensor networks. *International Journal of Ad Hoc and Ubiquitous Computing*, *3*(4), 209–217. doi:10.1504/IJAHUC.2013.055465

Rao, A., Marshall, S., Gubbi, J., Palaniswami, M., Sinnott, R., & Pettigrovet, V. (2013). Design of low-cost autonomous water quality monitoring system. *Proceedings of the 2nd International Conference on Advances in Computing, Communications and Informatics (ICACCI)*. 10.1109/ICACCI.2013.6637139

Rice, R., & Bales, R. C. (2010). Embedded-sensor network design for snow cover measurements around snow pillow and snow course sites in the Sierra Nevada of California. *Water Resources Research*, *46*(3), 35–37. doi:10.1029/2008WR007318

Riquelme, J. A.-L., Soto, F., Suardíaz, J., Sánchez, P., Iborra, A., & Vera, J. A. (2009). Wireless Sensor Networks for precision horticulture in Southern Spain. *Computers and Electronics in Agriculture*, *68*(1), 25–35. doi:10.1016/j.compag.2009.04.006

Rutishauser, M., Petkov, V., Boice, J., Obraczka, K., Mantey, P., Williams, T. M., & … . (2010). CARNIVORE: A Disruption-Tolerant System for Studying Wildlife. *EURASIP Journal on Wireless Communications and Networking*, *1*, 1–14.

Shaikh, Z. A., Shaikh, N. A., & Islam, N. (2010). An Integrated Framework to Develop Context-Aware Sensor Grid for Agriculture. *Australian Journal of Basic and Applied Sciences*, *4*(5), 922–931.

Shelby, Z., & Bormann, C. (2010). *6LoWPAN: The Wireless Embedded Internet*. New York: John Wiley & Sons, Inc.

Shepherd, R., Beirne, S., Lau, K., Corcoran, B., & Diamond, D. (2007). Monitoring chemical plumes in an environmental sensing chamber with a wireless chemical sensor network. *Sensors and Actuators. B, Chemical*, *121*(1), 142–149. doi:10.1016/j.snb.2006.09.045

Somov, A., Baranov, A., Spirjakin, D., Spirjakin, A., Sleptsov, V., & Passerone, R. (2013). Deployment and evaluation of a Wireless Sensor Network for methane leak detection. *Sensors and Actuators. A, Physical*, *202*, 217–225. doi:10.1016/j.sna.2012.11.047

Song, W.-Z., Huang, R., Xu, M., Ma, A., Shirazi, B., & Lahusen, R. (2009). Air-dropped Sensor Network for Real-time High-fidelity Volcano Monitoring. *Proceedings of the 7th International Conference on Mobile systems, applications, and services (MobiSys)*. 10.1145/1555816.1555847

Stephen, B., Michie, C., & Andonovic, I. (2013). Remote sensing in agricultural livestock welfare monitoring: Practical consideration. In S. Mukhopadhyay & J.-A. Jiang (Eds.), *Wireless Sensor Networks and Ecological Monitoring* (pp. 179–193). Berlin: Springer-Verlag Berlin Heidelberg. doi:10.1007/978-3-642-36365-8_7

Strazdins, G., Elsts, A., & Selavo, L. (2010). MansOS: easy to use, portable and resource efficient operating system for networked embedded devices. *Proceedings of the 8th ACM Conference on Embedded Networked Sensor Systems (SenSys)*. 10.1145/1869983.1870057

Su, X., Shao, G., Vause, J., & Tang, L. (2013). An integrated system for urban environmental monitoring and management based on the environmental Internet of Things. *International Journal of Sustainable Development and World Ecology*, *20*(3), 205–209. doi:10.1080/13504509.2013.782580

Szewczyk, R., Mainwaring, A., Polastre, J., Anderson, J., & Culler, D. (2004). An analysis of a large scale habitat monitoring application. *Proceedings of the 2nd international conference on Embedded networked sensor systems (SenSys)*. 10.1145/1031495.1031521

Talzi, I., Hasler, A., Gruber, S., & Tschudin, C. (2007). PermaSense: Investigating permafrost with a WSN in the Swiss Alps. *Proceedings of the 4th International Workshop on Embedded Networked Sensors (EmNets)*. 10.1145/1278972.1278974

Tan, R., Xing, G., Chen, J., Song, W.-Z., & Huang, R. (2013). Fusion-based volcanic earthquake detection and timing in wireless sensor networks. *ACM Trans. Sen. Netw., 9*(2), 17:1–17:25.

Terzis, A. R., Musaloiu, E., Cogan, J., Szlavecz, K., Szalay, A., Gray, J., et al. Wireless sensor networks for soil science. *International Journal of Sensor Networks, 7*(2), 53-70.

Texas Instruments Microcontrollers. (1980). Retrieved from http://www.ti.com/lsds/ti/microcontroller/overview.page

The Arduino platforms. (2009). Retrieved from http://arduino.cc/en/Main/Products

The BeagleBoard technology. (2008). Retrieved from http://beagleboard.org

The GlacsWeb project. (2012). Retrieved from http://glacsweb.org/technology/

The Pandaboard technology. (2010). Retrieved from http://pandaboard.org

TI CC2420 Datasheet. (2010). Retrieved from http://inst.eecs.berkeley.edu/~cs150/Documents/CC2420.pdf

Tik, L. B., Khuan, C. T., & Palaniappan, S. (2009). Monitoring of an Aeroponic Greenhouse with a Sensor Network. *International Journal of Computer Science and Network Security, 9*(3), 240–246.

Tolle, G., Polastre, J., Szewczyk, R., Culler, D., Turner, N., & Tu, K. (2005). A macroscope in the redwoods. *Proceedings of the 3rd International Conference on Embedded networked sensor systems (SenSys)*. 10.1145/1098918.1098925

Tseng, C.-L., Jiang, J.-A., Yang, E.-C., Lu, F.-M., Liao, C.-S., & Lin, C. (2008). An Automatic Counting Trap with Wireless Transmission Capability for the Oriental Fruit Fly. *Proceedings of the 4th International Symposium on Machinery and Mechatronics for Agriculture and Biosystems Engineering (ISMAB)*, 14-20.

Vasseur, J., & Dunkels, A. (2010). *Interconnecting Smart Objects with IP: The Next Internet*. San Francisco, CA, USA: Morgan Kaufmann Publishers Inc.

Vellidis, G., Tucker, M., Perry, C., Kvien, C., & Bednarz, C. (2008). Real-time Smart Sensor Array for Scheduling Irrigation. *Computers and Electronics in Agriculture, 61*(1), 44–50. doi:10.1016/j.compag.2007.05.009

Wang, G., Zhang, J., Li, W., Cui, D., & Jing, Y. (2010). A forest fire monitoring system based on GPRS and Zigbee wireless sensor network. *Proceedings of the 5th IEEE Conference on Industrial Electronics and Applications (ICIEA)*. doi: 10.1109/ICIEA.2010.5515417

Wang, J., Ren, X., Shen, Y., & Liu, S. (2010). A Remote Wireless Sensor Networks for Water Quality Monitoring. *Proceedings of the International Asia-Pacific Conf on Information Technology & Ocean Engineering (CICC-ITOE)*. 10.1109/CICC-ITOE.2010.9

Wang, Y., Wang, Y., Qi, X., Xu, L., Chen, J., & Wang, G. (2010). L3SN: A Level-Based, Large-Scale, Longevous Sensor Network for Agriculture Information Monitoring. *Wireless Sensor Network*, *2*(1), 655–660. doi:10.4236/wsn.2010.29078

Wark, T., Crossman, C., Hu, W., Guo, Y., Valencia, P., & Sikka, P. (2007). The design and evaluation of a mobile sensor/actuator network for autonomous animal control. *Proceedings of the 6th International Conference on Information Processing in Sensor Networks (IPSN)*. doi: 10.1109/IPSN.2007.4379680

Wark, T., Hu, W., Corke, P., Hodge, J., Keto, A., & Mackey, B. (2008). Springbrook: Challenges in developing a long-term, rainforest wireless sensor network. *Proceedings of the Internaional Conference on Intelligent Sensors, Sensor Networks and Information Processing (ISSNIP)*. 10.1109/ISSNIP.2008.4762055

Wen, T.-H., Jiang, J.-A., Sun, C.-H., Juang, J.-Y., & Lin, T.-S. (2013). Monitoring street-level spatial-temporal variations of carbon monoxide in urban settings using a wireless sensor network (WSN) framework. *International Journal of Environmental Research and Public Health*, *10*(12), 6380–6396. doi:10.3390/ijerph10126380 PMID:24287859

Werner-Allen, G., Lorincz, K., Johnson, J., Lees, J., & Welsh, M. (2006). Fidelity and Yield in a Volcano Monitoring Sensor Network. *Proceedings of the 7th symposium on Operating systems design and implementation (OSDI)*, 381-396.

Werner-Allen, G., Lorincz, K., Welsh, M., Marcillo, O., Johnson, J., Ruiz, M., & ... . (2006). Deploying a wireless sensor network on an active volcano. *IEEE Internet Computing*, *10*(2), 18–25. doi:10.1109/MIC.2006.26

Wijesinghe, L., Siriwardena, P., & Dias, D. (2013). A wireless sensor network for early warning of elephant intrusions. In F. Shaikh (Ed.), *Wireless Sensor Networks for Developing Countries* (pp. 38–50). Berlin: Springer-Verlag Berlin Heidelberg. doi:10.1007/978-3-642-41054-3_4

Woo, A., Tong, T., & Culler, D. (2003). Taming the underlying challenges of reliable multihop routing in sensor networks. *Proceedings of the 1st international conference on Embedded networked sensor systems (SenSys)*. 10.1145/958491.958494

Xia, J., Tang, Z., Shi, X., Fan, L., & Li, H. (2011). An environment monitoring system for precise agriculture based on wireless sensor networks. *Proceedings of the 7th International Conference on Mobile Ad-hoc and Sensor Networks (MSN)*. 10.1109/MSN.2011.16

Yaacoub, E., Kadri, A., Mushtaha, M., & Abu-Dayya, A. (2013). Air quality monitoring and analysis in Qatar using a Wireless Sensor Network deployment. *Proceedings of the 9th International Wireless Communications and Mobile Computing Conference (IWCMC)*. 10.1109/IWCMC.2013.6583625

Yang, J., Zhang, C., Li, X., Huang, Y., Fu, S., & Acevedo, M. F. (2010). Integration of wireless sensor networks in environmental monitoring cyber infrastructure. *Wireless Networks*, *16*(4), 1091–1108. doi:10.100711276-009-0190-1

Yoon, I., Noh, D. K., Lee, D., Teguh, R., Honma, T., & Shin, H. (2012). Reliable wildfire monitoring with sparsely deployed wireless sensor networks. *Proceedings of the 26th IEEE International Conference on Advanced Information Networking and Applications (AINA)*. 10.1109/AINA.2012.107

Zhang, L., Yuan, M., Tai, D., Ding, J., Xu, X., & Zhen, X. (2010). Design and implementation of granary monitoring system based on wireless sensor network node. *Proceedings of the International Conference on Measuring Technology and Mechatronics Automation (ICMTMA)*. 10.1109/ICMTMA.2010.296

Zheng, X.-Y., Liu, C.-Y., Chen, P.-T., Chuang, C.-L., Chen, C.-P., & Jiang, J.-A. (2013). Development of an embedded system-based gateway for environmental monitoring in wild fields. In S. Mukhopadhyay, K. Jayasundera, & A. Fuchs (Eds.), *Advancement in Sensing Technology* (pp. 23–42). Berlin: Springer-Verlag Berlin Heidelberg. doi:10.1007/978-3-642-32180-1_2

Zhu, Y., Dong, F., & Yuan, T. (2012). Automatic monitoring system for forest fire based on wireless sensor network. *Proceedings of the 2nd International Conference on Electric Information and Control Engineering (ICEICE)*. doi:10.1109/ICEICE.2012.936

Zia, H., Harris, N. R., Merrett, G. V., Rivers, M., & Coles, N. (2013). The impact of agricultural activities on water quality: A case for collaborative catchment-scale management using integrated Wireless Sensor Networks. *Computers and Electronics in Agriculture*, *96*, 126–138. doi:10.1016/j.compag.2013.05.001

Zoller, S., Reinhardt, A., Wachtel, M., & Steinmetz, R. (2013). Integrating wireless sensor nodes and smartphones for energy-efficient data exchange in smart environments. *Proceedings of the IEEE International Conference on Pervasive Computing and Communications Workshops (PERCOM Workshops)*. 10.1109/PerComW.2013.6529574

Zviedris, R., Elsts, A., Strazdins, G., Mednis, A., & Selavo, L. (2010). LynxNet: Wild animal monitoring using sensor networks. *Proceedings of the 4th International Conference on Real-world wireless sensor networks*. 10.1007/978-3-642-17520-6_18

ENDNOTE

1 MintRoute uses a shortest-path algorithm to route packets to the base station on the basis of a definable routing metric.

Chapter 10
Introducing Activity-Based Costing in Farm Management:
The Design of the FarmBO System

Giacomo Carli
Open University, UK

Maurizio Canavari
University of Bologna, Italy

Alessandro Grandi
University of Bologna, Italy

ABSTRACT

Recent research indicates that farm managers do not rely on adequate informative support in their decision-making processes. The authors propose a model of a farm management information system which integrates the activity-based costing approach. In describing the design and development of the "FarmBO" system, the authors provide a detailed functional requirement definition and the description of a working system prototype. The solution is designed to show the impact of general costs on the different crops, allocating them on the basis of the production cycle complexity. It includes a report section directly linked to the database which provides crop balance sheets and simulations in terms of what-if analyses. The system allows farm managers to 1) analyze deviations between budgeted and actual costs, 2) compare crop balance sheets across different years, and 3) perform sensitivity analyses. The authors account for prototype validation in two farms and discuss results and possible developments.

DOI: 10.4018/978-1-5225-5978-8.ch010

INTRODUCTION

In the last few years, new technologies applied on machines and equipment, new Web-based services, and new solutions from Precision Agriculture have proved able to generate large amounts of data that could improve farm management activities (Nikkilä, Seilonen, & Koskinen, 2010). Research on Farm Management Information Systems (FMIS) has proposed many models of information systems oriented towards the integration of multiple data sources, benefiting from the support of new methodologies (Sørensen, Pesonen, et al., 2010) and languages (Papajorgji, Pinet, Miralles, Jallas, & Pardalos, 2010). One of the efforts of this pervasive data collection activity is to enable cost analysis, which is a core part of the managerial decision-making activity. Nowadays, farmers are required to select not only the most profitable crops, but also the right level of investment in machines and the proper use of external services. All these decisions relate with cost analysis. Although agricultural practice can seem simple, farms are complex organizations which produce several products and a large part of the costs are indirect with respect to products. A long-standing problem is connected to the use of different procedures for the allocation of indirect costs to products and their impact on how the economic performance of products is reported and interpreted.

Surprisingly, current commercial FMIS present highly customized approaches towards product costing, and the existing literature has dedicated less attention to the design of cost analysis procedures. Furthermore, the great availability of data is not complemented by new developments in the elaboration phase (Sørensen, Fountas, et al., 2010): FMIS research remains focused on connecting new devices and stakeholders rather than on transforming heterogeneous data into useful information for farmers. In particular, cost analyses appear not particularly developed in FMIS. The current approaches tend to rely on parametric estimations of costs or on very specific approaches not validated in common managerial research and practice. Since indirect costs (e.g.: machine depreciation) are becoming the most important part of total costs in agricultural practice, their allocation plays a pivotal role. This is a classic problem in industrial accounting, and Activity-Based Costing is a well-known approach for allocating indirect costs to final cost objects, be they products, services or clients. Activity-Based Costing allows a part of the indirect costs proportional to their real use of the resources which originated those costs to be allocated to the final cost objects. Nevertheless, in agricultural research less attention has been focussed on this topic, and the few existing studies cover a limited range of applications and are related to a punctual use of Activity-Based Costing approach rather than to a broad definition of a systematic approach supported by an FMIS.

Hence, the possibility of integrating Activity-Based Costing procedures in an FMIS model is still questioned. The aim of this paper is to propose a model of FMIS which integrates Activity-Based Costing procedures. In this study, we present the development of the FarmBO system (the name comes from a contraction of Farm and University of Bologna), starting from the collection of functional requirements; we then describe the design aspects; finally, we show the reports produced by the system. FarmBO was tested in two validation cases, where it provided detailed support in understanding the cost of final products, comparing crops and performing crop choices in a farm.

This paper is structured as follows: first, the relevant literature about FMIS and Activity-Based Costing applied in farm management is reviewed. Second, the methodology applied in designing the FarmBO system, the functional requirements and the database model are presented. Then, the potentialities of the report section are shown. Finally, the advantages of introducing Activity-Based Costing procedure in FMIS are discussed.

THEORETICAL BACKGROUND

In recent years, the development of new technologies on machines and equipment has led to the generation of a large amount of data from the field. The information processing workload is increasing (Sørensen, Fountas, et al., 2010), and this stimulates the adoption of IT solutions in farms. Moreover, the introduction of Precision Agriculture technologies poses new challenges, requiring FMIS to process large amounts of raw data from multiple and dispersed sources (McBratney, Whelan, Ancev, & Bouma, 2005; Zhang, Wang, & Wang, 2002). The panorama remains very articulated: a vast group of farmers do not rely on IT technologies either for their day-by-day activities or for their decisional processes, but another smaller group is more oriented towards new technologies, not only related to machinery or equipment, but also devoted to information processing and decision support. While many software houses are incorporating cost analysis functionalities in their products, in many cases their solutions do not propose validated accounting approaches. Research about FMIS has developed a rich framework to address the data management issues of modern agriculture and PA applications. In this section, we review the core findings of this literature and we then focus on the specific applications of Activity-Based Costing in agriculture.

FMIS Development

Due to the increasing need to process large amounts of data, FMIS literature has shown a considerable growth in the number of researchers dealing with this issue. Research efforts have addressed conceptual models and functional requirements for future FMISs (Sørensen, Pesonen, Bochtis, Vougioukas, & Suomi, 2011; Sørensen, Fountas, et al., 2010; Sørensen, Pesonen, et al., 2010), architectural designs for the information systems (Nikkilä et al., 2010; Voulodimos, Patrikakis, Sideridis, Ntafis, & Xylouri, 2010), information flows (Fountas, Wulfsohn, Blackmore, Jacobsen, & Pedersen, 2006), data flows related to different processes (Nash, Dreger, Schwarz, Bill, & Werner, 2009).

FMIS represent a particular class of information systems which combine the specific needs of farms with database architectures and information management technologies. The presence of biological processes, a fixed supply of land, small company size, weather forecast and perfect competition are some of the specific features that differentiate farms from other companies (Kay, Edwards, & Duffy, 2011; Sørensen, Pesonen, et al., 2010). Considerable efforts have been made to evolve the FMIS, leading to new approaches to machinery performance monitoring, and collection of site specific data (Fountas et al., 2006). Furthermore, the specific needs of Precision Agriculture pose new challenges to FMIS. For instance, Geographic Information Systems (GIS) require appropriate designs (Nikkilä et al., 2010) and some new developments converging with this direction, as shown in the case of the vineyard zone definition proposed by Acevedo-Opazo et al. (2008). Nevertheless, as reported by Sørensen, Fountas, et al. (2010), many scholars point out that "automatically collected data or data by manual registration is not used due to data logistic problems, leaving a gap between the acquiring of such data and the efficient use of this in agricultural management decisions".

Our contribution is aimed at addressing that gap, proposing an FMIS prototype which supports Activity-Based Costing procedures in order to improve farmers' decision-making processes.

Activity-Based Costing in Farm Management

The introduction of the Activity-Based Costing approach can lead to improvements of managerial activities and decisions in farms. Allocation of fixed costs to products is complex and, as mentioned before, may be tricky, leading to major errors in product cost assessment, profitability analysis and in other areas of management decision making. Activity-Based Costing is a sophisticated methodology for allocating fixed costs to final cost objects, for instance products. The introduction of Activity-Based Costing was due to the increasing level of fixed costs in modern companies (Cooper & Kaplan, 1988; Johnson & Kaplan, 1987). Activity-Based Costing "*measures costs and performances of activities, resources and cost objects, assigns resources to activities and activities to cost objects based on their use, and recognizes causal relationships of cost drivers to activities*" (Dierks & Cokins, 2000). The Activity-Based Costing approach is based on the idea that the development of final products requires the use of resources (e.g.: machines, human resources, materials). Usually, resources generate fixed costs, which need to be charged to products. Activity-Based Costing requires costs to be allocated first to the activities that generated them; second, to the products that required those activities. According to Ferreira (2004), an Activity-Based Costing system is composed of two critical processes. The first is the Resource-Activity Assignment Process, which measures the resource consumption generated by the different activities performed in the company, with a high level of granularity. The second is the Activity-Cost-Object Tracing Process, which measures which activities are required by products (or final cost objects) and allocates the corresponding portion of costs.

In conventional approaches, fixed costs are allocated to final products using an allocation base such as crop extension, or the value of final products, which can lead to significant misrepresentations of final product costs. While traditional approaches may be useful to evaluate costs where land is the main cost driver, they may lead to significant evaluation errors when there are strong differences in the complexity of the products: simple products realized in large quantities receive a large portion of costs in comparison with complex products realized in small quantities. Therefore, simple products could appear less convenient than complex products.

The adoption of Activity-Based Costing systems is still particularly limited due to the high quantity of data needed for their application. Anthony, Hawkins, & Merchant (2010) propose some data about Activity-Based Costing adoption in the world: 6% of companies in the United Kingdom have adopted Activity-Based Costing systems; 36% of USA companies; and 12% in Italy. Abusalama (2008) classifies implementation issues or difficulties into three main types: "technical", "behavioral" and "systems" barriers. Examples of "technical" barriers are: difficulties in identifying activity centers, identifying cost drivers, and assigning costs to activities. Examples of "behavioral" barriers are individual issues, such as lack of senior management support, lack of suitable accounting staff, and internal resistance. Examples of "systems" barriers are inadequate hardware and software, and data collection difficulties. Finally, Cinquini et al. (2008) point out some critical aspects related to Activity-Based Costing implementation: the introduction of an Activity-Based Costing system requires (1) the change of the cost structure in which the overheads allocation requires sophisticated processes, (2) the design of a more accurate cost measurement system, (3) the design more analytic product costing tools. In other words, the size of the company and its organizational structure can affect the feasibility of the introduction of an Activity-Based Costing system, because it requires a considerable set up effort.

Research studies on Activity-Based Costing applications in the farming and food processing contexts are limited: fish processing in Finland (Setala & Gunasekaran, 1996), fish markets in Taiwan (Lee & Kao, 2001), sawmilling in Finland (Korpunen,

Mochan, & Uusitalo, 2010), winemaking in Spain (González-Gómez & Morini, 2006), ornamental plant cultivation in Spain (González-Gómez & Morini, 2009). Finally, Chrenková (2011) proposes a complete framework, without reference to a specific business, but her analysis based on a Microsoft Excel sheet is not supported by an FMIS.

Despite the paucity of research and applications, Activity-Based Costing offers the considerable advantage of improving decision making processes, providing a reliable method to process the wealth of data collected from the field, also with the aid of Precision Agriculture technologies. Farmers face many complex decisions, ranging from crop choices to machinery renewal and the use of external services. In many cases, indirect costs play a pivotal role in a decision, as more and more activities are completed using expensive machinery. If cost allocation procedures return a misleading representation, the company is more exposed to risks.

Applying Activity-Based Costing, farm managers can achieve greater control on the consumption of resources and improve their decision-making processes. Furthermore, the accounting process itself can be positively influenced. The scope of the Activity-Based Costing system allows fine and accurate adjustments to be made to the company cost structure, reducing overhead costs (Cooper & Kaplan, 1988).

The possibility of improving farm management activity providing a solution which integrates Activity-Based Costing triggered the development of the FarmBO system, as detailed in the next sections.

Design and Development of the FarmBO System

The development of the FarmBO system started with a detailed functional requirements definition phase. We then proceeded to develop a working prototype of the system. Finally, we validated it in two farms. The next subsections present this development process.

Functional Requirements Analysis

In this stage, the typical procedure suggested in accounting literature for the introduction of Activity-Based Costing systems (Anthony, Hawkins, & Merchant, 2010; Rafiq & Garg, 2002) was combined with the usual steps of IT systems development and adapted to the case of farming activity. We conducted interviews with different key users and stakeholders to collect their opinions about the managerial decisions farmers need to take in their activity and how they can be supported by a structured economic analysis tool. Since the likelihood of adoption is influenced by the coherence of the tool with the users' needs (Sørensen, Pesonen, et al., 2010), we conducted two rounds of interviews with a group of four farmers, who operate in Northern Italy, and we designed the initial data flows referring to the production of potatoes. We then traced the core functions of the system and its main elements. Therefore, we expanded our analysis by collecting the opinions of twelve farmers and technicians from three countries (Italy, Greece and Turkey). We focused on the specific outputs users are going to expect from the system. Furthermore, we compared the data provided manually by users with the data that could have been automatically collected. Finally, we developed four use case groups to describe the development phase of the system prototype according to common guidelines about use case modelling in UML language (Cockburn, 2001; Phillips, Kemp, & Kek, 2001; Rosenberg, & Stephens, 2007). The first use case group, reported in Figure 1, presents the planning activities which farmers usually conduct to decide which crops they are going to cultivate in their fields. These activities span from a simple crop selection and assignment (use case 1.1) to more advanced decisions, which can be anticipated at this stage: definition of crop cycles (use case 1.2), assignment of resources to activities (use case 1.3), and definition of budget costs related to the activities which are going to be conducted on each crop (use case 1.4). More

Figure 1. Use cases about initial set up of the system

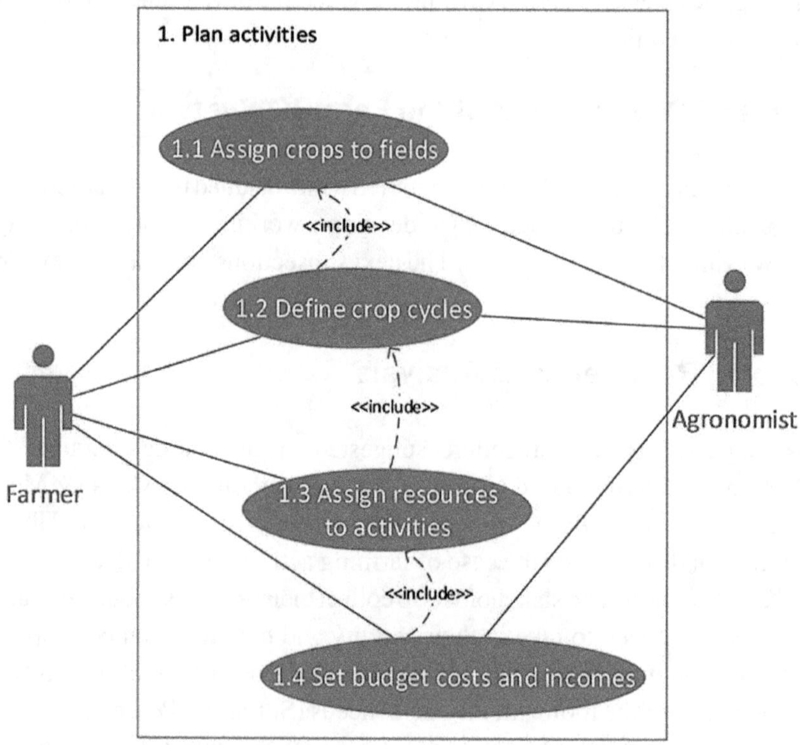

complex activities always include simpler activities from use case 1.4 to 1.1. In these preparatory activities, the farmer can be supported by an agronomist who helps in selecting the best crops, setting up the production cycles, and defining the budget costs. The cases are represented in grey background colour because they require human interaction and decision-making.

Figure 2 shows the use cases regarding the day-by-day activities conducted by the farmer, who is required to record the use of resources. The general case 2.1 can be specialized showing how its "children" are differentiated. Four different types of resources have been identified: human resources, machines and equipment, materials, and external services. The main task of the farmer is to record the consumption that activities make of resources. In general, it is a time-based consumption (in use cases 2.1a, 2.1c, 2.1d), which is quite easily measurable (Kaplan & Anderson, 2007). Use of materials is based on quantity variations as in use case 2.1b, e.g.: level of tanks or other containers. Moreover, machine use generates fuel costs, which is a case linked with an Include relationship because it can be performed when a machine is linked with a specific activity (use case 2.2). An automatic system could be particularly useful to measure fuel consumption. Moreover, all the use cases are in blue background colour to show that they could be supported by automatic systems for recording time or material consumption.

During the on field activities, the planned activities may, for some reason, have to be changed or cancelled, or some new activity may be needed. Therefore, the system should support these unexpected procedures in which a technician or an agronomist is likely to advise the farmer (use case 3.1, 3.2, 3.3). In the three conditions shown in Figure 3 (insert, modify, and delete an activity), the (re)definition of resources

Figure 2. Use cases regarding day-by-day activities management

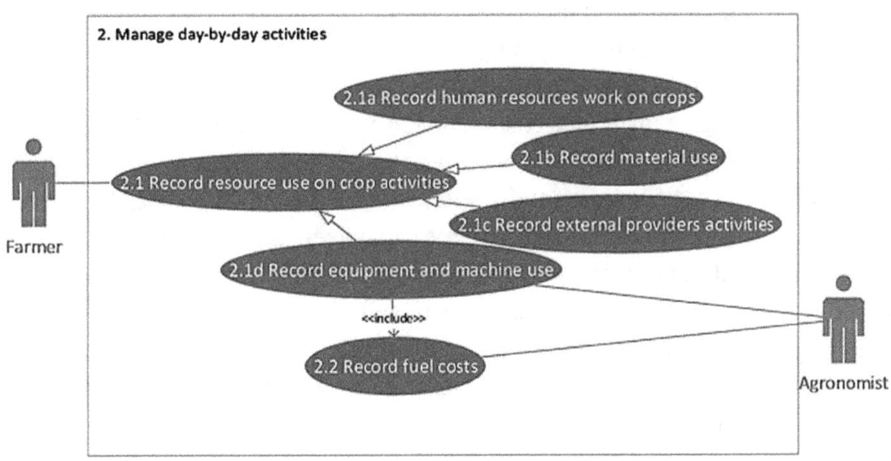

Figure 3. Use cases regarding activity updating

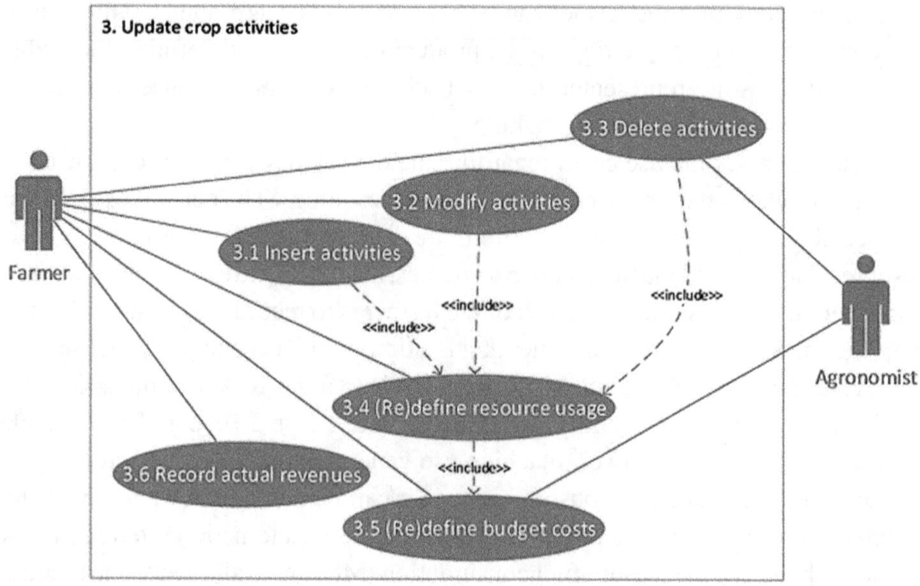

associated with the activity (use case 3.4) is required, along with a budget reformulation (use case 3.5). Moreover, use cases 3.4 and 3.5 can happen independently from the previous conditions, for different reasons (e.g.: the decision to switch from internal machines and human resources to an external provider of services, or vice versa). Use case 3.6 is about actual cost recording.

Figure 4 proposes the use case regarding the decisional phase, which is the most complex functional area supported by the FarmBO system. With the support of an agronomist, the farmer allocates indirect costs (use case 4.1) to activities and crops, according to an Activity-Based Costing procedure, described in the next pages. The allocation encompasses different types of costs: depreciation (use case 4.2), maintenance (use case 4.3), and all the other general costs of the farm (use case 4.4), for instance insurance costs. The farmer evaluates the results of the work carried out during the year and compares the actual outcomes with the budget forecasts (use case 4.5). Costs and revenues variances make it possible to focus attention on the performance in resources management. Comparisons between crops of different years facilitate the choice of future productions (use case 4.6). The farmer can assess crop balance sheets and crop sensitivity analyses, which are presented in the subsection dedicated to reports. While the generation of these reports is an automated process (use cases 4.7 and 4.8, with blue background), an agronomist is required to validate the output to check their consistency (e.g.: check

Figure 4. Use cases regarding decision-making phase

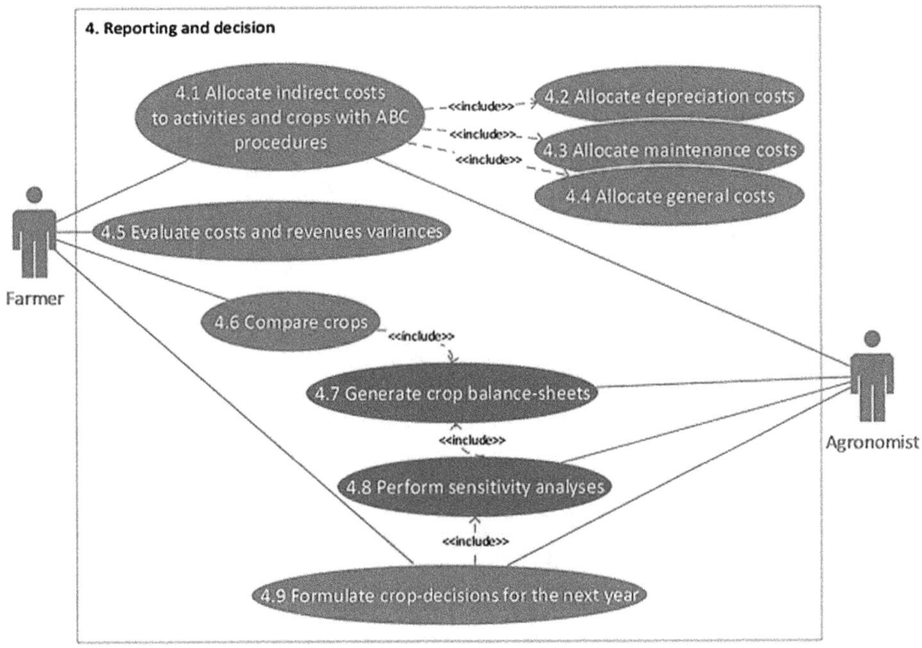

that depreciation procedures have been set up carefully by the farmer) before making it available to farmers. Decision-making may then become based on structured data from sensitivity analyses (use case 4.9).

Design of the System

In designing the system, the requirements of Activity-Based Costing systems were integrated in internal data management procedures. The relational database was defined adopting a flexible structure. First, all main entities were identified: crops, fields, activities resources (human, machines and equipment, materials, external services). Second, the system was modelled around these main entities, developing the most inclusive solution as possible, to allow extended compatibility with as many data structures as possible, including the automatic outputs from machines. Figure 5 presents the Entity-Relationship diagram with the key entities according to the guidelines and notation reported by Davis and Yen (1998).

Direct costs are recorded in the specific tables which match an activity with a resource per record. Adopting this solution, every cost is linked to both activity and resource. An activity generates the cost due to resource consumption which is then charged to the crop which required that activity. Indirect costs are only linked to

Figure 5. Entity-Relationship Diagram of the FarmBO system

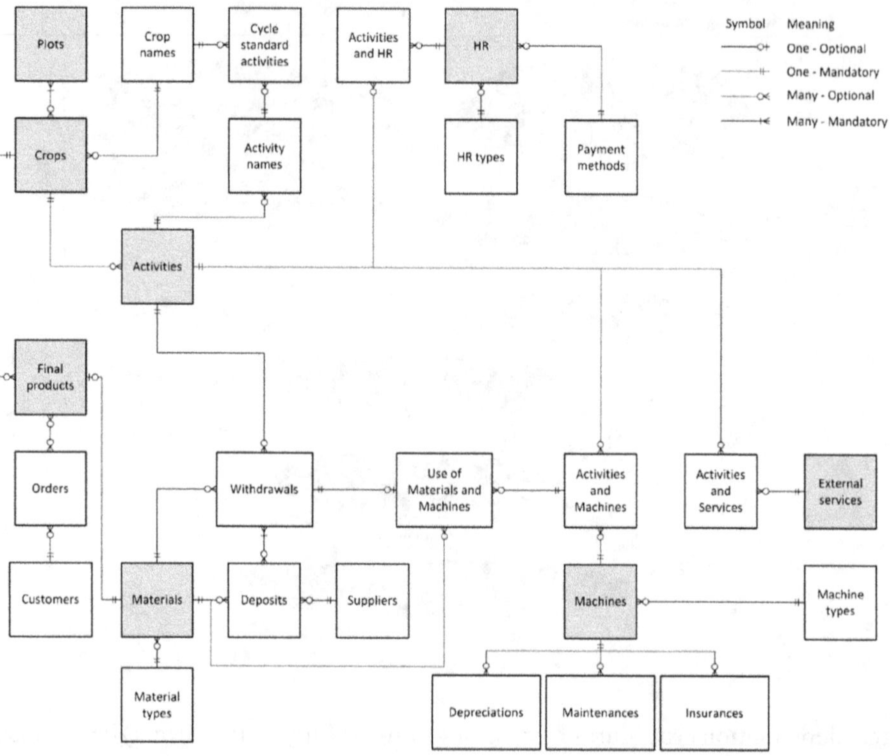

resources (e.g.: depreciation), so they are allocated to activities and then to crops using an Activity-Based Costing procedure.

In order to allocate general costs to crops and final products, after the identification of the activities, an activity driver needs to be identified for each activity. The choice of the activity drivers has to be consistent with the use of the underlying resource. Every time the activity is performed, it should generate the same consumption of the underlying resource and the same amount of costs (Anthony et al., 2010; Cooper & Kaplan, 1988).

While we refer to the accounting literature for the details of the Activity-Based Costing procedure, we present the information requirements of Activity-Based Costing implementation and how they are met in the design of the database of the FarmBO system. Figure 5 shows how indirect costs are allocated. Being indirect, they are recorded in tables linked only to the resources (in green in Figure 6).

Nevertheless, when a resource is associated with an activity, this latter uses the resource for a defined amount of time (Kaplan & Anderson, 2007) or according to a different driver. Therefore, in the allocation procedure depicted at the bottom

Figure 6. Allocation of indirect costs in FarmBO system using an Activity-Based Costing procedure

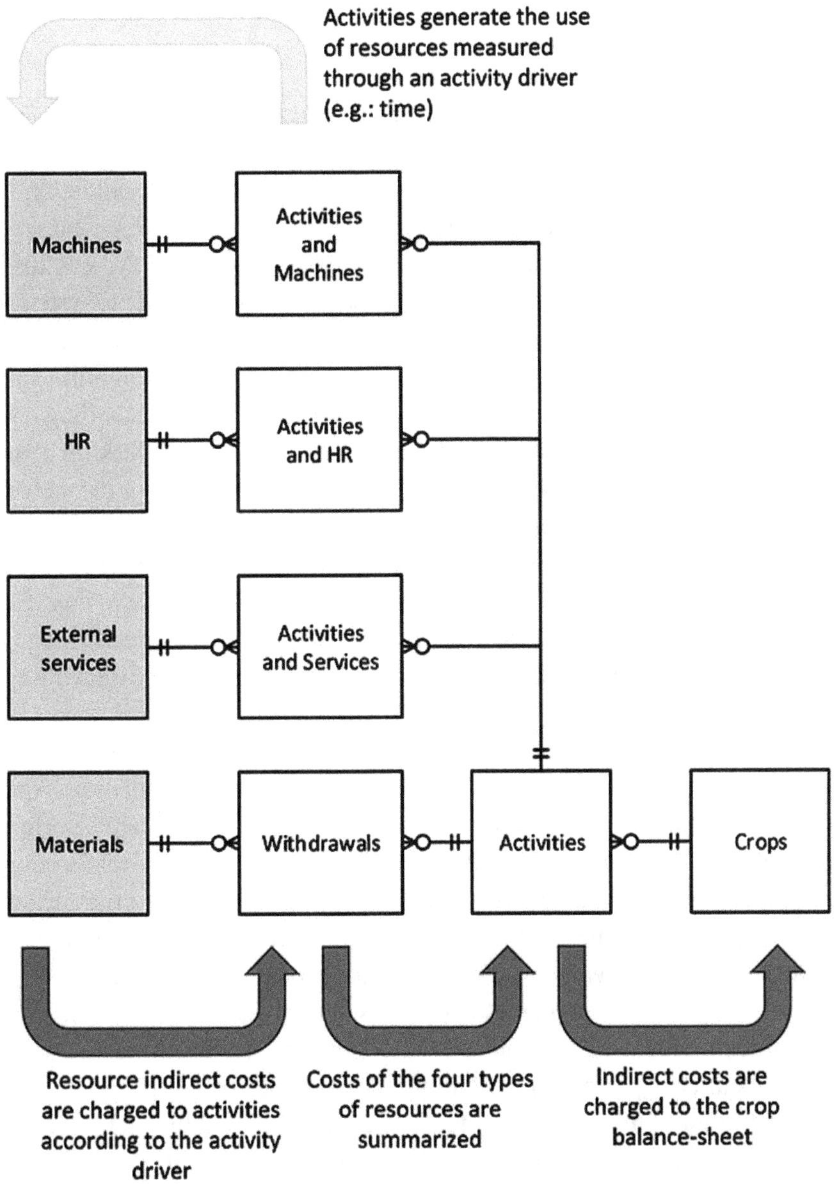

**For a more accurate representation see the electronic version.*

of the picture, the costs linked only to resources are moved to activities and then summed for the four classes of resources which can be used in performing a single activity. This process makes it possible to create a crop balance sheet, which is the topic of the next subsection.

System Prototype: Interfaces and Reports

We developed a prototype of the system and we tested it in two validation cases. In this subsection, we present the main results focusing on interfaces and reports.

The prototype of the database was developed both in Microsoft Access 2013 and Microsoft SQL Server 2012. To responsively adapt the interfaces of the system to the user requirements, we designed them in the Access prototype. This choice allowed fast cycles of user-tests and changes of the interfaces. Finally, a mobile solution for data entry was explored adopting the Microsoft LightSwitch technology, which enables the creation of data entry interfaces for both mobile and desktop websites.

The system design also supports automatic acquisition of data or the use of very structured approaches. Modern tractors and machines are equipped with sensors and communication systems. During the activities, sensors record a large amount of data, which can be transferred to the FarmBO system after preprocessing. Human resource activities are still very pervasive in the Italian, Greek, and Turkish farms, as suggested by our informants, therefore we opted for a simple time tracking system which can be semi-automated through the use of badges.

The FarmBO system can be easily converted in a cloud-based solution with the support of modern LightSwitch interfaces. During our interviews, many respondents pointed out that having the possibility to record data in the system while performing the activity on the crop is a key element that can facilitate the system's adoption. LightSwitch language makes it possible to design web interfaces for mobile devices without compatibility issues related to different mobile operating systems, because the system is accessible via a browser. The final model of the cloud-based system is depicted in Figure 7: machines and operators upload data in the system; operators receive guidelines for managing their activities; farmers and agronomists have access to reports about crop performances and what-if analyses.

In the definition of reports, we devoted particular attention to the user-requirements that emerged during interviews. Informants asked for a cost control interface which shows the different crops of their farms. The key point was related to the impact of general costs on the different crops, allocating them not on the basis of the land surface area only, but also on the basis of the complexity of the production cycle. Our informants also requested the possibility to perform simple simulations in terms of what-if analyses related to variations in selling prices, cultivated extensions, and production per Ha. Combining these requirements, we designed a reporting system

Figure 7. Model of the cloud-based system

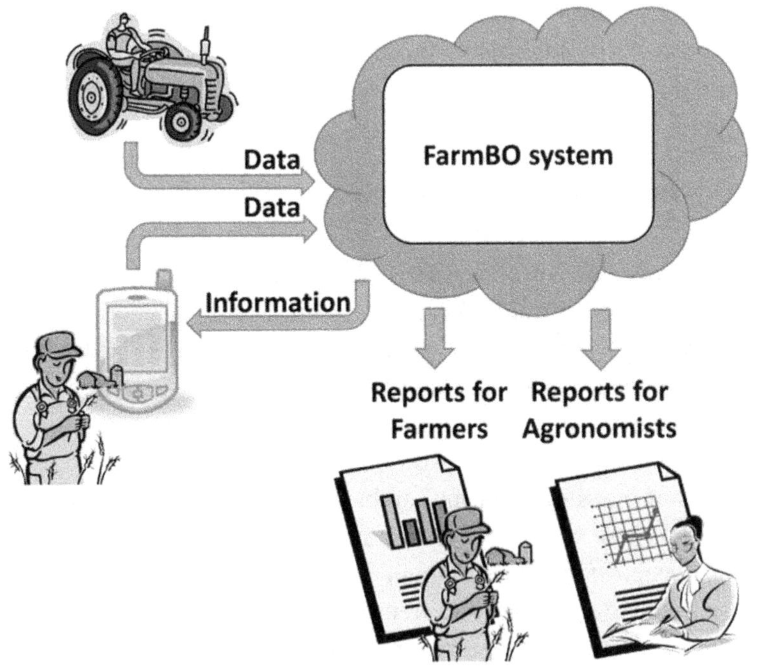

in Microsoft Excel linked to the database to reduce complexity and compatibility issues and leave the possibility of personal customization on reports. We combined a balance sheet designed to offer detailed information data per single crop, with a set of what-if analyses. The report section is designed in Microsoft Excel to enable farmers and agronomists to work on data. The system offers three main types of analysis. First, it makes it possible to analyze variances between budget and actual costs and identify if they are due to a change in the unit costs or to a change in volumes of activities (e.g.: number of hours, quantity of materials). Second, it traces the crop balance sheet for each crop, supporting comparisons between different crop varieties and different years. Third, it performs sensitivity analyses which are a specific class of what-if analyses which shows how a variable (e.g.: net profit) changes accordingly to the variation of two other variables (e.g.: unity price and produced quantity). This analysis was considered particularly important by our respondents because it is able to convey straightforward information about the positions of profit and loss and the break-even point at different levels of two exogenous variables.

The FarmBO system was validated in two farms. The first produces four different varieties of potatoes in the countryside near Bologna (Italy), in a land extension of 35 Ha. The second cultivates grain sorghum and wheat in a land extension of 7 Ha.

In both cases, the companies tested FarmBO in 2013, recording production cycles and costs on the platform. They evaluated the report section. Figure 8 reports the example of a sensitivity analysis for the production of a potato variety. It shows how the net profit per Ha changes according to variations in unit prices and in land use.

DISCUSSION AND CONCLUSION

From both the perspective of research and practice, there is a growing call for improving managerial decision making processes in farms. This contribution addresses these needs by proposing a system which supports Activity-Based Costing and can significantly improve the management of farms. In this section, we discuss the contribution to Farm Management Information Systems literature and we evaluate the benefits of introducing Activity-Based Costing in farms.

Information processing in farms can be significantly improved by introducing structured techniques, which rely on solid managerial approaches and accounting methods. The design of a module that supports managerial decision making processes complements existing solutions (Sørensen, Fountas, et al., 2010), offering the possibility to integrate data from multiple sources.

Key users and informants provided rich insights about how farms can benefit from clear and specific information based on real data, but filtered and summarized with proven and reliable approaches. One main point raised by interviewees was having

Figure 8. Sensitivity analysis of net profits per Ha related to production of potatoes

Sensitivity Analysis: Net Profits per Ha of Potato Primura (€/Ha)

Extension (Ha)	Unit Price (€/q) 350	375	400	425	450	475	500	525	550	575	600
15	-3653	-3365	-3076	-2787	-2498	-2210	-1921	-1632	-1343	-1055	-766
16	-3384	-3076	-2768	-2460	-2152	-1844	-1536	-1228	-920	-612	-304
17	-3114	-2787	-2460	-2133	-1805	-1478	-1151	-824	-496	-169	158
18	-2845	-2498	-2152	-1805	-1459	-1112	-766	-419	-73	274	620
19	-2575	-2210	-1844	-1478	-1112	-747	-381	-15	351	716	1082
20	-2306	-1921	-1536	-1151	-766	-381	4	389	774	1159	1544
21	-2036	-1632	-1228	-824	-419	-15	389	793	1198	1602	2006
22	-1767	-1343	-920	-496	-73	351	774	1198	1621	2045	2468
23	-1497	-1055	-612	-169	274	716	1159	1602	2045	2487	2930
24	-1228	-766	-304	158	620	1082	1544	2006	2468	2930	3392
25	-958	-477	4	485	967	1448	1929	2410	2892	3373	3854
26	-689	-188	312	813	1313	1814	2314	2815	3315	3816	4316
27	-419	100	620	1140	1660	2179	2699	3219	3739	4258	4778
28	-150	389	928	1467	2006	2545	3084	3623	4162	4701	5240
29	120	678	1236	1794	2353	2911	3469	4027	4586	5144	5702
30	389	967	1544	2122	2699	3277	3854	4432	5009	5587	6164

a system accessible from mobile applications and not hosted on local servers, but in a cloud based environment, to avoid the risks of losing data and the maintenance costs of the server infrastructure in their farms. FamrBO was created to be easily migrated on the cloud: it can be accessed through web browsers also from mobile devices with the same scalable interfaces designed with Microsoft LightSwitch technology. This solution is geared toward offering a flexible FMIS hosted in a cloud-based environment accessible through simple web interfaces, which can be designed using the same language for both desktop and mobile devices.

While many software houses are offering systems which integrate cost analysis, the introduction of highly customized methodologies could expose farmers to the risk that their decisions are not based on reliable approaches. The FarmBO prototype is designed to grant a high level of coherence with Activity-Based Costing methodology.

Traditional accounting and cost calculation systems are sufficiently accurate when most of the cost items can be easily assigned to a single activity or product; they offer the advantage that they need a relatively low input of information and if the share of overheads is low the allocation errors deriving from a lack of more specific information may be negligible.

However, Activity-Based Costing in farm management is potentially able to provide many benefits, mainly because of the intrinsic nature of farming as a multi-activity business. This is especially true in situations where the relative importance of durable multi-purpose assets and fixed costs burden is high.

A more reliable, accurate, and timely costing method based on Activity-Based Costing, therefore, would be beneficial to farm managers who need financial information to effectively support strategic and tactical decision-making in several farm management areas, such as crop rotations, introduction of new crops, farming process improvement, investment and disinvestment decisions. Another important area could be product pricing, although for most agricultural firms price is an exogenous variable rather than a variable decided by the manager. However, by comparing market price information and forecasts with a more accurately quantified cost of the single product, the farm manager increases the chances of making the right decision.

Another relevant area where Activity-Based Costing could impact the agricultural supply chain is the possibility for associated farmers organizations (such as cooperatives, consortia, etc.) to better plan their production and marketing activities and to provide more insightful advice to their associates.

A more accurate costing method would also be helpful in supporting the negotiations and decision-making when defining inter-professional agreements between farmers associations and processors and/or distributors. The knowledge of detailed and accurate cost data from the associates through an organization-wide database would support not only marketing and supply chain management, but could also provide background information for the identification of inefficiencies and weaknesses that need to be improved and help the farmer's association to plan extension and support actions addressed to those who need them the most.

An important element which makes Activity-Based Costing on one hand more useful and on the other hand more accessible/feasible to farms and farming-related companies is that agriculture-based businesses are making increasingly intensive use of information in several production and management processes (e.g., for quality management or food safety management purposes).

In addition, farm management is increasingly requiring greater flexibility and adaptation to rapidly changing market conditions; this makes it necessary to avoid under and over-costing situations due to inaccurate allocation of overheads and common costs with concurrent activities, which is a typical problem of traditional costing methods.

As stated before, the high degree of data intensity required by Activity-Based Costing has been (and still is) an important barrier to adoption. However, the evolution of technologies applied to farming can contribute to allow farm managers to automatically generate or recover existing data that may be the source of information necessary to define the cost drivers and to easily and effectively use the Activity-Based Costing approach. There could be many sources of extra-accountancy data, e.g.: software applications designed to support farmers and farm managers to control quality assurance parameters; digital controllers used to verify and record the working conditions of equipment such as tractors, pumps, sprayers and combines; robots and other automatic equipment with portable sensors as well as on-site sensors which are used in precision farming to verify and record several parameters referring to soil, crops, animals and the environment in which they live.

The design of a new farm management information system, therefore, should also include a management oriented accountancy/cost analysis module because this would allow the farm manager to consider key economic variables together with the technical aspects of his activity. In an inverse way, it could be hypothesized that the adoption of precision agriculture tools and techniques could be facilitated if these tools also provide guidance to the farmers in performing an economic evaluation of their activity.

A first direction for future research involves the possibility to introduce a linear programming module in FMIS, which evaluates different combinations of crops, taking into account the boundaries imposed by climate, costs, machines and other resources, and maximizing profits.

A second possibility is the design of an integrated solution between FMIS and machines, defining the interfaces and the communication protocols to support a completely automatic process of data collection. The variety of tools, machines, and equipment along with the differences between standards requires the use of a flexible staging area where data is consolidated and validated before being used in the FMIS. The integration process between data sources and FMIS is a long pathway where standards play a key role in shaping technological developments. In the current embryonic situation, the emergence of new unifying paradigms is still possible. In this context, different business models can be created. One could mirror the android business ecosystem where a flexible and open source platform is coupled with a market where small and big software houses propose their applications.

ACKNOWLEDGMENT

This research is included in the research project "RoboFarm" and financially supported by the Italian Ministry of Agriculture, Food and Forestry Policies in the framework of the EU FP7 ICT-AGRI ERA-NET. We thank Prof. Spyros Fountas, and the participants to the EFITA 2013 and HAICTA 2013 conferences for their useful feedback. We also thank the "Confederazione Italiana Agricoltori", the Institution of Agronomists and Foresters of Bologna, Beatrice Bolognini and Matteo Bertacchini for their precious help in data collection.

REFERENCES

Abusalama, F. A. (2008). *Barriers to adopting activity-based costing systems (ABC): an empirical investigation using cluster analysis* (Doctoral dissertation). Dublin Institute of Technology. Retrieved from http://arrow.dit.ie/cgi/viewcontent.cgi?article=1010&context=busdoc

Acevedo-Opazo, C., Tisseyre, B., Guillaume, S., & Ojeda, H. (2008). The potential of high spatial resolution information to define within-vineyard zones related to vine water status. *Precision Agriculture*, *9*(5), 285–302. doi:10.100711119-008-9073-1

Anthony, R. N., Hawkins, D. F., & Merchant, K. A. (2010). *Accounting Texts & Cases, 13E by Robert N. Anthony* (13th ed.). McGraw-Hill.

Chrenková, I. (2011). Controlling in the Conditions of Czech Republic. *AGRIS On-Line Papers in Economics and Informatics*, *3*, 3–14.

Cinquini, L., Collini, P., Marelli, A., & Tenucci, A. (2008). *An exploration of the factors affecting the diffusion of Advanced Costing techniques: a comparative analysis of two surveys (1996-2005)*. Retrieved from http://mpra.ub.uni-muenchen.de/10911/

Cockburn, A. (2001). *Writing effective use cases*. Pearson Education.

Cooper, R., & Kaplan, R. S. (1988). Measure costs right: Make the right decisions. *Harvard Business Review*, *66*, 96–103.

Davis, W., & Yen, D. (1998). Entity-relationship diagrams. In W. Davis & D. Yen (Eds.), The Information System Consultant's Handbook. CRC Press. doi:10.1201/9781420049107.ch26

Dierks, P., & Cokins, G. (2000). *The CAM-I Glossary of Activity-based Management*. CAM-I.

Ferreira, W. N. (2004). *Is Activity Based Costing a Blockbuster?* Extension Economic Reports.

Fountas, S., Wulfsohn, D., Blackmore, B. S., Jacobsen, H. L., & Pedersen, S. M. (2006). A model of decision-making and information flows for information-intensive agriculture. *Agricultural Systems*, *87*(2), 192–210. doi:10.1016/j.agsy.2004.12.003

González-Gómez, J. I., González-Gómez, J. I., & Morini-Marrero, S. (2009). A model for cost calculation and management in a multiproduct agricultural framework. The case for ornamental plants. *Spanish Journal of Agricultural Research*, *7*(1), 12–23. doi:10.5424jar/2009071-394

González-Gómez, J. I., & Morini, S. (2006). An Activity-Based Costing of Wine. *Journal of Wine Research*, *17*(3), 195–203. doi:10.1080/09571260701286650

Johnson, H. T., & Kaplan, R. S. (1987). *Relevance Lost: The Rise and Fall of Management Accounting*. Harvard Business Press.

Kaplan, R. S., & Anderson, S. R. (2007). *Time-Driven Activity-Based Costing: A Simpler and More Powerful Path to Higher Profits*. Harvard Business Press.

Kay, R., Edwards, W., & Duffy, P. (2011). *Farm Management*. McGraw-Hill Education.

Korpunen, H., Mochan, S., & Uusitalo, J. (2010). An Activity-Based Costing Method for Sawmilling. *Forest Products Journal*, *60*(5), 420–431. doi:10.13073/0015-7473-60.5.420

Lee, T.-R., & Kao, J.-S. (2001). Application of simulation technique to activity-based costing of agricultural systems: A case study. *Agricultural Systems*, *67*(2), 71–82. doi:10.1016/S0308-521X(00)00042-1

McBratney, A., Whelan, B., Ancev, T., & Bouma, J. (2005). Future Directions of Precision Agriculture. *Precision Agriculture*, *6*(1), 7–23. doi:10.100711119-005-0681-8

Nash, E., Dreger, F., Schwarz, J., Bill, R., & Werner, A. (2009). Development of a model of data-flows for precision agriculture based on a collaborative research project. *Computers and Electronics in Agriculture*, *66*(1), 25–37. doi:10.1016/j.compag.2008.11.005

Nikkilä, R., Seilonen, I., & Koskinen, K. (2010). Software architecture for farm management information systems in precision agriculture. *Computers and Electronics in Agriculture*, *70*(2), 328–336. doi:10.1016/j.compag.2009.08.013

Papajorgji, P., Pinet, F., Miralles, A., Jallas, E., & Pardalos, P. M. (2010). Modeling: A Central Activity for Flexible Information Systems Development in Agriculture and Environment. *International Journal of Agricultural and Environmental Information Systems*, *1*(1), 1–25. doi:10.4018/jaeis.2010101301

Phillips, C., Kemp, E., & Kek, S. M. (2001). Extending UML use case modelling to support graphical user interface design. In *Software Engineering Conference, 2001. Proceedings. 2001 Australian* (pp. 48–57). IEEE. 10.1109/ASWEC.2001.948497

Rafiq, A., & Garg, A. (2002). Activity based costing and financial institutions: Old wine in new bottles or corporate panacea? *The Journal of Bank Cost & Management Accounting*, *15*, 12–29.

Rosenberg, D., & Stephens, M. (2007). *Use Case Driven Object Modeling with UML*. Berkeley, CA: Apress. Retrieved from http://www.springerlink.com/index/10.1007/978-1-4302-0369-8

Setala, J., & Gunasekaran, A. (1996). Activity-based costing and management--a way to improve the profitability of fish processing? *Production and Inventory Management Journal*, *37*, 63–69.

Sørensen, C. G., Fountas, S., Nash, E., Pesonen, L., Bochtis, D., Pedersen, S. M., ... Blackmore, S. B. (2010). Conceptual model of a future farm management information system. *Computers and Electronics in Agriculture*, *72*(1), 37–47. doi:10.1016/j.compag.2010.02.003

Sørensen, C. G., Pesonen, L., Bochtis, D. D., Vougioukas, S. G., & Suomi, P. (2011). Functional requirements for a future farm management information system. *Computers and Electronics in Agriculture*, *76*(2), 266–276. doi:10.1016/j.compag.2011.02.005

Sørensen, C. G., Pesonen, L., Fountas, S., Suomi, P., Bochtis, D., Bildsøe, P., & Pedersen, S. M. (2010). A user-centric approach for information modelling in arable farming. *Computers and Electronics in Agriculture*, *73*(1), 44–55. doi:10.1016/j.compag.2010.04.003

Voulodimos, A. S., Patrikakis, C. Z., Sideridis, A. B., Ntafis, V. A., & Xylouri, E. M. (2010). A complete farm management system based on animal identification using RFID technology. *Computers and Electronics in Agriculture*, *70*(2), 380–388. doi:10.1016/j.compag.2009.07.009

Zhang, N., Wang, M., & Wang, N. (2002). Precision agriculture—a worldwide overview. *Computers and Electronics in Agriculture*, *36*(2-3), 113–132. doi:10.1016/S0168-1699(02)00096-0

Compilation of References

Abusalama, F. A. (2008). *Barriers to adopting activity-based costing systems (ABC): an empirical investigation using cluster analysis* (Doctoral dissertation). Dublin Institute of Technology. Retrieved from http://arrow.dit.ie/cgi/viewcontent.cgi?article=1010&context=busdoc

Acevedo-Opazo, C., Tisseyre, B., Guillaume, S., & Ojeda, H. (2008). The potential of high spatial resolution information to define within-vineyard zones related to vine water status. *Precision Agriculture*, *9*(5), 285–302. doi:10.100711119-008-9073-1

Adamo, F., Attivissimo, F., Carducci, C. G. C., & Lanzolla, A. M. L. (2015). A smart sensor network for sea water quality monitoring. *IEEE Sensors Journal*, *15*(5), 2514–2522. doi:10.1109/JSEN.2014.2360816

Adu-Manu, K., Tapparello, K. S., Heinzelman, C., Katsriku, W., & Abdulai, J. D. (2017). Water Quality Monitoring Using Wireless Sensor Networks: Current Trends and Future Research Directions. *ACM Transactions on Sensor Networks*, *13*(1), 4. doi:10.1145/3005719

Ahas, R., & Mark, U. (2005). Location based services - new challenges for planning and public administrations? *Futures*, *37*(6), 547–561. doi:10.1016/j.futures.2004.10.012

Ahmedi, L., Jajaga, E., & Ahmedi, F. (2013). An Ontology Framework for Water Quality Management. In C. H. O. Corcho (Ed.), *Proceedings of the 6th International Workshop on Semantic Sensor Networks co-located with the 12th International Semantic Web Conference* (vol. 1063, pp. 35-50). Sydney, Australia: CEUR Workshop Proceedings.

Ahmedi, L., Sejdiu, B., Bytyçi, E., & Ahmedi, F. (2015). An Integrated Web Portal for Water Quality Monitoring through Wireless Sensor Networks. *International Journal of Web Portals*, *7*(1), 28–46. doi:10.4018/IJWP.2015010102

Ahonen, T., Virrankoski, R., & Elmusrati, M. (2008). Greenhouse Monitoring with Wireless Sensor Network. In *Proceedings of the IEEE/ASME International Conference on Mechtronic and Embedded Systems and Applications* (pp. 1-6). Beijing, China: IEEE.

Akyildiz, I. F., Su, W., Sankarasubramaniam, Y., & Cayirci, E. (2002). Wireless Sensor Networks: A Survey. *Computer Networks*, *38*(4), 393–422. doi:10.1016/S1389-1286(01)00302-4

Alippi, C., Boracchi, G., Camplani, R., & Roveri, M. (2012). Wireless sensor networks for monitoring vineyards. In G. Anastasi, B. E., E. Di Nitto, C. Ghezzi, L. Tanca, & E. Zimeo (Eds.), Methodologies and Technologies for Networked Enterprises (pp. 295–310). Berlin: Springer-Verlag Heidelberg. doi:10.1007/978-3-642-31739-2_15

Alippi, C., Camplani, R., Galperti, C., & Roveri, M. (2011). A Robust, Adaptive, Solar-Powered WSN Framework for Aquatic Environmental Monitoring. *Sensors Journal, IEEE*, *11*(1), 45–55. doi:10.1109/JSEN.2010.2051539

Alippi, C., Camplani, R., Marullo, A., & Roveri, M. (2012). A real-time monitoring framework for landslide and rock-collapse forecasting. In S. Mukhopadhyay (Ed.), *Smart Sensing Technology for Agriculture and Environmental Monitoring* (pp. 285–302). Berlin: Springer-Verlag Berlin Heidelberg. doi:10.1007/978-3-642-27638-5_14

Alkandari, A., Alnasheet, M., Alabduljader, Y., & Moein, S. (2012). Water monitoring system using wireless sensor network (wsn): Case study of kuwait beaches. *Proceedings of the Second International Conference on Digital Information Processing and Communications (ICDIPC)*. 10.1109/ICDIPC.2012.6257270

Al-Karaki, J. N., & Kamal, A. E. (2004). Routing techniques in Wireless Sensor Networks: A Survey. *Wireless Communications*, *11*(6), 6–28. doi:10.1109/MWC.2004.1368893

Allan, I. J., Vrana, B., Greenwood, R., Mills, G. A., Knutsson, J., Holmberg, A., ... Laschi, S. (2006). Strategic monitoring for the European water framework directive. *TrAC Trends in Analytical Chemistry*, *25*(7), 704–715. doi:10.1016/j.trac.2006.05.009

Anastasiadis, A. D., Magoulas, G. D., & Vrahatis, M. N. (2005). New globally convergent training scheme based on the resiliente propagation Algorithm. *Neurocomputing*, *64*, 253–270. doi:10.1016/j.neucom.2004.11.016

Anderson, N. H. (2001). *Empirical direction in design and analysis*. London: Taylor and Francis. Retrieved from http://cds.cern.ch/record/2123907

Anthony, D., Bennett, W. P., Vuran, M. C., Dwyer, M. B., Elbaum, S., & Lacy, A. (2011). Sensing through the continent: Towards monitoring migratory birds using cellular sensor networks. *Proceedings of the 11th International Conference on Information Processing in Sensor Networks*. doi: 10.1145/2185677.2185747

Anthony, R. N., Hawkins, D. F., & Merchant, K. A. (2010). *Accounting Texts & Cases, 13E by Robert N. Anthony* (13th ed.). McGraw-Hill.

Antoin, L., Javier, T., Brendan, O., John, W., & Fiona, R. (2012). Deploy: A long term deployment of a water quality sensor monitoring system. *Sensor Review*, *1*(12), 29–38.

AOBDR. (2016). *android-obd-reader and obd-server, Github Repository Android OBD-II Reader Project*. Retrieved from https://github.com/pires/

Aqeel-ur-Rehman, Abbasi, A. Z., Islam, N., & Shaikh, Z. A. (2014). A review of wireless sensors and networks' applications in agriculture. *Computer Standards & Interfaces*, *36*(2), 263–270. doi:10.1016/j.csi.2011.03.004

Aquino-Santos, R., Apolinar, G.-P., Edwards-Block, A., & Virgen-Ortiz, R. A. (2011). Developing a New Wireless Sensor Network Platform and Its Application in Precision Agriculture. *Sensors (Basel)*, *11*(1), 1193–1211. doi:10.3390110101192 PMID:22346622

Asche, H., & Engemaier, R. (2012). From Concept to Implementation: Web-based Cartographic Visualisation with CartoService. In *Computational Science and its Application – ICCSA 2012* (Vol. 2, pp. 414–424). Springer International Publishing. doi:10.1007/978-3-642-31075-1_31

Ascher, F. (2004). Les sens du mouvement. Modernité et mobilités. In S. Allemand, F. Ascher, & J. Lévy (Eds.), Les sens du mouvement. Modernité et mobilités dans les sociétés urbaines contemporaines (pp. 20–34). Paris: Belin.

Atmel AVR Microcontrollers. (1984). Retrieved from http://www.atmel.com/products/microcontrollers/avr/default.aspx

Aubert, B. A., Schroeder, A., & Grimaudo, J. (2012). IT as enabler of sustainable farming: An empirical analysis of farmers' adoption decision of precision agriculture technology. *Decision Support Systems*, *54*(1), 510–520. doi:10.1016/j.dss.2012.07.002

Ayday, C., & Safak, S. (2009). Application of Wireless Sensor Networks with GIS on the Soil Moisture Distribution Mapping. *Proceedings of the 16th International Symposium GIS Ostrava.*

Baggio, A. (2009). Wireless Sensor Networks in Precision Agriculture. *Proceedings of the ACM Workshop on Real-World Wireless Sensor Networks.* Retrieved from www.sics.se/realwsn05/papers/baggio05wireless.pdf

Balducci, A., Fedeli, V., & Pasqui, G. (2010). *Strategic planning for contemporary urban regions: city of cities : a project for Milan.* Burlington, VT: Ashgate.

Bansal, N., & Malik, S. K. (2011). A framework for agriculture ontology development in semantic web. *Int. Conf on Communication Systems and Network Technologies*, 283 – 286. 10.1109/CSNT.2011.68

Barrenetxea, G., Ingelrest, F., Schaefer, G., & Vetterli, M. (2008). The Hitchhiker's Guide to Successful Wireless Sensor Network Deployments. *Proceedings of the 6th ACM Conference on Embedded Network Sensor Systems, (SenSys).* 10.1145/1460412.1460418

Becker, R. A., Caceres, R., Hanson, K., Loh, J. M., Urbanek, S., Varshavsky, A., & Volinsky, C. (2011). A Tale of One City: Using Cellular Network Data for Urban Planning. *IEEE Pervasive Computing*, *10*(4), 18–26. doi:10.1109/MPRV.2011.44

Bédard, Y., Proulx, M. J., & Rivest, S. (2005). *Enrichissement du OLAP pour l'analyse géographique: exemples de réalisation et différentes possibilités technologiques. In Entrepôts de Données et Analyse en ligne, RNTI B_1* (pp. 1–20). Paris: Cépaduès.

Bejaoui, L. (2009). *Qualitative topological relationships for objects with possibly vague shapes: implications on the specification of topological integrity constraints in transactional spatial databases and in spatial data warehouses*. Université Blaise Pascal.

Bejaoui, L., Pinet, F., Bedard, Y., & Schneider, M. (2009). Qualified topological relations between spatial objects with possible vague shape. *International Journal of Geographical Information Science*, *23*(7), 877–921. doi:10.1080/13658810802022814

Bekhor, S., Cohen, Y., & Solomon, C. (2011). Evaluating long-distance travel patterns in Israel by tracking cellular phone positions. *Journal of Advanced Transportation*. doi:10.1002/atr.170

Bencini, L. D. (2010). Wireless sensor networks for on-field agricultural management process. In G. V. Merrett & Y. K. Tan (Eds.), *Wireless Sensor Networks: application-centric design* (pp. 1–18). InTech - Open Access Company. doi:10.5772/13001

Bernareggi, G. M. (2013). L'istituzionalizzazione della città metropolitana di Milano – Aspetti economici. In A. Ceriani (Ed.), *Aggiornamento della ricerca "Gli enti locali nella transizione verso il federalismo-Effetti ordinamentali della Spending review in Lombardia"* (pp. 75–84). Éupolis Lombardia.

Berners-Lee, T., Hendler, J., & Lassila, O. (2001). The semantic web. *Scientific American*, *284*(5), 34–43. doi:10.1038cientificamerican0501-34 PMID:11396337

Bhatti, S., Carlson, J., Dai, H., Deng, J., Rose, J., Sheth, A., ... Han, R. (2005). MANTIS OS: An embedded multithreaded operating system for wireless micro sensor platforms. *Mobile Networks and Applications*, *10*(4), 563–579. doi:10.100711036-005-1567-8

Białobrzewski, I. (2008). Neural modeling of relative air humidity. *Computers and Electronics in Agriculture*, *60*(1), 1–7. doi:10.1016/j.compag.2007.02.009

Bimonte, S., Nazih, H., Kang, M.-A., Edoh-Alove, E., & Rizzi, S. (2013). *ProtOLAP: Rapid OLAP Prototyping with On-Demand Data Supply*. Paper presented at the DOLAP'13, San Francisco, CA, USA. 10.1145/2513190.2513199

Blackmore, S. (1994). *Precision Farming: An Introduction* (Vol. 23). Outlook on Agriculture Journal.

Blank, S., Bartolein, C., Meyer, A., Ostermeier, R., & Rostanin, O. (2013). iGreen: A ubiquitous dynamic network to enable manufacturer independent data exchange in future precision farming. *Computers and Electronics in Agriculture*, *98*, 109–116. doi:10.1016/j.compag.2013.08.001

Bolla, R., & Davoli, F. (2000). Road traffic estimation from location tracking data in the mobile cellular network. *2000 IEEE Wireless Communications and Networking Conference. Conference Record (Cat. No.00TH8540), 3*. doi:10.1109/WCNC.2000.904783

Bonifati, A., Cattaneo, F., Ceri, S., Fuggetta, A., & Paraboschi, S. (2001). Designing data marts for data warehouses. *ACM Transactions on Software Engineering and Methodology*, *10*(4), 452–483. doi:10.1145/384189.384190

Boulil, K., Bimonte, S., & Pinet, F. (2011). Un modèle UML et des contraintes OCL pour les entrepôts de données spatiales. De la représentation conceptuelle à l'implémentation. *Ingénierie des Systèmes d'Information, 16*(6), 11–39. doi:10.3166/isi.16.6.11-39

Bourdin, A. (2005). Les mobilités et le programme de la sociologie. *Cahiers Internationaux de Sociologie, 118*(1), 5–21. doi:10.3917/cis.118.0005

Boyes, E., Stanisstreet, M., & Yongling, Z. (2008). Combating global warming: The ideas of high school students in the growing economy of South East China. *The International Journal of Environmental Studies, 65*(2), 233–245. doi:10.1080/00207230701284543

Braga, A. P., Carvalho, A. P. L., & Ludemir, T. B. (2005). Redes Neurais Artificiais. In S. O. Rezende (Ed.), *Sistemas Inteligentes: Fundamentos e aplicações* (pp. 141–168). Barueri: Manole.

Bredemeir, C., Variani, C., Almeida, D., & Rosa, A. T. (2013). Estimativa do potencial produtivo em trigo utilizando sensor óptico ativo para adubação nitrogenada em taxa variável. *Ciência Rural, 43*(7), 1147–1154. doi:10.1590/S0103-84782013005000080

Breitenberger, S. (2004). Extended Floating Car Data – Potenziale für die Verkehrsinformation und notwendige Durchdringungsraten. In M. Rohleder (Ed.), *Straßenverkehrstechnik, 10/2014* (pp. 522–531). Bonn: Kirschbaum Verlag.

Brewer, E. (2012). CAP twelve years later: How the "rules" have changed. *Computer, 45*(2), 23–29. doi:10.1109/MC.2012.37

Bri, D., Garcia, M., Lloret, J., & Dini, P. (2009). Real Deployments of Wireless Sensor Networks. *Proceedings of the third International Conference on Sensor Technologies and Applications (SENSORCOMM)*. 10.1109/SENSORCOMM.2009.69

Bryan, E., Deressa, T. T., Gbetibouo, G. A., & Ringler, C. (2009). Adaptation to climate change in Ethiopia and South Africa: Options and constraints. *Environmental Science & Policy, 12*(4), 413–426. doi:10.1016/j.envsci.2008.11.002

Burrough, P. A., & Frank, A. U. (1996). *Geographic Objects with Indeterminate Boundaries*. Academic Press.

Butler, Z., Corke, P., Peterson, R., & Rus, D. (2004). Virtual Fences for Controlling Cows. *Proceedings of the IEEE International Conference on Robotics and Automation, (ICRA)*. doi: 10.1109/ROBOT.2004.1302415

Bytyçi, E., Ahmedi, L., & Kurti, A. (2016). Association rule mining with context ontologies: An application to mobile sensing of water quality. *Communications in Computer and Information Science, 672*, 67–78. doi:10.1007/978-3-319-49157-8_6

Caceres, N., Romero, L. M., Benitez, F. G., & del Castillo, J. M. (2012). Traffic Flow Estimation Models Using Cellular Phone Data. *IEEE Transactions on Intelligent Transportation Systems, 13*(3), 1430–1441. doi:10.1109/TITS.2012.2189006

Caceres, N., Wideberg, J. P., & Benitez, F. G. (2008). Review of traffic data estimations extracted from cellular networks. *Intelligent Transport Systems, IET*, *2*(3), 179–192. doi:10.1049/iet-its:20080003

Calabrese, F., Liu, L., Lorenzo, G. D., & Ratti, C. (2011). Estimating Origin-Destination flows using opportunistically collected mobile phone location data from one million users in Boston Metropolitan Area. *IEEE Pervasive Computing*. doi:10.1109/MPRV.2011.41

Cao, X., Chena, J., Zhangb, Y., & Suna, Y. (2008). Development of an integrated wireless sensor network micro-environmental monitoring system. *ISA Transactions*, *47*(3), 247–255. doi:10.1016/j.isatra.2008.02.001 PMID:18355827

Capella, J., Bonastre, A., Ors, R., & Peris, M. (2010). A Wireless Sensor Network approach for distributed in-line chemical analysis of water. *Talanta*, *80*(5), 1789–1798. doi:10.1016/j.talanta.2009.10.025 PMID:20152412

Cardell-Oliver, R., Smettem, K., Kranz, M., & Mayer, K. (2005). Field Testing a Wireless Sensor Network for Reactive Environmental Monitoring. *International Journal of Distributed Sensor Networks*, *1*, 149–162. doi:10.1080/15501320590966422

Carto, D. B. (2016). *CartoDB Github Repository*. Vizzuality Inc. Retrieved from https://github.com/CartoDB/cartodb/

Cattell, R. (2011). Scalable SQL and NoSQL Data Stores. *SIGMOD Record*, *39*(4), 12–27. doi:10.1145/1978915.1978919

Cayford, R., & Johnson, T. (2003). Operational parameters affecting use of anonymous cell phone tracking for generating traffic information. In *Institute of transportation studies for the 82th TRB Annual Meeting* (Vol. 1). Academic Press.

Chaamwe, N. (2010). Wireless sensor networks for water quality monitoring: A case of Zambia. In *Proceedings of the 4th International Conference on Bioinformatics and Biomedical Engineering*. IEEE. 10.1109/ICBBE.2010.5515792

Chang, M., & Bonnet, P. (2010). Monitoring in a High-Arctic Environment: Some Lessons from MANA. *Pervasive Computing, IEEE*, *9*(4), 16–23. doi:10.1109/MPRV.2010.53

Chaudhary, D., Nayse, S., & Waghmare, L. (2011). Application of Wireless Sensor Networks for Greenhouse parameter Control. *Precision Agriculture International Journal of Wireless & Mobile Networks*, *3*(1), 140–148. doi:10.5121/ijwmn.2011.3113

Chave, P. (2001). *The EU water framework directive*. IWA Publishing.

Chen, P., & Lu, Z. (2013). A web-based indoor environment monitoring system using wireless sensor networks. *Proceedings of the Fifth International Conference on Computational and Information Sciences (ICCIS)*. 10.1109/ICCIS.2013.529

Chen, X., Xun, Y., Li, W., & Zhang, J. (2010). Combining discriminant analysis and neural networks for corn variety identification. *Computers and Electronics in Agriculture*, *71*(1), 48–53. doi:10.1016/j.compag.2009.09.003

Choi, S., Kim, N., Cha, H., & Ha, R. (2009). Micro Sensor Node for Air Pollutant Monitoring: Hardware and Software Issues. *Sensors (Basel)*, *9*(10), 7970–7987. doi:10.339091007970 PMID:22408489

Chrenková, I. (2011). Controlling in the Conditions of Czech Republic. *AGRIS On-Line Papers in Economics and Informatics*, *3*, 3–14.

Christin, D., Reinhardt, A., Mogre, P., & Steinmetz, R. (2009). Wireless sensor networks and the Internet of Things: Selected Challenges. *Proceedings of the 8th GI/ITG KuVS Fachgespräch Drahtlose Sensornetze*, 31-34.

Cinquini, L., Collini, P., Marelli, A., & Tenucci, A. (2008). *An exploration of the factors affecting the diffusion of Advanced Costing techniques: a comparative analysis of two surveys (1996-2005)*. Retrieved from http://mpra.ub.uni-muenchen.de/10911/

CIRCA. (2003). *Monitoring under the Water Framework Directive, Policy Summary to Guidance Document 7, Produced by Working Group 2.7 – Monitoring, Common Implementation Strategy for the Water Framework Directive (2000/60/EC), Published Guidance Documents*. CIRCA Library EPA Ireland - Water Framework Directive Monitoring Programme. Retrieved from http://www.epa.ie/pubs/reports/water/other/wfd/

Cockburn, A. (2001). *Writing effective use cases*. Pearson Education.

Codd, E. F. (1990). *The Relational Model for Database Management: Version 2*. Boston, MA: Addison-Wesley Longman Publishing Co., Inc.

Cohn, A. G., & Gotts, N. M. (1996). *Geographic objects with indeterminate boundaries*. Academic Press.

Collins, F., Orpen, D., Maher, D., Cleary, J., Fay, C., & Diamond, D. (2011). Distributed Chemical Sensor Networks for Environmental Sensing. *Proceedings of the 2nd International Conference on Sensor Device Technologies and Applications*, 58-62.

Contiki: The Open Source OS. (2006). Retrieved from http://www.contiki-os.org

Cooper, R., & Kaplan, R. S. (1988). Measure costs right: Make the right decisions. *Harvard Business Review*, *66*, 96–103.

Corke, P., Wark, T., Jurdak, R., Hu, W., Valencia, P., & Moore, D. (2010). Environmental Wireless Sensor Networks. *Proceedings of the IEEE*, *98*(11), 1903–1917. doi:10.1109/JPROC.2010.2068530

Dasgupta, S., Patel, R., Padia, A., & Shah, K. (2013). *Description Logics based formalization of Wh-Queries*. CoRR abs/1312.6948

Davis, W., & Yen, D. (1998). Entity-relationship diagrams. In W. Davis & D. Yen (Eds.), The Information System Consultant's Handbook. CRC Press. doi:10.1201/9781420049107.ch26

Davis, A. P., Gole, T. W., Baena, S., & Moat, J. (2012). The Impact of Climate Change on Indigenous Arabica Coffee (Coffea arabica): Predicting Future Trends and Identifying Priorities. *PLoS One*, *7*(11), 1–13. doi:10.1371/journal.pone.0047981 PMID:23144840

Delgado, M., Molina, C., Sánchez, D., Ariza, L. R., & Vila, M. A. (2004). *A flexible approach to the multidimensional model: The fuzzy datacube. In Current Topics in Artificial Intelligence* (pp. 26–36). Springer. doi:10.1007/978-3-540-25945-9_3

Demirbas, M., Chow, K. Y., & Wan, C. S. (2006). INSIGHT: Internet-Sensor Integration for Habitat Monitoring. *Proceedings of the IEEE International Symposium on a World of Wireless, Mobile and Multimedia Networks (WoWMoM)*, 553-558.

Demirkol, I., Ersoy, C., & Alagoz, F. (2006). MAC Protocols for Wireless Sensor Networks: A Survey. *Communications Magazine, IEEE*, *44*(4), 115–121. doi:10.1109/MCOM.2006.1632658

Devillers, R., Bédard, Y., & Jeansoulin, R. (2005). Multidimensional Management of Geospatial Data Quality Information for its Dynamic Use Within Geographical Information Systems. *American Society for Photogrammetry and Remote Sensing (PE&RS)*, *71*(2), 205–215. doi:10.14358/PERS.71.2.205

Devillers, R., Bedard, Y., Jeansoulin, R., & Moulin, B. (2007). Towards spatial data quality information analysis tools for experts assessing the fitness for use of spatial data. *International Journal of Geographical Information Science*, *21*(3), 261–282. doi:10.1080/13658810600911879

Di Tria, F., Lefons, E., & Tangorra, F. (2012). Hybrid methodology for data warehouse conceptual design by UML schemas. *Information and Software Technology*, *54*(4), 360–379. doi:10.1016/j.infsof.2011.11.004

Dierks, P., & Cokins, G. (2000). *The CAM-I Glossary of Activity-based Management*. CAM-I.

Dilo, A., de By, R. A., & Stein, A. (2007). A system of types and operators for handling vague spatial objects. *International Journal of Geographical Information Science*, *21*(4), 397–426. doi:10.1080/13658810601037096

Doctorow, C. (2011). *The Coming War on General Computation*. Presented at 28C3 conference in Berlin, Germany. Retrieved from https://github.com/jwise/28c3-doctorow/blob/master/transcript.md

Doolin, D. M., & Sitar, N. (2005). *Wireless sensors for wildfire monitoring. In Sensors and Smart Structures Technologies for Civil, Mechanical, and Aerospace Systems*. SPIE. doi:10.1117/12.605655

Dubois-Ferriere, H., Meier, R., Fabre, L., & Metrailler, P. (2006). TinyNode: a comprehensive platform for wireless sensor network applications. *Proceedings of the Fifth International Conference on Information Processing in Sensor Networks (IPSN)*. 10.1145/1127777.1127831

Dunbabin, M., Alistair, G., & James, U. (2009). An autonomous surface vehicle for water quality monitoring. *Proceedings of the Australasian Conference on Robotics & Automation*, 1-6.

Dunbabin, M., & Corke, P. (2010). A framework for marine sensor network & autonomous vehicle interaction. *Proceedings of the OCEANS 2010 IEEE*. 10.1109/OCEANSSYD.2010.5603677

Dunkels, A., Gronvall, B., & Voigt, T. (2004). Contiki-A Lightweight and Flexible Operating System for Tiny Networked Sensors. *Proceedings of the 29th Annual IEEE Conference on Local Computer Networks*. 10.1109/LCN.2004.38

Dursun, M., & Ozden, S. (2011). A wireless application of drip irrigation automationsupported by soil moisture sensors. *Scientific Research and Essays*, *6*, 1573–1582.

Dyo, V., Ellwood, S. A., Macdonald, D. W., Markham, A., Scellato, S., & Trigoni, N. (2010). Evolution and Sustainability of a Wildlife Monitoring Sensor Network. *Proceedings of the 8th ACM Conference on Embedded Networked Sensor Systems (SenSys)*, 127-140. 10.1145/1869983.1869997

Ehrenberg, A. (1995). *L'individu incertain*. Paris: Calmann-Lévy.

Ehsan, S., Bradford, K., Brugger, M., Hamdaoui, B., Kovchegov, Y., & Johnson, D. (2012). Design and analysis of delay-tolerant sensor networks for monitoring and tracking free-roaming animals. *Wireless Communications. IEEE Transactions on*, *11*(3), 1220–1227.

Elsts, A., Balass, R., Judvaitis, J., Zviedris, R., Strazdins, G., & Mednis, A. (2012). SADmote: a robust and cost-effective device for environmental monitoring. *Proceedings of the 25th International Conference on Architecture of Computing Systems (ARCS)*. 10.1007/978-3-642-28293-5_19

EPA. (2013). *Frequently asked questions about global warming and climate change: back to basics*. Retrieved from http://www.epa.gov/climatechange/basics/

Erdle, K., Mistele, B., & Schimidhalter, U. (2011). Comparison of active and passive spectral sensors in discriminating biomass parameters and nitrogen status in wheat cultivars. *Field Crops Research*, *124*(1), 74–84. doi:10.1016/j.fcr.2011.06.007

Estebe, P. (2008). *Gouverner la ville mobile*. Paris: PUF. doi:10.3917/puf.esteb.2008.01

Exposito, E., & Diop, C. (2014). *Smart SOA Platforms in Cloud Computing Architectures*. Hoboken, NJ: John Wiley and Sons Inc.; doi:10.1002/9781118761489

FAO, IFAD, & WFP. (2015). The State of Food Insecurity in the World 2015. Meeting the 2015 international hunger targets: taking stock of uneven progress. Rome: Authors.

FAO. (2011). *Global food losses and food waste – Extent, causes and prevention*. Retrieved from http://www.fao.org/docrep/014/mb060e/mb060e00.pdf

Fasel, D., & Zumstein, D. (2009). *A fuzzy data warehouse approach for web analytics*. Springer. doi:10.1007/978-3-642-04754-1_29

Fay, C., Doherty, A., Beirne, S., Collins, F., Foley, C., Healy, J., ... Diamond, D. (2011). Remote Real-Time Monitoring of Subsurface Landfill Gas Migration. *Sensors (Basel)*, *11*(7), 6603–6628. doi:10.3390110706603 PMID:22163975

Feng, W., Yao, X., Zhu, Y., Tian, Y. C., & Cao, W. X. (2008a). Monitoring leaf nitrogen status with hyperspectral reflectance in wheat. *European Journal of Agronomy*, *28*(3), 394–404. doi:10.1016/j.eja.2007.11.005

Feng, W., Zhu, Y., Tian, Y., Cao, W., Yao, X., & Li, Y. (2008b). Monitoring leaf nitrogen accumulation in wheat with hyper spectral remote sensing. *Acta Ecologica Sinica*, *28*(1), 23–32. doi:10.1016/S1872-2032(08)60018-9

Ferreira, W. N. (2004). *Is Activity Based Costing a Blockbuster?* Extension Economic Reports.

Flecks Datasheet. (2011). Retrieved from http://www.snm.ethz.ch/snmwiki/pub/uploads/Projects/Flecks.pdf

Fleischmann, N., Staubmann, K., & Langergraber, G. (2002). Management of sensible water uses with real-time measurements. *Water Science and Technology*, *46*(3), 33–40. PMID:12227601

Fountas, S., Carli, G., Sørensen, C. G., Tsiropoulos, Z., Cavalaris, C., Vatsanidou, A., ... Tisserye, B. (2015). Farm management information systems: Current situation and future perspectives. *Computers and Electronics in Agriculture*, *115*, 40–50. doi:10.1016/j.compag.2015.05.011

Fountas, S., Wulfsohn, D., Blackmore, B. S., Jacobsen, H. L., & Pedersen, S. M. (2006). A model of decision-making and information flows for information-intensive agriculture. *Agricultural Systems*, *87*(2), 192–210. doi:10.1016/j.agsy.2004.12.003

Freitas, A. A. C., Silva, I. N., & Souza, A. N. (2002). Aplicação de redes neurais na estimação da temperatura interna de transformadores de distribuição imersos em óleo. *Controle & Automação*, *13*(3), 266–274. doi:10.1590/S0103-17592002000300005

Garcia-Sanchez, A.-J., Garcia-Sanchez, F., & Garcia-Haro, J. (2011). Wireless sensor network deployment for integrating video-surveillance and data-monitoring in precision agriculture over distributed crops. *Computers and Electronics in Agriculture*, *75*(2), 288–303. doi:10.1016/j.compag.2010.12.005

Gere, A., Székely, G., Kovács, S., Kókai, Z., & Sipos, L. (2017). Readiness to adopt insects in Hungary: A case study. *Food Quality and Preference*, *59*, 81–86. doi:10.1016/j.foodqual.2017.02.005

Gervais, M., Bédard, Y., Levesque, M.-A., Bernier, E., & Devillers, R. (2009). Data Quality Issues and Geographic Knowledge Discovery. *Geographic Data Mining and Knowledge Discovery*, 99-115.

Ghidini, C., & Giunchiglia, F. A. (2004). Semantics for abstraction. In *Proceedings of ECAI* (pp. 343-347). Academic Press.

Giannopoulos, N., Goumopoulos, C., & Kameas, A. (2009). Design Guidelines for Building a Wireless Sensor Network for Environmental Monitoring. *Proceedings of the 13th Panhellenic Conference on Informatics*. 10.1109/PCI.2009.17

Giorgini, P., Rizzi, S., & Garzetti, M. (2005). Goal-oriented requirement analysis for data warehouse design. *Proceedings of the 8th ACM international workshop on Data warehousing and OLAP*. 10.1145/1097002.1097011

Girard, L., Beutel, J., Gruber, S., Hunziker, J., Lim, R., & Weber, S. (2012). A custom acoustic emission monitoring system for harsh environments: Application to freezing-induced damage in alpine rock walls. *Geoscientific Instrumentation. Methods and Data Systems*, *1*(2), 155–167.

Gitelson, A. A. (2004). Wide Dynamic Range Vegetation Index for Remote Quantification of Biophysical Characteristics of Vegetation. *Journal of Plant Physiology*, *161*(2), 165–173. doi:10.1078/0176-1617-01176 PMID:15022830

Glorio, O., & Trujillo, J. (2008). An MDA Approach for the Development of Spatial Data Warehouses. In I.-Y. Song, J. Eder & T. Nguyen (Eds.), Data Warehousing and Knowledge Discovery (Vol. 5182, pp. 23-32). Springer Berlin / Heidelberg. doi:10.1007/978-3-540-85836-2_3

Godfray, H. C. J., Beddington, J. R., Crute, I. R., Haddad, L., Lawrence, D., Muir, J. F., ... Toulmin, C. (2010). Food Security: The Challenge of Feeding 9 Billion People. *Science*, *327*(5967), 812–818. doi:10.1126cience.1185383 PMID:20110467

González-Gómez, J. I., González-Gómez, J. I., & Morini-Marrero, S. (2009). A model for cost calculation and management in a multiproduct agricultural framework. The case for ornamental plants. *Spanish Journal of Agricultural Research*, *7*(1), 12–23. doi:10.5424jar/2009071-394

González-Gómez, J. I., & Morini, S. (2006). An Activity-Based Costing of Wine. *Journal of Wine Research*, *17*(3), 195–203. doi:10.1080/09571260701286650

Gonzalez, M. C., Hidalgo, C. A., & Barabasi, A.-L. (2008). Understanding individual human mobility patterns. *Nature*, *453*(7196), 779–782. doi:10.1038/nature06958 PMID:18528393

Goumopoulos, C., Kameas, A., & O'Flynn, B. (2007, July). Proactive agriculture: an integrated framework for developing distributed hybrid systems. In *International Conference on Ubiquitous Intelligence and Computing* (pp. 214-224). Springer Berlin Heidelberg. 10.1007/978-3-540-73549-6_22

Goumopoulos, C., O'Flynn, B., & Kameas, A. (2014). Automated zone-specific irrigation with wireless sensor/actuator network and adaptable decision support. *Computers and Electronics in Agriculture*, *105*, 20–33. doi:10.1016/j.compag.2014.03.012

Grath, J., Ward, R., Scheidleder, A., & Quevauviller, P. (2007). Report on EU guidance on groundwater monitoring developed under the common implementation strategy of the water framework directive. *Journal of Environmental Monitoring*, *9*(11), 1162–1175. doi:10.1039/b710665n PMID:17968442

Green, P. E., & Krieger, A. M. (1991). Segmenting Markets with Conjoint Analysis. *Journal of Marketing*, *55*(4), 20–31. doi:10.2307/1251954

Green, P. E., & Srinivasan, V. (1990). Conjoint Analysis in Marketing: New Developments with Implications for Research and Practice. *Journal of Marketing*, *54*(4), 3–19. doi:10.2307/1251756

Greenwood, R., Webster, J., & Regan, F. (2008). *Sustainable water: chemical science priorities*. RSC Report, 1-18.

Grira, J., Bédard, Y., & Roche, S. (2013). *Revisiting the Concept of Risk Analysis within the Context of Geospatial Database Design: A Collaborative Framework.* Paper presented at the World Academy of Science, Engineering and Technology 75, Madrid, Spain.

Grohs, D. S., Bredemeier, C., Mundstock, C. M., & Polleto, N. (2009). Modelo para estimativa do potencial produtivo em trigo e cevada por meio do sensor GreenSeeker. *Engenharia Agrícola*, *29*(1), 101–112. doi:10.1590/S0100-69162009000100011

Grohs, D. S., Bredemeier, C., Polleto, N., & Mundstock, C. M. (2011). Validação de *modelo* para predição do potencial produtivo de trigo com sensor óptico ativo. *Pesquisa Agropecuária Brasileira*, *46*(4), 446–449. doi:10.1590/S0100-204X2011000400015

Gruber, T. (1993). A translation approach to portable ontology specifications. *Knowledge Acquisition*, *5*(2), 199–220. doi:10.1006/knac.1993.1008

Guimond, L.-E. (2005). *Conception d'un environnement de découvertes des besoins pour le développement de solutions SOLAP*. Québec, Canada: Master, Université Laval.

Günther, W. A., Mehrizi, M. H. R., Huysman, M., & Feldberg, F. (2017). Debating big data: A literature review on realizing value from big data. *The Journal of Strategic Information Systems*, *26*(3), 191–209. doi:10.1016/j.jsis.2017.07.003

Guo, Y., Corke, P., Poulton, G., Wark, T., Bishop-Hurley, G., & Swain, D. (2006). Animal Behaviour Understanding using Wireless Sensor Networks. *Proceedings of the 31st IEEE Conference on Local Computer Networks (LCN)*. 10.1109/LCN.2006.322023

Haber, R. B., & McNabb, D. (1990). Visualization Idioms: A Conceptual Model for Scientific Visualization Systems. In G. Nielson & B. Shriver (Eds.), *Visualization in Scientific Computing*. Los Alamitos, CA: IEEE Computer Society Press.

Haerder, T., & Reuter, A. (1983). Principles of transaction-oriented database recovery. *ACM Computing Surveys*, *15*(4), 287–317. doi:10.1145/289.291

Hakala, I., Tikkakoski, M., & Kivelä, I. (2008). Wireless Sensor Network in Environmental Monitoring - Case Foxhouse. *Proceedings of the 2nd International Conference on Sensor Technologies and Applications (SENSORCOMM)*. 10.1109/SENSORCOMM.2008.27

Halbritter, G. (2008). Strategien für Verkehrsinnovationen: Umsetzungsbedingungen, Verkehrstelematik, internationale Erfahrungen. Edition Sigma.

Hansen, P. M., & Schjoerring, J. K. (2003). Reflectance measurement of canopy biomass and nitrogen status in wheat crops using normalized difference vegetation indices and partial least squares regression. *Remote Sensing of Environment, 86*(4), 542–553. doi:10.1016/S0034-4257(03)00131-7

Hartung, C., Han, R., Seielstad, C., & Holbrook, S. (2006). FireWxNet: a multi-tiered portable wireless system for monitoring weather conditions in wildland fire environments. *Proceedings of the 4th International Conference on Mobile systems, applications and services.* 10.1145/1134680.1134685

Hastie, T., Tibshirani, R., & Friedman, J. (2011). *The Elements of Statistical Learning* (2nd ed.). New York, NY: Springer.

Haykin, S. (2001). *Redes Neurais: princípios e práticas* (2nd ed.). Porto Alegre: Bookman.

Hazarika, S., & Cohn, A. (2001). *Qualitative Spatio-Temporal Continuity Spatial Information Theory.* Springer Berlin/Heidelberg.

Heidmann, F. (2013). Interaktive Karten und Geovisualisierungen. In W. Weber, H. Burmeister, & R. Tille (Eds.), *Interaktive Infografiken* (pp. 66–67). Berlin: Springer-Verlag; doi:10.1007/978-3-642-15453-9_3

Hill, J. (2003). *System Architecture for Wireless Sensor Networks.* Berkeley, CA: University of California.

Hiu, L., Meng, Z., & Wang, M. (2009). A Wireless Sensor Network for Cropland Environmental Monitoring. *Proceedings of the International Conference on Network Security, Wireless Communications and Trusted Computing (NSWCTC).* doi: 10.1109/NSWCTC.2009.306

Hongye, C. (2010). Study of agricultural knowledge service system model based on semantic web. *International Forum on Information Technology and Applications*, 381 – 383.

Hou, J. C., Li, N., & Stojmenović, I. (2005). *Topology Construction and Maintenance in Wireless Sensor Networks.* New York, NJ: John Wiley & Sons, Inc. doi:10.1002/047174414X.ch10

Hou, K., De Sousa, G., Zhou, H., Chanet, J., Kara, M., Amamra, A., & ... . (2007). LiveNode: LIMOS versatile embedded wireless sensor node. *Journal of Harbin Institute of Technology, 7*, 140–144.

Huang, R., Song, W.-Z., Xu, M., Peterson, N., Shirazi, B., & LaHusen, R. (2012). Real-World Sensor Network for Long-Term Volcano Monitoring: Design and Findings. *IEEE Transactions on Parallel and Distributed Systems, 23*(1), 321–329. doi:10.1109/TPDS.2011.170

Huete, A. R. (1988). A Soil-Adjusted Vegetation Index (SAVI). *Remote Sensing of Environment, 25*(3), 295–309. doi:10.1016/0034-4257(88)90106-X

Hui, J. W., & Culler, D. (2004). The dynamic behavior of a data dissemination protocol for network programming at scale. *Proceedings of the 2nd international conference on Embedded networked sensor systems (SenSys).* 10.1145/1031495.1031506

Hu, J., Shen, L., Yang, Y., & Lv, R. (2010). Design and Implementation of Wireless Sensor and Actor Network for Precision Agriculture. *Proceedings of the IEEE International Conference on Wireless Communications, Networking and Information Security (WCNIS)*. 10.1109/WCINS.2010.5544151

Hu, W., Bulusu, N., Chou, T. C., Jha, S., Taylor, A., & Tran, V. N. (2009). Design and evaluation of a hybrid sensor network for cane toad monitoring. *ACM Transactions on Sensor Networks*, *5*(1), 1–28. doi:10.1145/1464420.1464424

Hu, W., Dinh, L., Corke, P., & Jha, S. (2012). Outdoor sensornet design and deployment: Experiences from a sugar farm. *IEEE Pervasive Computing*, *11*(2), 82–91. doi:10.1109/MPRV.2010.70

IDS Monitoring. (n.d.). Retrieved from https://www.idsmonitoring.com

IEEE-802.15.4. (2011). *IEEE Standard for Local and Metropolitan Area Networks-Part 15.4: Low-Rate Wireless Personal Area Networks (LR-WPANs)*. Retrieved from http://ieeexplore.ieee.org/xpl/articleDetails.jsp?reload=true&arnumber=6185525

Iftikhar, N., & Pedersen, T. B. (2011). Flexible exchange of farming device data. *Computers and Electronics in Agriculture*, *75*(1), 52–63. doi:10.1016/j.compag.2010.09.010

Imote2 Datasheet. (2008). Retrieved from http://web.univ-pau.fr/~cpham/ENSEIGNEMENT/PAU-UPPA/RESA-M2/DOC/Imote2_Datasheet.pdf

Ingelrest, F., Barrenetxea, G., Schaefer, G., Vetterli, M., Couach, O., & Parlange, M. (2010). SensorScope: Application-specific sensor network for environmental monitoring. *ACM Transactions on Sensor Networks*, *6*(2), 1–32. doi:10.1145/1689239.1689247

Inman, D., Khosla, R., Westfall, D. G., & Reich, R. (2005). Nitrogen uptake across site specific management zones in irrigated maize production systems. *Agronomy Journal*, *97*(1), 169–176. doi:10.2134/agronj2005.0169

IPCC. (2001). *The Regional Impacts of Climate Change: An Assessment of Vulnerability*. Geneva: IPCC.

IPPC. (2013). *CLIMATE CHANGE 2013. The Physical Science Basis. Intergovernmental Panel on Climate Change*. Retrieved from https://www.ipcc.ch/pdf/assessment-report/ar5/wg1/WGIAR5_SPM_brochure_en.pdf

Irvine, K., Boelens, R., Fitzsimmons, J., Kemp, A., & Johnston, P. (2002). *Review of Monitoring and Research to meet the needs of the EU Water Framework Directive (200-DS-5-M1)*. Academic Press.

ISO/IEC-51. (1999). ISO/IEC 51 Safety aspects - Guidelines for their inclusion in standards.

Jacobs, R. (1988). Increased rates of convergence through learning rate adaptation. *Neural Networks*, *1*(4), 295–307. doi:10.1016/0893-6080(88)90003-2

Jadidi, A., Mostafavi, M. A., Bédard, Y., & Long, B. (2012). *Towards an Integrated Spatial Decision Support System to Improve Coastal Erosion Risk Assessment: Modeling and Representation of Risk Zones*. Paper presented at the FIG Working Week 2012, Rome, Italy.

Jain, A. K. (2010). Data clustering: 50 years beyond K-means. *Pattern Recognition Letters, 31*(8), 651–666. doi:10.1016/j.patrec.2009.09.011

Jain, A., & Dubes, R. (1998). *Algorithms for Clustering Data*. Upper Saddle River, NJ: Prentice-Hall.

Jajaga, E., & Ahmedi, L. (2017). C-SWRL: SWRL for Reasoning over Stream Data information. *The 11th IEEE International Conference on Semantic Computing (ICSC 2017)*.

Jajaga, E., Ahmedi, L., & Ahmedi, F. (2016a). StreamJess: Enabling Jess for Stream Data Reasoning and the Water Domain Case. *20th International Conference on Knowledge Engineering and Knowledge Management*. Bologna, Italy: Springer Lecture Notes in Computer Science.

Jajaga, E., Ahmedi, L., & Ahmedi, F. (2015). An Expert System for Water Quality Monitoring based on Ontology. *Communications in Computer and Information Science, 544*, 89–100. doi:10.1007/978-3-319-24129-6_8

Jajaga, E., Ahmedi, L., & Ahmedi, F. (2016b). StreamJess: Stream Data Reasoning System for Water Quality Monitoring. *International Journal of Metadata, Semantics and Ontologies, 11*(4), 207. doi:10.1504/IJMSO.2016.083507

Jelicic, V., Razov, T., Oletic, D., Kuri, M., & Bilas, V. (2011). MasliNET: A Wireless Sensor Network based environmental monitoring system. *Proceedings of the 34th International Convention (MIPRO)*, 150-156.

Jiang, J., Li, A., Deng, W., & Bian, J. (2010). Construction of a New Classifier Integrated Multiple Sources and Multi-temporal Remote Sensing Data for wetlands. *Procedia Environmental Sciences, 2*, 302–314. doi:10.1016/j.proenv.2010.10.036

Jiang, P., Xia, H., He, Z., & Wang, Z. (2009). Design of a Water Environment Monitoring System Based on Wireless Sensor Networks. *Sensors (Basel), 9*(8), 6411–6434. doi:10.339090806411 PMID:22454592

Jiang, X., Zhou, G., Liu, Y., & Wang, Y. (2010). Wireless sensor networks for forest environmental monitoring. *Proceedings of the 2nd International Conference on Information Science and Engineerrng (ICISE)*. 10.1109/ICISE.2010.5691483

Johnson, H. T., & Kaplan, R. S. (1987). *Relevance Lost: The Rise and Fall of Management Accounting*. Harvard Business Press.

Johnson, K. S., Needoba, J. A., Riser, S. C., & Showers, W. J. (2007). Chemical sensor networks for the aquatic environment. *Chemical Reviews, 107*(2), 623–640. doi:10.1021/cr050354e PMID:17249737

Jordan, C. F. (1969). Derivation of leaf-area index from quality of light on the forest floor. *Ecology, 50*(4), 633–666. doi:10.2307/1936256

Junfeng, S., & Yong, L. (2016). Semantic similarity algorithm based on agricultural ontology and its application on crop ontology. *Nongye Gongcheng Xuebao (Beijing)*, *32*(16), 175–182.

Jurdak, R., Sommer, P., Kusy, B., Kottege, N., Crossman, C., & Mckeown, A. (2013). Camazotz: multimodal activity-based GPS sampling. *Proceedings of the 12th ACM International conference on Information processing in sensor networks (IPSN)*. doi: 10.1145/2461381.2461393

Kadri, A., Yaacoub, E., Mushtaha, M., & Abu-Dayya, A. (2013). Wireless sensor network for real-time air pollution monitoring. *Proceedings of the 1st International Conference on Communications, Signal Processing, and their Applications (ICCSPA)*. 10.1109/ICCSPA.2013.6487323

Kahneman, D. (2011). *Thinking, Fast and Slow*. New York, NY: Farrar, Straus and Giroux. Retrieved from https://books.google.hu/books?id=ZuKTvERuPG8C

Kaladevi, R., Geetha, G., & Narayanasamy, P. (2016). Ontological based interoperability and integration framework for heterogeneous weather systems. *Rev. Téc. Ing. Univ. Zulia*, *39*(1), 185–192.

Kaladevi, R., & Mirnalinee, T. T. (2015). Heterogeneous information management using ontology mapping. *Journal of Engineering and Applied Sciences (Asian Research Publishing Network)*, *10*(5), 2078–2081.

Kaloxylos, A., Eigenmann, R., Teye, F., Politopoulou, Z., Wolfert, S., Shrank, C., ... Kormentzas, G. (2012). Farm management systems and the Future Internet era. *Computers and Electronics in Agriculture*, *89*, 130–144. doi:10.1016/j.compag.2012.09.002

Kaplan, R. S., & Anderson, S. R. (2007). *Time-Driven Activity-Based Costing: A Simpler and More Powerful Path to Higher Profits*. Harvard Business Press.

Kashyap, V., & Sheth, A. (1996). Semantic and schematic similarities between database objects: A context-based approach. *The VLDB Journal*, *5*(4), 276–304. doi:10.1007007780050029

Kaufmann, V. (2000). *Re-thinking mobility*. Aldershot, UK: Ashgate.

Kay, R., Edwards, W., & Duffy, P. (2011). *Farm Management*. McGraw-Hill Education.

Kenny, G. J., & Shao, J. (1992). An assessment of a latitude-temperature index for predicting climate suitability for grapes in Europe. *Journal of Horticultural Science*, *67*(2), 239–246. doi: 10.1080/00221589.1992.11516243

Kerzner, H. (2006). *Project Management: a systems approach to planning, scheduling, and controlling* (9th ed.). Academic Press.

Kiernan, B. M., Beirne, S., Fay, C., & Diamond, D. (2008). Landfill gas monitoring at borehole wells using an autonomous environmental monitoring system. *Proceedings of the 5th International Conference on Climate Change and Global Warming*, 166-171.

Kim, Y.-D., Yang, Y.-M., Kang, W.-S., & Kim, D.-K. (2014). On the design of beacon based wireless sensor network for agricultural emergency monitoring systems. *Computer Standards & Interfaces, 36*(2), 288–299. doi:10.1016/j.csi.2011.05.004

Kitchen, N. R. (2008). Emerging technologies for real-time and integrated agriculture decisions. *Computers and Electronics in Agriculture, 61*(1), 1–3. doi:10.1016/j.compag.2007.06.007

Koch, B., Khosla, R., Frasier, M., Westfall, D. G., & Inman, D. (2004). Economic feasibility of variable rate nitrogen application utilizing site-specific management zones. *Agronomy Journal, 96*(6), 1572–1580. doi:10.2134/agronj2004.1572

Könnölä, T. (2008). *Future outlook on water quality monitoring technologies.* VTT Technical Research Centre of Finland Report.

Korpunen, H., Mochan, S., & Uusitalo, J. (2010). An Activity-Based Costing Method for Sawmilling. *Forest Products Journal, 60*(5), 420–431. doi:10.13073/0015-7473-60.5.420

Kotamäki, N., Thessler, S., Koskiaho, J., Hannukkala, A. O., Huitu, H., Huttula, T., ... Järvenpää, M. (2009). Wireless in-situ Sensor Network for Agriculture and Water Monitoring on a River Basin Scale in Southern Finland: Evaluation from a Data User's Perspective. *Sensors (Basel), 9*(4), 2863–2883. doi:10.339090402862 PMID:22574050

Kotis, K., & Vouros, G. A. (2004). The HCONE approach to ontology merging. In ESWS (pp. 137–151). Academic Press.

Kraak, M.-J. (2002). Cartographic Animation and legends for temporal maps: exploration and interaction. In *Multimedia Cartography* (2nd ed.). Berlin: Springer-Verlag.

Kumar, S., Iyengar, S. S., Lochan, R., Wiggins, U., Sekhon, K., Chakraborty, P., & .... (2009). Application of Sensor Networks for Monitoring of Rice Plants: A Case Study. *Proceedings of the 4th International Symposium on Innovations and Real-time Applications of Distributed Sensor Networks (IRADSN),* 1-7.

Kwong, K. H., Wu, T., Goh, H. G., Stephen, B., Gilroy, M. P., Michie, C., & Andonovic, I. (2009). Wireless sensor networks in agriculture: Cattle monitoring for farming industries. *PIERS Online, 5*(1), 31–35. doi:10.2529/PIERS081001110156

Kwong, K. H., Wu, T.-T., Goh, H. G., Sasloglou, K., Stephen, B., Glover, I., ... Andonovic, I. (2012). Practical considerations for wireless sensor networks in cattle monitoring applications. *Computers and Electronics in Agriculture, 81*, 33–44. doi:10.1016/j.compag.2011.10.013

Lagacherie, P., Andrieux, P., & Bouzigues, R. (1996). Fuzzyness and uncertainty of soil boundaries: from reality to coding in GIS. In P. A. Burrough & A. U. Frank (Eds.), *Geographic Objects with Indeterminate Boundaries, GISDATA series* (Vol. 2, pp. 275–286). London: Taylor & Francis.

Lambrix, P., & Tan, H. (2006). SAMBO– A system for aligning and merging biomedical ontologies. *Journal of Web Semantics, 4*(1), 196–206. doi:10.1016/j.websem.2006.05.003

Langendoen, K., Baggio, A., & Visser, O. (2006). Murphy Loves Potatoes. Experiences from a Pilot Sensor Network Deployment in Precision Agriculture. *Proceedings of the 20th IEEE International Parallel Distributed Processing Symposium (IPDPS)*. 10.1109/IPDPS.2006.1639412

Larios, D. F., Barbancho, J., Sevillano, J. L., Rodrıguez, G., Molina, F. J., Gasull, V. G., et al. (2013). Five years of designing wireless sensor networks in the doñana biological reserve (Spain): An applications approach. *Sensors, 13*(9), 12:044–12:069.

Laurent, A. (2002). Bases de données multidimensionnelles floues et leur utilisation pour la fouille de données. Paris: Academic Press.

Lawlor, A., Torres, J., O'Flynn, B., Wallace, J., & Regan, F. (2012). DEPLOY: A long term deployment of a water quality sensor monitoring system. *Sensor Review, 32*(1), 29–38. doi:10.1108/02602281211197125

Le Dinh, T., Hu, W., Sikka, P., Corke, P., Overs, L., & Brosnan, S. (2007). Design and Deployment of a Remote Robust Sensor Network: Experiences from an outdoor Water Quality Monitoring. *Proceedings of the IEEE Conference on Local Computer Networks*. doi: 10.1109/LCN.2007.39

Lee, T.-R., & Kao, J.-S. (2001). Application of simulation technique to activity-based costing of agricultural systems: A case study. *Agricultural Systems, 67*(2), 71–82. doi:10.1016/S0308-521X(00)00042-1

Lee, W., Alchanatis, V., Yang, C., Hirafuji, M., Moshou, D., & Li, C. (2010). Sensing technologies for precision specialty crop production. *Computers and Electronics in Agriculture, 74*(1), 2–33. doi:10.1016/j.compag.2010.08.005

Lenka, S., & Mohapatra, A. (2012). A novel approach of cattle health monitoring system based on wireless sensor network for early detection of disease. *Journal of Grid and Distributed Computing, 2*(1), 2249–7056.

Lévesque, M.-A. (2008). *Formal Approach for a better identification and management of risks of inappropriate use of geodecisional data (Master of Science)*. Laval University.

Levis, P., Madden, S., Polastre, J., Szewczyk, R., Whitehouse, K., Woo, A., ... Culler, D. (2005). Tinyos: An operating system for sensor networks. *Ambient Intelligence, 35*, 115–148. doi:10.1007/3-540-27139-2_7

Liao, J., Li, L., & Liu, X. (2015). An integrated, ontology based agricultural information system. *Information Development, 31*(2), 150–163. doi:10.1177/0266666913510716

Libelium Waspmote. (2013). Retrieved from http://www.libelium.com/products/waspmote/

LiteOS operating System. (2011). Retrieved from https://www.linuxliteos.com

Liu, M., Liu, X., Li, M., Fang, M., & Chi, W. (2010). Neural-network model for estimating leaf chlorophyll concentration in rice under stress from heavy metals using four spectral indices. *Biosystems Engineering, 106*(3), 223–233. doi:10.1016/j.biosystemseng.2009.12.008

Liu, X., Duan, X., & Zhang, H. (2012). Application of ontology in classification of agricultural information. *IEEE Symposium on Robotics and Applications (ISRA)*, 451 – 454. 10.1109/ISRA.2012.6219222

Liu, Y., Zhang, J., & Richards, M. (2009). Towards Continuous Surveillance of Fruit Flies Using Sensor Networks and Machine Vision. *Proceedings of the 5th International Conference on Wireless Communications, Networking and Mobile Computing*. 10.1109/WICOM.2009.5303034

Lloret, J., Garcia, M., Bri, D., & Sendra, S. (2009). A Wireless Sensor Network Deployment for Rural and Forest Fire Detection and Verification. *Sensors (Basel)*, *9*(11), 8722–8747. doi:10.339091108722 PMID:22291533

Lofton, J., Tubana, B. S., Kanke, Y., Teboh, J., Viator, W., & Dalen, M. (2012). Estimating Sugarcane Yield Potential Using an In-Season Determination of Normalized Difference Vegetative Index. *Sensors (Basel)*, *12*(6), 7529–7547. doi:10.3390120607529 PMID:22969359

Lorkowski, S. (2003). Erste Mobilitätsdienste auf Basis von "Floating Car Data". In *Stadt, Region, Land – 4. Aachener Kolloqium "Mobilität und Stadt" (AMUS)*. Aachen: Institut für Stadtbauwesen der RWTH.

Luce, R. D., & Tukey, J. W. (1964). Simultaneous conjoint measurement: A new type of fundamental measurement. *Journal of Mathematical Psychology*, *1*(1), 1–27. doi:10.1016/0022-2496(64)90015-X

Mafuta, M., Zennaro, M., Bagula, A., Ault, G., Gombachika, H., & Chadza, T. (2012). Successful deployment of a wireless sensor network for precision agriculture in Malawi. *Proceedings of the 3rd IEEE International Conference on Networked Embedded Systems for Every Application (NESEA)*. 10.1109/NESEA.2012.6474009

Mainwaring, A. C. (2002). Wireless Sensor Networks for Habitat Monitoring. *Proceedings of the 1st ACM International Workshop on Wireless Sensor Networks and Applications, (WSNA)*. 10.1145/570738.570751

Malavolta, E., Vitti, G. C., & Oliveira, S. A. (1997). *Avaliação do estado nutricional das plantas: princípios e aplicações* (2nd ed.). Piracicaba: Potafos.

Malinowski, E., & Zimányi, E. (2008). *Designing Spatial and Temporal Data Warehouses. In Advanced Data Warehouse Design* (pp. 307–343). Berlin: Springer - Verlag Berlin Heidelberg.

Mallikarjuna, A., & Kunz, T. (2011). Operating Systems for Wireless Sensor Networks: A Survey. *Sensors, 11(6)*, 5900-5930.

Mancuso, M., & Bustaffa, F. (2006). Wireless Sensors Network For Monitoring Environmental Variables in a Tomato Greenhouse. *Proceedings of the IEEE International Workshop on Factory Communication Systems*. 10.1109/WFCS.2006.1704135

Manfredini, F., Pucci, P., Secchi, P., Tagliolato, P., Vantini, S., & Vitelli, V. (2012). *Treelet decomposition of mobile phone data for deriving city usage and mobility pattern in the Milan urban region*. Academic Press.

Manikandan, N., Sakthiganesh, M., & Kumar, P. J., & SenthilKumar, M. (2012). Web based farmers bulletin for agricultural development using PAP Approach. *International Journal of Soft Computing and Engineering*, *2*(1), 171–174.

Mao, M. (2008). *Ontology Mapping: Towards semantic interoperability in distributed and heterogeneous environments (Ph.D. dissertation)*. Pittsburgh, PA: Pittsburgh Univ.

Markham, A., Trigoni, N., Macdonald, D. W., & Ellwood, S. A. (2012). Underground localization in 3-D using magneto-inductive tracking. *Sensors Journal, IEEE*, *12*(6), 1809–1816. doi:10.1109/JSEN.2011.2178064

Martinez, K., Basford, P., De Jager, D., & Hart, J. (2012). A wireless sensor network system deployment for detecting stick slip motion in glaciers. *Proceedings of the IET Conference on Wireless Sensor Systems (WSS)*. 10.1049/cp.2012.0580

Martinez, K., Hart, J. K., & Ong, R. (2009). Deploying a Wireless Sensor Network in Iceland. *Proceedings of the 3rd International Conference on Geosensor Networks*. 10.1007/978-3-642-02903-5_13

Martinez, K., Hart, J., & Ong, R. (2004). Environmental sensor networks. *Computer*, *37*(8), 50–56. doi:10.1109/MC.2004.91

Martinotti, G. (1993). *Metropoli: la nuova morfologia sociale della città*. Bologna: Il Mulino.

Martin, V., Moisan, S., Paris, B., & Nicolas, O. (2008). Towards a Video Camera Network for Early Pest Detection in Greenhouses. *Proceedings of the Endure International Conference Diversifying Crop Protection*, 1-5.

Matese, A., Vaccari, F. P., Tomasi, D., Di Gennaro, S. F., Primicerio, J., Sabatini, F., & Guidoni, S. (2013). Crossvit: Enhancing canopy monitoring management practices in viticulture. *Sensors (Basel)*, *13*(6), 7652–7667. doi:10.3390130607652 PMID:23765273

Mazón, J.-N., Trujillo, J., & Lechtenbörger, J. (2007). Reconciling requirement-driven data warehouses with data sources via multidimensional normal forms. *Data & Knowledge Engineering*, *63*(3), 725–751. doi:10.1016/j.datak.2007.04.004

McBratney, A., Whelan, B., Ancev, T., & Bouma, J. (2005). Future Directions of Precision Agriculture. *Precision Agriculture*, *6*(1), 7–23. doi:10.100711119-005-0681-8

McGuinness, D. L., Fikes, R., Rice, J., & Wilder, S. (2000). An environment for merging and testing large ontologies. Academic Press.

McKinsey. (2011). *Big data: The next frontier for innovation, competition, and productivity*. New York: McKinsey & Company Inc. Retrieved from http://www.mckinsey.com/business-functions/business-technology/our-insights/big-data-the-next-frontier-for-innovation

Meenakshi, N., Verma, K., Sheth, A., & Miller, J. A. (2007). Ontology driven data mediation in web service. *International Journal of Web Services Research*, *4*(4), 104–126. doi:10.4018/jwsr.2007100105

Mica2 Datasheet. (2005). Retrieved from http://www.eol.ucar.edu/isf/facilities/isa/internal/CrossBow/DataSheets/mica2.pdf

MICAz Datasheet. (2005). Retrieved from http://www.openautomation.net/uploadsproductos/micaz_datasheet.pdf

Milutinovic, V., & Salom, J. (2016). *Mind Genomics - A Guide to Data-Driven Marketing Strategy*. Springer International Publishing; doi:10.1007/978-3-319-39733-7

Miluzzo, E., Lu, H., Peebles, D., Choudhury, T., & Campbell, A. (2010). A survey of mobile phone sensing. *Communications Magazine, IEEE*, *48*(9), 140–150. doi:10.1109/MCOM.2010.5560598

Mingli, Y., & Yihai, S. (2013). *Marine data collection based on embedded system with wired and wireless transmission*. University of Agder.

Mishra, A., Karimi, D., Ehsani, R., & Albrigo, L. G. (2011). Evaluation of an active optical sensor for detection of Huanglongbing (HLB) disease. *Biosystems Engineering*, *110*(3), 302–309. doi:10.1016/j.biosystemseng.2011.09.003

Mitra, P., & Wiederhold, G. (2002). Resolving terminological heterogeneity in ontologies. *Workshop on Ontologies and Semantic Interoperability*.

Mittal, A., Chetan, K., Jayaraman, S., Jagyasi, B., Pande, A., & Balamuralidhar, P. (2012). mKRISHI: Wireless sensor network platform for precision agriculture. *Proceedings of the 6th International Conference on Sensing Technology (ICST)*. 10.1109/ICSensT.2012.6461755

Monmonier, M. (1990). Strategies for the visualization of geographic time-series data. *Cartographica*, *27*(1), 30–45. doi:10.3138/U558-H737-6577-8U31

Morais, R., Fernandes, M. A., Matos, S., Serôdio, C., Ferreira, P., & Reis, M. J. (2008). A ZigBee multi-powered wireless acquisition device for remote sensing applications in precision viticulture. *Computers and Electronics in Agriculture*, *62*(2), 94–106. doi:10.1016/j.compag.2007.12.004

Moteiv Tmote Sky. (2004). Retrieved from http://www.snm.ethz.ch/snmwiki/pub/uploads/Projects/tmote_sky_datasheet.pdf

Mozell, M. R., & Thach, L. (2014). The impact of climate change on the global wine industry: Challenges & solutions. *Wine Economics and Policy*, *3*(2), 81–89. doi:10.1016/j.wep.2014.08.001

Mukhopadhyay, D., Banik, A., Mukherjee, S., & Bhattacharya, J. (2011). A domain specific ontology based semantic web search engine. *7th International Workshop on MSPT*, 1-9.

Mullen, R. W., Freman, K. W., Raun, W. R., Johnson, G. V., Stone, M. L., & Solie, J. B. (2003). Identifying an In-Season Response Index and the Potencial to Increase wheat Yield with nitrogen. *Agronomy Journal*, *95*(2), 347–351. doi:10.2134/agronj2003.0347

Mun, M., Reddy, S., Shilton, K., Yau, N., Burke, J., & Estrin, D. (2009). PEIR: the personal environmental impact report, as a platform for participatory sensing systems research. *Proceedings of the 7th international conference on Mobile systems, applications, and services (MobiSys)*. 10.1145/1555816.1555823

Muriithi, A. G., Bett, E., & Ogaleh, S. A. (2009). Information technology for agriculture and rural development in africa: experiences from Kenya. In *Conference on International Research on Food Security, Natural Resource Management and Rural Development*. University of Hamburg.

Nadareishvili, I. (2016). *Microservice Architecture – Aligning Principles, Practices, and Culture*. Sebastopol, CA: O'Reilly Media Inc.

Nadimi, E. S., Jørgensen, R. N., Blanes-Vidal, V., & Christensen, S. (2012). Monitoring and classifying animal behavior using zigbee-based mobile ad hoc wireless sensor networks and artificial neural networks. *Computers and Electronics in Agriculture*, *82*, 44–54. doi:10.1016/j.compag.2011.12.008

Nance, C., Losser, T., Iype, R., & Harmon, G. (2013). Nosql vs rdbms-why there is room for both. *Proceedings of the Southern Association for Information Systems Conference*, 111–116.

Naouali, S., & Missaoui, R. (2006). *Approximation des cubes OLAP et génération de règles dans les entrepôts de données*. Academic Press.

Naouali, S., & Missaoui, R. (2005). *Flexible query answering in data cubes. In Data Warehousing and Knowledge Discovery* (pp. 221–232). Springer. doi:10.1007/11546849_22

Nash, E., Dreger, F., Schwarz, J., Bill, R., & Werner, A. (2009). Development of a model of data-flows for precision agriculture based on a collaborative research project. *Computers and Electronics in Agriculture*, *66*(1), 25–37. doi:10.1016/j.compag.2008.11.005

Nash, E., Korduan, P., & Bill, R. (2009). Applications of open geospatial web services in precision agriculture: A review. *Precision Agriculture*, *10*(6), 546–560. doi:10.100711119-009-9134-0

Nasser, N., Ali, A., Karim, L., & Belhaouari, S. (2013). An efficient wireless sensor network-based water quality monitoring system. In *Proceedings of the 2013 ACS International Conference on Computer Systems and Applications*. IEEE. 10.1109/AICCSA.2013.6616432

Neha. (2012). *Building crop ontology for farmers* (M.Tech thesis). Banasthali University, Rajasthan, India.

Newman, S. (2015). *Building Microservices – Designing fine-grained systems*. Sebastopol, CA: O'Reilly Media Inc.

Neyamadpour, A., Taib, S., & Abdullah, W. (2009). Using artificial neural networks to invert 2D DC resistivity imaging data for high resistivity contrast regions: A MATLAB application. *Computers & Geosciences*, *35*(11), 2268–2274. doi:10.1016/j.cageo.2009.04.004

Nikander, J., Koistinen, M., Laajalahti, M., Pesonen, L., Ronkainen, A., & Suomi, P. (2015). Farm Information Management Infrastructures in the Future. In *2015 26th International Workshop on Database and Expert Systems Applications (DEXA)* (pp. 104–107). IEEE. 10.1109/DEXA.2015.38

Nikkilä, R., Seilonen, I., & Koskinen, K. (2010). Software architecture for farm management information systems in precision agriculture. *Computers and Electronics in Agriculture*, *70*(2), 328–336. doi:10.1016/j.compag.2009.08.013

Noh, H., & Zang, Q. (2012). Shadow effect on multi-spectral image for detection of nitrogen deficiency in corn. *Computers and Electronics in Agriculture*, *83*, 52–57. doi:10.1016/j.compag.2012.01.014

Noy, N. F., & Musen, M. A. (2000). PROMPT: Algorithm and Tool for Automated Ontology Merging andAlignment. In AAAI/IAAI (pp. 450–455). AAAI.

Nuvolati, G. (2003). Resident and Non-resident Populations: Quality of Life, Mobility and Time Policy. *The Journal of Regional Analysis & Policy*, *33*(2), 67–83.

O'Flynn, B., Martinez, R., Cleary, J., Slater, C., Regan, F., Diamond, D., & Murphy, H. (2007). SmartCoast: a wireless sensor network for water quality monitoring. In *Local Computer Networks, 2007. LCN 2007. 32nd IEEE Conference on* (pp. 815-816). IEEE. 10.1109/LCN.2007.34

O'Flynn, B., Regan, F., Lawlor, A., Wallace, J., Torres, J., & O'Mathuna, C. (2010). Experiences and recommendations in deploying a real-time, water quality monitoring system. *Measurement Science & Technology*, *21*(12), 124004. doi:10.1088/0957-0233/21/12/124004

Ogbuene, E. B. (2010). Impact of meteorological parameters on rice yield: An approach for environmental resource sustainability in Ebonyi rice farmland, Nigeria. *Journal of Environmental Issues and Agriculture in Developing Countries*, 2(2& 3), 103–116.

Oksanen, T., Öhman, M., Miettinen, M., & Visala, A. (2005). ISO 11783 –STANDARD AND ITS IMPLEMENTATION. In *IFAC Proceedings Volumes* (Vol. 38, pp. 69–74). Academic Press. 10.3182/20050703-6-CZ-1902.02102

Oliveira, A. C. S., Souza, A. A., Lacerda, W. S., & Gonçalves, L. R. (2010). Aplicação de redes neurais artificiais na previsão da produção de álcool. *Ciência e Agrotecnologia*, *34*(2), 279–284. doi:10.1590/S1413-70542010000200002

Oliveira, L. M., & Rodrigues, J. J. (2011). Wireless Sensor Networks: A Survey on Environmental Monitoring. *JCM*, *6*(2), 143–151. doi:10.4304/jcm.6.2.143-151

Oliveira, L. M., & Rodrigues, J. J. (2011). Wireless Sensor Networks: A Survey on Environmental Monitoring. *Journal of Communication*, *6*, 143–151.

Ong, K. G., Yang, X., Mukherjee, N., Wang, H., Surender, S., & Grimes, C. A. (2004). A wireless sensor network for long-term monitoring of aquatic environments: Design and implementation. *Sensor Letters*, *2*(1), 48–57. doi:10.1166l.2004.023

Padhy, P., Martinez, K., Riddoch, A., Ong, H. L., & Hart, J. K. (2005). Glacial Environment Monitoring using Sensor Networks. *Proceedings of the International Conference on Real-World Wireless Sensor Networks*, 10-14.

Pahuja, R., Verma, H., & Uddin, M. (2013). A wireless sensor network for greenhouse climate control. *Pervasive Computing, IEEE, 12*(2), 49–58. doi:10.1109/MPRV.2013.26

Panda, S. S., Ames, D. P., & Panigrahi, S. (2010). Application of Vegetation Indices for Agricultural Crop Yield Prediction Using Neural Network Techniques. *Remote Sensing, 2*(3), 673–696. doi:10.3390/rs2030673

Papajorgji, P., Pinet, F., Miralles, A., Jallas, E., & Pardalos, P. M. (2010). Modeling: A Central Activity for Flexible Information Systems Development in Agriculture and Environment. *International Journal of Agricultural and Environmental Information Systems, 1*(1), 1–25. doi:10.4018/jaeis.2010101301

Pardillo, J., Mazon, J.-N., & Trujillo, J. (2010). Designing OLAP schemata for data warehouses from conceptual models with MDA. *Decision Support Systems*, 1–13.

Patel, M., Koch, T., Doerr, M., & Sinaraki, C. (2005). *Semantic interoperability in digital library systems*. UKOLN, University of Bat.

Pathak, R. R., Lochab, S., & Raghuram, N. (2011). Improving Plant Nitrogen-Use Efficiency. *Agricultural and Related Biotechnologies, 4*, 209–218.

Pauly, A., & Schneider, M. (2010). VASA: An algebra for vague spatial data in databases. *Information Systems, 35*(1), 111–138. doi:10.1016/j.is.2009.05.003

Peets, S., Mouazen, A. M., Blackburn, K., Kuang, B., & Wiebensohn, J. (2012). Methods and procedures for automatic collection and management of data acquired from on-the-go sensors with application to on-the-go soil sensors. *Computers and Electronics in Agriculture, 81*, 104–112. doi:10.1016/j.compag.2011.11.011

Perez, D., Somodevilla, M. J., & Pineda, I. H. (2007). Fuzzy Spatial Data Warehouse: A Multidimensional Model. *Proceedings of the Eighth Mexican International Conference on Current Trends in Computer Science*. 10.1109/ENC.2007.15

Perry, E. M., & Davenport, J. R. (2007). Spectral and spatial differences in response of vegetation indices to nitrogen treatments on apple. *Computers and Electronics in Agriculture, 59*(1-2), 56–65. doi:10.1016/j.compag.2007.05.002

Petersen, W. (1999). *Malthus: Founder of Modern Demography*. Routledge.

Pfoser, D. (2008a). Floating Car Data. In S. Shekhar & H. Xiong (Eds.), *Encyclopedia of Geographical Information Sciences* (p. 321). New York: Springer-Verlag.

Pfoser, D. (2008b). Map-Matching. In S. Shekhar & H. Xiong (Eds.), *Encyclopedia of Geographical Information Sciences* (pp. 632–633). New York: Springer-Verlag.

Phillips, C., Kemp, E., & Kek, S. M. (2001). Extending UML use case modelling to support graphical user interface design. In *Software Engineering Conference, 2001. Proceedings. 2001 Australian* (pp. 48–57). IEEE. 10.1109/ASWEC.2001.948497

Pierce, F. J., & Elliott, T. V. (2008). Regional and on-farm wireless sensor networks for agricultural systems in Eastern Washington. *Computers and Electronics in Agriculture*, *61*(1), 32–43. doi:10.1016/j.compag.2007.05.007

Pitarch, Y., Favre, C., Laurent, A., & Poncelet, P. (2012). *Enhancing flexibility and expressivity of contextual hierarchies.* Paper presented at the Fuzzy Systems (FUZZ-IEEE), 2012 IEEE International Conference on. 10.1109/FUZZ-IEEE.2012.6251176

Polastre, J. (2003). *Design and Implementation of Wireless Sensor Networks for Habitat Monitoring*. Berkeley, CA: University of California.

Porter, C. H., Villalobos, C., Holzworth, D., Nelson, R., White, J. W., Athanasiadis, I. N., ... Jones, J. W. (2014). Harmonization and translation of crop modeling data to ensure interoperability. *Environmental Modelling & Software*, *62*, 495–508. doi:10.1016/j.envsoft.2014.09.004

Portz, G., Molin, J. P., & Jasper, J. (2012). Active crop sensor to detect variability of nitrogen supply and biomass on sugarcane fields. *Precision Agriculture*, 11–12.

Prat, N., Akoka, J., & Comyn-Wattiau, I. (2006). A UML-based data warehouse design method. *Decision Support Systems*, *42*(3), 1449–1473. doi:10.1016/j.dss.2005.12.001

Puccinelli, D., & Haenggi, M. (2005). Wireless sensor networks: Applications and challenges of ubiquitous sensing. *Circuits and Systems Magazine, IEEE*, *5*(3), 19–31. doi:10.1109/MCAS.2005.1507522

Quevauviller, P. (2006). Chemical monitoring activity under the common implementation strategy of the WFD. *Journal of Soils and Sediments*, *6*(1), 2–3. doi:10.1065/jss2006.02.002

Rafiq, A., & Garg, A. (2002). Activity based costing and financial institutions: Old wine in new bottles or corporate panacea? *The Journal of Bank Cost & Management Accounting*, *15*, 12–29.

Ramos, M., Foster, A., Castell, S. F., Fos, V. G., & Solano, J. J. P. (2013). Gatherer: An environmental monitoring application based on IPv6 using wireless sensor networks. *International Journal of Ad Hoc and Ubiquitous Computing*, *3*(4), 209–217. doi:10.1504/IJAHUC.2013.055465

Rao, A., Marshall, S., Gubbi, J., Palaniswami, M., Sinnott, R., & Pettigrovet, V. (2013). Design of low-cost autonomous water quality monitoring system. *Proceedings of the 2nd International Conference on Advances in Computing, Communications and Informatics (ICACCI)*. 10.1109/ICACCI.2013.6637139

Rasin, Z., & Abdullah, M. R. (2009). Water quality monitoring system using ZigBee based wireless sensor network. *IACSIT International Journal of Engineering and Technology*, *9*(10), 24–28.

Ratti, C., Pulselli, R. M., Williams, S., & Frenchman, D. (2006). Mobile Landscapes: Using location data from cell phones for urban analysis. *Environment and Planning. B, Planning & Design, 33*(5), 727–748. doi:10.1068/b32047

Raunich, S., & Rahm, E. (2011). ATOM: Automatic Target-driven Ontology Merging. *Proc. 27th International Conference on Data Engineering (ICDE)*, 1276–1279.

Raun, W. R., Solie, J. B., Johnson, G. B., Stone, M. L., Mullen, R. W., Freman, K. W., ... Lukina, E. V. (2002). Improving nitrogen use efficiency in cereal grain production with optical sensing and variable rate application. *Agronomy Journal, 94*(4), 815–820. doi:10.2134/agronj2002.8150

Reades, J., Calabrese, F., Sevtsuk, A., & Ratti, C. (2007). Cellular Census: Explorations in Urban Data Collection. *IEEE Pervasive Computing, 6*(3), 30–38. doi:10.1109/MPRV.2007.53

Regione Lombardia, Direzione Generale Infrastrutture e mobilità. (2002). *Indagine Origine/Destinazione regionale 2002 - sintesi*. Author.

Rice, R., & Bales, R. C. (2010). Embedded-sensor network design for snow cover measurements around snow pillow and snow course sites in the Sierra Nevada of California. *Water Resources Research, 46*(3), 35–37. doi:10.1029/2008WR007318

Riedmiller, M., & Braun, H. (1993). A direct adaptive meted for faster backpropagation learning: the RPROP algorithm. *Internacional Conference on Neural Network (ICNN), San Francisco: Proceedings*, 586-591.

Riquelme, J. A.-L., Soto, F., Suardíaz, J., Sánchez, P., Iborra, A., & Vera, J. A. (2009). Wireless Sensor Networks for precision horticulture in Southern Spain. *Computers and Electronics in Agriculture, 68*(1), 25–35. doi:10.1016/j.compag.2009.04.006

Romero, O., & Abelló, A. (2009). *A Survey of Multidimensional Modeling Methodologies*. IGI Global.

Romero, O., & Abelló, A. (2011). Data-Driven Multidimensional Design for OLAP. In J. Bayard Cushing, J. French, & S. Bowers (Eds.), *Scientific and Statistical Database Management* (Vol. 6809, pp. 594–595). Springer Berlin Heidelberg. doi:10.1007/978-3-642-22351-8_51

Rosenberg, D., & Stephens, M. (2007). *Use Case Driven Object Modeling with UML*. Berkeley, CA: Apress. Retrieved from http://www.springerlink.com/index/10.1007/978-1-4302-0369-8

Rossato, O. B., Aandrade-Sanchez, P., Guerra, S. P. S., & Crusciol, C. A. C. (2012). Sensores de refletância e fluorescência na avaliação de teores de nitrogênio, produção de biomassa e produtividade do algodoeiro. *Pesquisa Agropecuária Brasileira, 47*(8), 1133–1141. doi:10.1590/S0100-204X2012000800014

Rouse, J. W. J., Haas, H. R., Schell, A. J., & Deering, W. D. (1974). Monitoring vegetation systems in the great plains with ERTS. *NASA Special Publication, 1*, 309–317.

Ruggaber, T. P., Talley, J. W., & Montestruque, L. A. (2007). Using embedded sensor networks to monitor, control, and reduce CSO events: A pilot study. *Environmental Engineering Science*, *24*(2), 172–182. doi:10.1089/ees.2006.0041

Rutishauser, M., Petkov, V., Boice, J., Obraczka, K., Mantey, P., Williams, T. M., & … . (2010). CARNIVORE: A Disruption-Tolerant System for Studying Wildlife. *EURASIP Journal on Wireless Communications and Networking*, *1*, 1–14.

Sadalage, P., & Fowler, M. (2012). *NoSQL Distilled: A Brief Guide to the Emerging World of Polyglot Persistence* (1st ed.). Addison-Wesley Professional.

Schäffer, F. (2012). *OBD-Fahrzeugdiagnose in der Praxis*. Franzis Verlag, Haar.

Scheller, M., & Urry, J. (2006). The new mobilities paradigm. *Environment & Planning A*, *38*(2), 207–226. doi:10.1068/a37268

Schmitz, M., Martini, D., Kunisch, M., & Mösinger, H.-J. (2009). agroXML Enabling Standardized, Platform-Independent Internet Data Exchange in Farm Management Information Systems. In Metadata and Semantics (pp. 463–468). Boston, MA: Springer US. doi:10.1007/978-0-387-77745-0_45

Schneider, M. (1999). Uncertainty Management for Spatial Datain Databases: Fuzzy Spatial Data Types Advances in Spatial Databases. Springer Berlin/Heidelberg.

Schneider, S. (2010). Extended Floating Car Data in Co-operative Traffic Management. In J. Barceló & M. Kuwahara (Eds.), *Traffic Data Collection and its Standardization* (pp. 161–170). New York: Springer-Verlag. doi:10.1007/978-1-4419-6070-2_11

Seders, L. A., Shea, C. A., Lemmon, M. D., Maurice, P. A., & Talley, J. W. (2007). LakeNet: An Integrated Sensor Network for Environmental Sensing in Lakes. *Environmental Engineering Science*, *24*(2), 183–191. doi:10.1089/ees.2006.0044

Setala, J., & Gunasekaran, A. (1996). Activity-based costing and management--a way to improve the profitability of fish processing? *Production and Inventory Management Journal*, *37*, 63–69.

Shaikh, Z. A., Shaikh, N. A., & Islam, N. (2010). An Integrated Framework to Develop Context-Aware Sensor Grid for Agriculture. *Australian Journal of Basic and Applied Sciences*, *4*(5), 922–931.

Shanahan, J. F., Kitchen, N. R., Raun, W. R., & Schepers, J. S. (2008). Responsive in-season nitrogen management for cereals. *Computers and Electronics in Agriculture*, *61*(1), 51–62. doi:10.1016/j.compag.2007.06.006

Sheiver, T. M., Khosla, R., & Westfall, D. G. (2011). Evaluation of two crop canopy sensors for nitrogen variability determination in irrigated maize. *Precision Agriculture*, *12*(6), 892–904. doi:10.100711119-011-9229-2

Shelby, Z., & Bormann, C. (2010). *6LoWPAN: The Wireless Embedded Internet*. New York: John Wiley & Sons, Inc.

Shepherd, R., Beirne, S., Lau, K., Corcoran, B., & Diamond, D. (2007). Monitoring chemical plumes in an environmental sensing chamber with a wireless chemical sensor network. *Sensors and Actuators. B, Chemical*, *121*(1), 142–149. doi:10.1016/j.snb.2006.09.045

Shoaib, M., & Basharat, A. (2010). Semantic web based integrated agriculture information framework. *IEEE Second International Conference on Computer Research and Development*, 285 – 289. 10.1109/ICCRD.2010.33

Silva, S., Nguyen, H. N., Tiporlini, V., & Alameh, K. (2011). Web based water quality monitoring with sensor network: Employing ZigBee and WiMax technologies. In *Proceedings of the Conference on High Capacity Optical Networks and Enabling Technologies*. IEEE. 10.1109/HONET.2011.6149804

Singh, T. N., & Verma, A. K. (2012). Comparative analysis of intelligent algorithms to correlate strength and petrographic properties of some schistose rocks. *Engineering with Computers*, *28*(1), 1–12. doi:10.100700366-011-0210-5

Siqueira, T. L. L., Aguiar Ciferri, C. D., Times, V. C., & Ciferri, R. R. (2012). Towards Vague Geographic Data Warehouses. In N. Xiao, M.-P. Kwan, M. Goodchild, & S. Shekhar (Eds.), *Geographic Information Science* (Vol. 7478, pp. 173–186). Springer Berlin Heidelberg.

Solari, F., Shanahan, J., Ferguson, R., Schepers, J., & Gitelson, A. (2008). Active sensor reflectance measurements of corn nitrogen status and yield potential. *Agronomy Journal*, *100*(3), 571–579. doi:10.2134/agronj2007.0244

Solid, I. T. (2017). *DB-Engines Ranking per database model category*. Retrieved August 29, 2017, from https://db-engines.com/en/ranking_categories

Somov, A., Baranov, A., Spirjakin, D., Spirjakin, A., Sleptsov, V., & Passerone, R. (2013). Deployment and evaluation of a Wireless Sensor Network for methane leak detection. *Sensors and Actuators. A, Physical*, *202*, 217–225. doi:10.1016/j.sna.2012.11.047

Song, W.-Z., Huang, R., Xu, M., Ma, A., Shirazi, B., & Lahusen, R. (2009). Air-dropped Sensor Network for Real-time High-fidelity Volcano Monitoring. *Proceedings of the 7th International Conference on Mobile systems, applications, and services (MobiSys)*. 10.1145/1555816.1555847

Sørensen, C. G., Fountas, S., Nash, E., Pesonen, L., Bochtis, D., Pedersen, S. M., ... Blackmore, S. B. (2010). Conceptual model of a future farm management information system. *Computers and Electronics in Agriculture*, *72*(1), 37–47. doi:10.1016/j.compag.2010.02.003

Sørensen, C. G., Pesonen, L., Bochtis, D. D., Vougioukas, S. G., & Suomi, P. (2011). Functional requirements for a future farm management information system. *Computers and Electronics in Agriculture*, *76*(2), 266–276. doi:10.1016/j.compag.2011.02.005

Sørensen, C. G., Pesonen, L., Fountas, S., Suomi, P., Bochtis, D., Bildsøe, P., & Pedersen, S. M. (2010). A user-centric approach for information modelling in arable farming. *Computers and Electronics in Agriculture*, *73*(1), 44–55. doi:10.1016/j.compag.2010.04.003

Soto, V., & Frías-Martínez, E. (2011). Automated land use identification using cell-phone records. In *Proceedings of the 3rd ACM international workshop on MobiArch* (pp. 17–22). Academic Press. 10.1145/2000172.2000179

Soto, V., & Frías-Martínez, E. (2011). Robust Land Use Characterization of Urban Landscapes using Cell Phone Data. *The first workshop on pervasive Urban Applications (PURBA).*

SPARQL Query Language. (n.d.). Available: http://www.w3.org/TR/rdf-sparql-query/

Srinivas, S., & Nair, A. (2015). Security maturity in NoSQL databases - are they secure enough to haul the modern IT applications? In *2015 International Conference on Advances in Computing, Communications and Informatics (ICACCI)* (pp. 739–744). IEEE. 10.1109/ICACCI.2015.7275699

Statutory Instruments. (2009). *European Communities Environmental Objectives (Surface Waters) Regulations 2015*. Retrieved from http://www.irishstatutebook.ie/eli/2015/si/386/made/en/pdf

Steinberger, G., Rothmund, M., & Auernhammer, H. (2009). Mobile farm equipment as a data source in an agricultural service architecture. *Computers and Electronics in Agriculture*, *65*(2), 238–246. doi:10.1016/j.compag.2008.10.005

Stemp-Morlock, G. (2007). Natural Resources: Australia's War on Drought. *Environmental Health Perspectives*, *115*(7), A348. doi:10.1289/ehp.115-a348

Stephen, B., Michie, C., & Andonovic, I. (2013). Remote sensing in agricultural livestock welfare monitoring: Practical consideration. In S. Mukhopadhyay & J.-A. Jiang (Eds.), *Wireless Sensor Networks and Ecological Monitoring* (pp. 179–193). Berlin: Springer-Verlag Berlin Heidelberg. doi:10.1007/978-3-642-36365-8_7

Stevens, S. S. (1946). On the Theory of Scales of Measurement. *Science*, *103*(2684), 677–680. doi:10.1126cience.103.2684.677

Stottan, T. (2013). XFCD als Basistechnologie für die Mobilität 3.0 - Entstehung, Entwicklung, Zukunftsanwendungen, Marktentwicklung. In *Schritte in die künftige Mobilität* (pp. 47–59). Wiesbaden: Springer Fachmedien. doi:10.1007/978-3-8349-4308-8_4

Strazdins, G., Elsts, A., & Selavo, L. (2010). MansOS: easy to use, portable and resource efficient operating system for networked embedded devices. *Proceedings of the 8th ACM Conference on Embedded Networked Sensor Systems (SenSys).* 10.1145/1869983.1870057

Sun, X., Zhu, H., Gu, J., Wu, H., & Feng, C. (2009). Research on the semantic web-based technology of knowledge integration for agricultural production. *Sixth International Conference on Fuzzy Systems and Knowledge Discovery*, 2, 361-366. 10.1109/FSKD.2009.58

Su, X., Shao, G., Vause, J., & Tang, L. (2013). An integrated system for urban environmental monitoring and management based on the environmental Internet of Things. *International Journal of Sustainable Development and World Ecology*, *20*(3), 205–209. doi:10.1080/13504509.2013.782580

Swaran, L., Bhaskar, S., Ela, K., Somnath, C., & Raghu, A. (2013). Semantic web query on e-Governance data and designing ontology for agriculture domain. *International Journal of Web & Semantic Technology*, *4*(3), 65–72. doi:10.5121/ijwest.2013.4307

Szewczyk, R., Mainwaring, A., Polastre, J., Anderson, J., & Culler, D. (2004). An analysis of a large scale habitat monitoring application. *Proceedings of the 2nd international conference on Embedded networked sensor systems (SenSys).* 10.1145/1031495.1031521

Talzi, I., Hasler, A., Gruber, S., & Tschudin, C. (2007). PermaSense: Investigating permafrost with a WSN in the Swiss Alps. *Proceedings of the 4th International Workshop on Embedded Networked Sensors (EmNets).* 10.1145/1278972.1278974

Tan, R., Xing, G., Chen, J., Song, W.-Z., & Huang, R. (2013). Fusion-based volcanic earthquake detection and timing in wireless sensor networks. *ACM Trans. Sen. Netw., 9*(2), 17:1–17:25.

Tateson, J., Roadknight, C., Gonzalez, A., Khan, T., Fitz, S., Henning, I., ... Marshall, I. (2005). Real world issues in deploying a wireless sensor network for oceanography. *Workshop on Real-World Wireless Sensor Networks REALWSN'05.*

Terzis, A. R., Musaloiu, E., Cogan, J., Szlavecz, K., Szalay, A., Gray, J., et al. Wireless sensor networks for soil science. *International Journal of Sensor Networks, 7*(2), 53-70.

Texas Instruments Microcontrollers. (1980). Retrieved from http://www.ti.com/lsds/ti/microcontroller/overview.page

The Arduino platforms. (2009). Retrieved from http://arduino.cc/en/Main/Products

The BeagleBoard technology. (2008). Retrieved from http://beagleboard.org

The GlacsWeb project. (2012). Retrieved from http://glacsweb.org/technology/

The Pandaboard technology. (2010). Retrieved from http://pandaboard.org

TI CC2420 Datasheet. (2010). Retrieved from http://inst.eecs.berkeley.edu/~cs150/Documents/CC2420.pdf

Tik, L. B., Khuan, C. T., & Palaniappan, S. (2009). Monitoring of an Aeroponic Greenhouse with a Sensor Network. *International Journal of Computer Science and Network Security*, *9*(3), 240–246.

Tolle, G., Polastre, J., Szewczyk, R., Culler, D., Turner, N., & Tu, K. (2005). A macroscope in the redwoods. *Proceedings of the 3rd International Conference on Embedded networked sensor systems (SenSys).* 10.1145/1098918.1098925

Tollenaere, T. (1990). Supersab: Fast adaptative backpropagation with good scaling properties. *Neural Networks*, *3*(5), 561–573. doi:10.1016/0893-6080(90)90006-7

Trotter, T. F., Frazier, P., Trotter, M. G., & Lamb, D. W. (2008). Objective biomass assessment using an active plant sensor (CROP CIRCLE) preliminary experiences on a variety of agricultural landscapes. In R. Koshla (Ed.), *9th Conference on Precision Agriculture*. Colorado State University.

Tseng, C.-L., Jiang, J.-A., Yang, E.-C., Lu, F.-M., Liao, C.-S., & Lin, C. (2008). An Automatic Counting Trap with Wireless Transmission Capability for the Oriental Fruit Fly. *Proceedings of the 4th International Symposium on Machinery and Mechatronics for Agriculture and Biosystems Engineering (ISMAB)*, 14-20.

Urry, J. (2002). *Sociology beyond societies: Mobilities for the twenty-first century*. Routledge.

Vantini, S., Vitelli, V., & Zanini, P. (2012). Treelet Analysis and Independent Component Analysis of Milan Mobile-Network Data: Investigating Population Mobility and Behavior. Analysis and Modeling of Complex Data in Behavioural and Social Sciences - Joint Meeting of of the Italian and the Japanese Statistical Societies.

Vasseur, J., & Dunkels, A. (2010). *Interconnecting Smart Objects with IP: The Next Internet*. San Francisco, CA, USA: Morgan Kaufmann Publishers Inc.

Vellidis, G., Tucker, M., Perry, C., Kvien, C., & Bednarz, C. (2008). Real-time Smart Sensor Array for Scheduling Irrigation. *Computers and Electronics in Agriculture*, *61*(1), 44–50. doi:10.1016/j.compag.2007.05.009

Verbeke, W. (2015). Profiling consumers who are ready to adopt insects as a meat substitute in a Western society. *Food Quality and Preference*, *39*, 147–155. doi:10.1016/j.foodqual.2014.07.008

Voigt, T., Osterlind, F., Finne, N., Tsiftes, N., He, Z., Eriksson, J., . . . Hjort, K. (2007). Sensor networking in aquatic environments-experiences and new challenges. In *Local Computer Networks, 2007. LCN 2007. 32nd IEEE Conference on* (pp. 793-798). IEEE. 10.1109/LCN.2007.23

Voland, P. (2014). *Webbasierte Visualisierung von Extended Floating Car Data (XFCD) - Ein Ansatz zur raumzeitlichen Visualisierung und technischen Implementierung mit Open Source Software unter spezieller Betrachtung des Umwelt- und Verkehrsmonitoring* (Master Thesis). University of Potsdam.

Voulodimos, A. S., Patrikakis, C. Z., Sideridis, A. B., Ntafis, V. A., & Xylouri, E. M. (2010). A complete farm management system based on animal identification using RFID technology. *Computers and Electronics in Agriculture*, *70*(2), 380–388. doi:10.1016/j.compag.2009.07.009

Wang, F., Brent Hall, G., & Subaryono. (1990). Fuzzy information representation and processing in conventional GIS software: Database design and application. *International Journal of Geographical Information Systems*, *4*(3), 261–283. doi:10.1080/02693799008941546

Wang, G., Zhang, J., Li, W., Cui, D., & Jing, Y. (2010). A forest fire monitoring system based on GPRS and Zigbee wireless sensor network. *Proceedings of the 5th IEEE Conference on Industrial Electronics and Applications (ICIEA)*. doi: 10.1109/ICIEA.2010.5515417

Wang, J., Ren, X., Shen, Y., & Liu, S. (2010). A Remote Wireless Sensor Networks for Water Quality Monitoring. In *Proceedings of the 2010 International Conference on Innovative Computing and Communication and 2010 Asia-Pacific Conference on Information Technology and Ocean Engineering*. IEEE Computer Society. 10.1109/CICC-ITOE.2010.9

Wang, Y., Wang, Y., Qi, X., Xu, L., Chen, J., & Wang, G. (2010). L3SN: A Level-Based, Large-Scale, Longevous Sensor Network for Agriculture Information Monitoring. *Wireless Sensor Network*, *2*(1), 655–660. doi:10.4236/wsn.2010.29078

Wang, Z., Wang, Q., & Hao, X. (2009). The design of the remote water quality monitoring system based on WSN. In *Proceedings of the 5th International Conference on Wireless Communications, Networking, and Mobile Computing*. IEEE. 10.1109/WICOM.2009.5303974

Wark, T., Crossman, C., Hu, W., Guo, Y., Valencia, P., & Sikka, P. (2007). The design and evaluation of a mobile sensor/actuator network for autonomous animal control. *Proceedings of the 6th International Conference on Information Processing in Sensor Networks (IPSN)*. doi: 10.1109/IPSN.2007.4379680

Wark, T., Hu, W., Corke, P., Hodge, J., Keto, A., & Mackey, B. (2008). Springbrook: Challenges in developing a long-term, rainforest wireless sensor network. *Proceedings of the Internaional Conference on Intelligent Sensors, Sensor Networks and Information Processing (ISSNIP)*. 10.1109/ISSNIP.2008.4762055

Web Ontology Language (OWL). (n.d.). Available: http://www.w3.org/TR/owl-ref/

Wen, T.-H., Jiang, J.-A., Sun, C.-H., Juang, J.-Y., & Lin, T.-S. (2013). Monitoring street-level spatial-temporal variations of carbon monoxide in urban settings using a wireless sensor network (WSN) framework. *International Journal of Environmental Research and Public Health*, *10*(12), 6380–6396. doi:10.3390/ijerph10126380 PMID:24287859

Werner-Allen, G., Lorincz, K., Johnson, J., Lees, J., & Welsh, M. (2006). Fidelity and Yield in a Volcano Monitoring Sensor Network. *Proceedings of the 7th symposium on Operating systems design and implementation (OSDI)*, 381-396.

Werner-Allen, G., Lorincz, K., Welsh, M., Marcillo, O., Johnson, J., Ruiz, M., & … . (2006). Deploying a wireless sensor network on an active volcano. *IEEE Internet Computing*, *10*(2), 18–25. doi:10.1109/MIC.2006.26

Wijesinghe, L., Siriwardena, P., & Dias, D. (2013). A wireless sensor network for early warning of elephant intrusions. In F. Shaikh (Ed.), *Wireless Sensor Networks for Developing Countries* (pp. 38–50). Berlin: Springer-Verlag Berlin Heidelberg. doi:10.1007/978-3-642-41054-3_4

Wolfert, S., Goense, D., & Sorensen, C. A. G. (2014). A Future Internet Collaboration Platform for Safe and Healthy Food from Farm to Fork. In *2014 Annual {SRII} Global Conference*. IEEE. 10.1109/SRII.2014.47

Wolfert, S., Ge, L., Verdouw, C., & Bogaardt, M.-J. (2017). Big Data in Smart Farming – A review. *Agricultural Systems*, *153*, 69–80. doi:10.1016/j.agsy.2017.01.023

Woo, A., Tong, T., & Culler, D. (2003). Taming the underlying challenges of reliable multihop routing in sensor networks. *Proceedings of the 1st international conference on Embedded networked sensor systems (SenSys)*. 10.1145/958491.958494

Worboys, M. (1998). Computation with imprecise geospatial data. *Computers, Environment and Urban Systems*, *22*(2), 85–106. doi:10.1016/S0198-9715(98)00023-4

WordNet. (n.d.). Retrieved from www.w3.org/TR/wordnet-rdf/

Xia, J., Tang, Z., Shi, X., Fan, L., & Li, H. (2011). An environment monitoring system for precise agriculture based on wireless sensor networks. *Proceedings of the 7th International Conference on Mobile Ad-hoc and Sensor Networks (MSN)*. 10.1109/MSN.2011.16

Xie, N., Wang, W., Ma, B., Zhang, X., Sun, W., & Guo, F. (2015). Research on an agricultural knowledge fusion method for big data. *Data Science Journal*, *14*(7), 1–9.

Yaacoub, E., Kadri, A., Mushtaha, M., & Abu-Dayya, A. (2013). Air quality monitoring and analysis in Qatar using a Wireless Sensor Network deployment. *Proceedings of the 9th International Wireless Communications and Mobile Computing Conference (IWCMC)*. 10.1109/IWCMC.2013.6583625

Yang, J., Zhang, C., Li, X., Huang, Y., Fu, S., & Acevedo, M. F. (2010). Integration of wireless sensor networks in environmental monitoring cyber infrastructure. *Wireless Networks*, *16*(4), 1091–1108. doi:10.100711276-009-0190-1

Yang, X., Ong, K. G., Dreschel, W. R., Zeng, K., Mungle, C. S., & Grimes, C. A. (2002). Design of a wireless sensor network for long-term, in-situ monitoring of an aqueous environment. *Sensors (Basel)*, *2*(11), 455–472. doi:10.339021100455

Yang, Y., Zhu, J., Zhao, C., Liu, S., & Tong, X. (2011). The spatial continuity study of NDVI based on Kriging and BPNN algorithm. *Mathematical and Computer Modelling in Agriculture.*, *54*(3-4), 1138–1144. doi:10.1016/j.mcm.2010.11.046

Yao, X., Zhu, Y., Tian, Y., Feng, W., & Cao, W. (2010). Exploring hyperspectral bands and estimation indices for leaf nitrogen accumulation in wheat. *International Journal of Applied Earth Observation and Geoinformation*, *12*(2), 89–100. doi:10.1016/j.jag.2009.11.008

Yoon, I., Noh, D. K., Lee, D., Teguh, R., Honma, T., & Shin, H. (2012). Reliable wildfire monitoring with sparsely deployed wireless sensor networks. *Proceedings of the 26th IEEE International Conference on Advanced Information Networking and Applications (AINA)*. 10.1109/AINA.2012.107

Yunpeng, C., Peng, T., Shihong, L., & Sufen, S. (2010). Study of agricultural search engine based on FAO Agrovoc ontology and Google API. *Conf on World Automation Congress (WAC)*, 439 – 444.

Zafar, R., Yafi, E., Zuhairi, M. F., & Dao, H. (2016). Big Data: The NoSQL and RDBMS review. In *2016 International Conference on Information and Communication Technology (ICICTM)* (pp. 120–126). Academic Press. 10.1109/ICICTM.2016.7890788

Zhang, L., Yuan, M., Tai, D., Ding, J., Xu, X., & Zhen, X. (2010). Design and implementation of granary monitoring system based on wireless sensor network node. *Proceedings of the International Conference on Measuring Technology and Mechatronics Automation (ICMTMA)*. 10.1109/ICMTMA.2010.296

Zhang, N., Wang, M., & Wang, N. (2002). Precision agriculture—a worldwide overview. *Computers and Electronics in Agriculture*, *36*(2-3), 113–132. doi:10.1016/S0168-1699(02)00096-0

Zheng, X.-Y., Liu, C.-Y., Chen, P.-T., Chuang, C.-L., Chen, C.-P., & Jiang, J.-A. (2013). Development of an embedded system-based gateway for environmental monitoring in wild fields. In S. Mukhopadhyay, K. Jayasundera, & A. Fuchs (Eds.), *Advancement in Sensing Technology* (pp. 23–42). Berlin: Springer-Verlag Berlin Heidelberg. doi:10.1007/978-3-642-32180-1_2

Zhu, Y., Dong, F., & Yuan, T. (2012). Automatic monitoring system for forest fire based on wireless sensor network. *Proceedings of the 2nd International Conference on Electric Information and Control Engineering (ICEICE)*. doi:10.1109/ICEICE.2012.936

Zia, H., Harris, N. R., Merrett, G. V., Rivers, M., & Coles, N. (2013). The impact of agricultural activities on water quality: A case for collaborative catchment-scale management using integrated Wireless Sensor Networks. *Computers and Electronics in Agriculture*, *96*, 126–138. doi:10.1016/j.compag.2013.05.001

Zoller, S., Reinhardt, A., Wachtel, M., & Steinmetz, R. (2013). Integrating wireless sensor nodes and smartphones for energy-efficient data exchange in smart environments. *Proceedings of the IEEE International Conference on Pervasive Computing and Communications Workshops (PERCOM Workshops)*. 10.1109/PerComW.2013.6529574

Zviedris, R., Elsts, A., Strazdins, G., Mednis, A., & Selavo, L. (2010). LynxNet: Wild animal monitoring using sensor networks. *Proceedings of the 4th International Conference on Real-world wireless sensor networks*. 10.1007/978-3-642-17520-6_18

About the Contributors

Figene Ahmedi is an Assistant Professor at the University of Prishtina, Kosova, where she joint the Faculty of Civil Engineering in 2000. She received her PhD in Wastewater Science at Civil Engineering Faculty, Skopje, Macedonia. In addition to long experience in teaching, she has published in renowned journals with high SCI impact factor in Wastewater Treatment, and currently focusing her research also in the areas of Small, On-site and Decentralized Wastewater Management Systems, and Water Protection. She was author and a senior researcher in an EU funded research project InWaterSense, as well as initiated and implemented another course development project granted by WUS Austria.

Lule Ahmedi is a Full Professor at the University of Prishtina, Kosova, Faculty of Electrical and Computer Engineering, where she joint in 2005. From 2005 to 2010, she was also affiliated to the South East European University, Macedonia, as visiting lecturer. She received her PhD in 2004 in computer science from University of Freiburg, Germany, where she was from 1999 to 2004 employed as teaching / research staff. She is, in recent years, repeatedly providing invited lectures to several universities in Europe in teaching, or as part of seminars or conferences. She is moreover experienced in designing curriculums in ICT. She has published papers in renowned international venues in the areas of Semantic Web, Social Network Analysis, XML and Databases, and Information Integration, and is currently also interested in Information Retrieval, Recommender Systems, and Stream Data Mining. She has reviewed for prestigious journals of ACM, Elsevier and Springer, and conferences like the International Semantic Web Conference. She is also serving as mentor or external evaluator in dissertation thesis at universities in Europe. Dr. Ahmedi has implemented over ten of research and capacity-building projects in ICT, mainly own-authored and as a manager in addition to serving as a lead researcher, to mention few highly competitive projects funded by EU (FP7 Programme, IPA, Tempus), DAAD Germany, HERD Norway, Linnaeus-Palme Sweden, the national research funding Kosova. She is author of several studies reflecting the state of the art and future perspectives in ICTs, and is continuously nominated to consultancy boards related to research in Kosovo. She is co-founder of the annual conference for open source software in Kosova.

Hartmut Asche is a professor of geoinformation science, remote sensing and cartography at the University of Potsdam, Germany. He received his PhD in geography from the University of Goettingen, Germany, in 1981. He has subsequently worked as an atlas editor with Georg Westermann Publishers (Brunswick, Germany) and professor of cartography at Karlsruhe and Berlin Universities of Applied Sciences. His research interests span geoinformation, visualisation/cartography, geodata management and remote sensing, with applications in atlases and atlas information systems, spatial analysis and decision making, environmental management, geospatial vulnerability and georisks, pedestrian navigation and planetary information systems (Mars). He has authored more than 40 peer-reviewed publications on cartography and geoinformation science. He has also co-edited books on web mapping and the Demographic Atlas of Albania.

Yvan Bedard was professor for 28 years in GIS and Spatial Database at Laval University Dept. Of Geomatics Sciences. He's now Senior Scientific and Strategic Advisor for Intelli3, a private company specialized in Geospatial Business Intelligence in Quebec City, Canada. While at University, he collaborated with many private and governmental organizations for research projects and strategic advising. He was the founding-director of the Centre for Research in Geomatics and has a multi-million dollars research record. Author of more than 160 full-referred papers, he's known as a pioneer in Spatial On-Line Analytical Processing (SOLAP).

Sandro Bimonte is a researcher at Irstea, and more exactly he is at TSCF. He received his PhD from INSA-Lyon, France (2004–2007). From 2007–2008, he carried out researches at IMAG, France. He is an editorial board member of International Journal of Decision Support System Technology, and International Journal of Data Mining, Modelling and Management and a member of the Commission on GeoVisualisation of the International Cartographic Association. His research activities concern spatial datawarehouses and spatial OLAP, visual languages, geographic information systems, spatio-temporal databases and geovisualisation.

Eliot Bytyçi is a PhD candidate of the Computer Science Program in the Faculty of Electrical and Computer Engineering, University of Prishtina, Kosovo. At the same time, he is a teaching assistant at the Department of Mathematics, Faculty of Mathematical and Natural Sciences, University of Prishtina. His current research lies in the area of Data Mining, where he uses different kinds of algorithms in relation to the Semantic Web, in order to gain new insights. Through ontology engineering and with the help of the context, interesting rules are outlined, providing domain experts new knowledge. He is active member of the research community in Kosovo, taking part in different projects, national and international, related to the ICT in general.

Eduardo Fávero Caires is an Associate Professor of Soil Science and Agricultural Engineering Department at State University of Ponta Grossa – Brazil. Dr. Caires teaches classes and guides students in Agronomy undergraduate and in Agronomy and Applied Computation graduate degrees. Dr. Caires has experience in Agriculture with emphasis in Soil Fertility and Plant Nutrition, acting on the following subjects: soil acidity control, liming, phosphogypsum use, and fertilization under no-till systems. Dr. Caires is researcher of CNPq (Conselho Nacional de Desenvolvimento Cientifico e Tecnol6gico), a Brazilian agency related to scientific development.

Maurizio Canavari holds a degree in Agricultural Sciences (Bologna, 1990) and a Doctorate in Appraisal and Land Economics (Padova, 1997); he is Associate Professor of Agricultural Economics and Appraisal at the University of Bologna, and he lectures on agri-food marketing and marketing research. Current research interests regard topics in agri-food marketing and economics of quality in the agri-food chains, such as innovation, trust and quality assurance and certification in food networks, marketing and consumer behaviour related to quality food products, such as organic, functional and unique speciality food.

Giacomo Carli is a Lecturer in Strategic Management at the Open University Business School. He was formerly a post-doc at the University of Bologna, where he obtained a MA in Engineering Management in 2007 and a PhD in Management in 2012, after visiting the Fuqua School of Business at Duke University. He served for 18 months as junior researcher and Project Manager of the EU ERA-NET funded project RoboFarm. His research interests include strategic management and organizational change. He devotes special attention to the agri-food industry, conducting research on precision agriculture adoption and Farm Management Information Systems.

Anastasios A. Economides is Full Professor at the University of Macedonia, Thessaloniki, Greece (http://www.uom.gr) and the deputy representative of the University of Macedonia at EL.ID.E.K. (the official administration for the National/ Greek Research Policy). He is Founder and Director of the CONTA (Computer Networks and Telematic Applications, http://conta.uom.gr) group and the SMILE (Smart & Mobile Interactive Learning Environments, http://smile.uom.gr) laboratory. He received the Ph.D. degrees in computer engineering from the University of Southern California, Los Angeles. His research interests include Internet of Things, Mobile Networks, Mobile Ubiquitous and Pervasive Learning, User Experience and Acceptance of Smart Systems and Services. He has published over two hundred peer-reviewed papers and has received over three thousand citations. He has been the Chairman of the Information Systems Postgraduate Program (2008-2014) ath the University of Macedonia (http://mis.uom.gr). He has been a Visiting Professor at several Universities (e.g. Univ. of Southern California, Univ. Oberta de Catalunya, Univ. Pompe Fabra). He has been the plenary speaker in International Conferences on Internet of Things as well on Computer-based Assessment. He has been the principal investigator and participated in many funded projects.

Elodie Edoh-Alove received her PhD in Geomatics (2015) from Laval University, Canada and PhD in Computer Science from Université Blaise Pascal - Clermont -Ferrand. Her thesis was set in a co-supervised context where the participating laboratories were the Centre for Research in Geomatics of Laval University and the French Research Institute for Agricultural and Environmental Engineering (Clermont-Ferrand, France). Her research is focused on the spatial vagueness issues in SOLAP systems and spatial datacubes design methods. Her activities now concern spatial data integration, geographic information systems, spatio-temporal databases, datawarehouses and geovisualisation. She is an author of several articles published in international conferences and journals.

Attila Gere is s sensory scientist received his Ph.D. in Sensory Analysis form Szent István University, Hungary in 2016. He started his career at Corvinus University of Budapest in 2015 and has been working as an assistant professor at Szent István University since 2017. He has been publishing research articles since 2014 and working in the Sensory Laboratory of SZIU. Attila currently lives in Budapest with his family, he enjoys hiking and reading.

Alessandro Grandi is full professor of Economics and Organization Theory at the School of Engineering and Architecture of the University of Bologna, where he also served as Chairman of the Engineering Management program from 2007 to 2013. He was visiting scholar at Graduate School of Business of the New York University. Prof. Grandi's current research interests include: technology strategy, management of new product development, organizational design and management of R&D. He is author or co-author of about 80 papers published on national and international scientific journals and books and on conference proceedings. He is currently chairman of AlmaCube srl, member of the board of Consorzio Centro ceramico; director of the Master of Management at the Bologna Business School. Prof. Grandi has also been teaching in business policy, organizational design and technology management at several major business schools in Italy.

Alaine Guimarães is Professor at Department of Computer Science, State University of Ponta Grossa (UEPG). Coordinator of the Master Program in Applied Computing at UEPG. Member of the Managing Board of the Brazilian Association of Computing Applied to Agriculture (SBIAgro). Graduated in Bachelor of Computing (1989) from State University of Ponta Grossa . Master in Computer Science (2000) from Federal University of Paraná (UFPR) and Ph.D. in Agronomy (2005) by the State Univeristy Paulista Júlio de Mesquita Filho (UNESP). Conducted part of her doctoral thesis at the University of Florida, USA (2005). Participates in the scientific and execution team of projects in which more than U$ 2,000,000.00 was raised on resources for research development, equipment acquirement and buildings construction for the University in which she Works. That resources were obtained from agencies fostering (FINEP, CNPq, Araucaria Foundation, CAPES). She served as Director of Research and Program Coordinator of Scientific Initiation UEPG during the period 2010-2013. Conducts researches related to Computing Applied to Agriculture, Machine Learning and Data Mining.

Claudio Kapp Jr. graduated in Business Administration and Accounting Sciences, Master in Computing Applied to Agriculture and PhD in Agronomy, all by the State University of Ponta Grossa. The autor is a substitute professor at Ponta Grossa State University and other institutions. The author works in the Graduation and Post-Graduation with the courses of Administration, Agronomy and Accounting Sciences. He is a Researcher in Rural Economy at ABC Technology Solutions for Agribusiness.

Eirini Karapistoli holds a PhD (2009) in Electrical Engineering from the Aristotle University of Thessaloniki, Greece. She was also the recipient of a post-doctoral research grant (2012-2015) from the Greek Secretariat of Research and Technology (GSRT) and has worked as a postdoctoral researcher at the University of Macedonia, Thessaloniki, Greece.. Irene has a broad expertise in wireless sensor networking technologu as well as in cyber security and privacy, decision-making based on cost-benefit analysis, network intrusion detection and mathematical optimisation. She has published several papers in international conferences and journals and she has attracted more than 200 citations to her papers. She is also competent in project management holding an MBA Degree (2013) from the Blekinge Institute of Technology, Sweden. In the past, Irene has participated in five FP7 and H2020 EU-funded projects, and has coordinated the preparation of several H2020 proposals.

Arianit Kurti, PhD, Associate Professor works at the Department of Computer Science at Linnaeus University (Lnu) in Sweden. He also holds the title of Docent in Computer Science awarded by South East European University (Macedonia). Dr. Kurti's research interests are within the areas of mobile systems and pervasive and ubiquitous computing, especially in exploring the use of contextual information for designing and developing innovative services and applications in specific application domains. He has also a strong interest in research cooperation with industry in the field of user engagement, requirements engineering, usability and accessibility studies, interaction design and visualization. Furthermore he has been the project leader for multiple nationally funded as well as EU funded research project in the past years. His research activities are funded by Swedish Knowledge Foundation (KKS), EU commission (EACEA), STINT, Vinnova etc. Furthermore Dr. Kurti is also a Director of RISE Interactive Institute in Norrköping.

Fabio Manfredini graduated in Environmental Sciences at the State University of Milan in 1999, is the responsible of the Data Analysis and Mapping Laboratory, Department of Architecture and Planning, Politecnico di Milano. His main areas of expertise are methods and techniques of urban and environmental analysis, design and management of geographical information systems, statistical and spatial analysis.

Howard R. Moskowitz founded Mind Genomics Associates in 2014, to promote the development and use of the new science of Mind Genomics. The science creates an archival and action-oriented database for knowledge about how people think about the aspects of their daily life. The science is used in business to drive 1:1 sales and marketing, and in social research to understand what is important, and what should be communicated to citizens of a country for socially relevant issues. Dr. Moskowitz is also chairman of Mind Genomics Advisors, a consulting firm which provides innovative, customized predictive analytics for a variety of organizations, using the Mind Genomics platform. Dr. Howard Moskowitz received awards for his innovative contributions to the field of marketing research. He received the "2012 Edison Award for Innovation" (Bronze Award - Applied Technology/Research Tools category) for the creation of Mind Genomics®, the aforementioned internationally-recognized science using experimental design and predictive analytics to uncover individual human preferences for products, services, ideas and social issues. Dr. Moskowitz has also received: the "2010 Walston Chubb Award for Innovation" across all sciences from Sigma Xi - The Scientific Research Society; the Advertising Research Foundation's first "Research Innovation Award" in 2006; the 2004 "David R. Peryam Award" from ASTM for outstanding contributions in basic and applied sensory science, and ESOMAR (European Society of Market Research) Awards in 2001, 2003, 2004 & 2006 for research innovation.

Brendan O'Flynn is a Senior Staff Researcher at the Tyndall National Institute, and is Head of Group for Wireless Sensor Networks Group. Since 2004 he has been defining and leading the research activities of the Wireless Sensor Networks (WSN) group. This multidisciplinary group consists of over 20 researchers of comprising of postdoctoral researchers, postgraduate students, engineers and scientists. Brendans key inputs to the Tyndall Institute in helping it to achieve its strategic objectives are in the areas of: Science, Technology and Innovation, Enterprise, Innovation and Growth, Education and Training, Support to Irish Academia, Communications, People and Sustainability – through the development, management & coordination of a world-class, multidisciplinary research group focused on delivering Smart Sensors, Circuits and Systems. The research focus of the WSN group is in the development of next generation embedded systems. These low-power consumption solutions are designed to meet the requirements associated with the indefinite lifetime deployment of low maintenance context aware sensing systems for deployment in the world around us so as to create information As a funded investigator on the SFI Centre CONNECT, Brendan is developing system level hardware and software for the "Internet of Things" and smart sensor integration. These are cross cutting research activities which address the requirements of the different applications the team is investigating – energy/building monitoring, environmental monitoring and health/motion monitoring. Brendan has an extensive track record in publishing this research in peer reviewed journals and conferences. Brendan has supervised to completion six Ph.D. students and seven research masters students and has published extensively in this field Brendan is a reviewer for multiple Journals and is a reviewer and committee member on several conference panels internationally.

Paola Pucci is a Full Professor in Urban planning and Research Director of the PhD course in Urban Planning Design and Policy (UPDP) at the Politecnico di Milano. She has taken part, also with roles of team coordinator, in national and international research projects, dealing with the following research topics: Mobility policy and transport planning, mobile phone data and territorial transformations. Her most recent publications are Mapping urban practices through mobile phone data, PoliMI SpringerBriefs Series (with Manfredini F., Tagliolato P.); La desserte ferroviaire des territoires périurbains, Éditions Recherches, Paris (with Grosjean B., Leloutre G., Grillet-Aubert A., Bowie K.); Large infrastructures and conflicts in Italy: searching "boundary objects", in Gualini E. (ed) Planning and Conflict. Routledge, London. She is co-editing Understanding Mobilities for Designing Contemporary Cities, Springer, with Matteo Colleoni.

Dalma Radványi is a food scientist received her Ph.D. in Analytical Chemistry from Szent István University, Hungary in 2016. She has been working as lecturer at Szent István University since 2016 and is a postdoctoral research fellow at Hungarian Academy of Sciences. She has been a speaker of conferences and published several research articles in highly respected journals. Dalma currently lives in Budapest with her family, she enjoys traveling, reading and working on publications.

Giancarlo Rodrigues graduated in Systems Analysis and Development at Federal Technological University of Paraná - UTFPR (2015). Currently is Applied Computing masters student at State University of Ponta Grossa - UEPG.

Astrit Salihu is Master of technical Sciences – Civil Engineering. Research visit at the Institute for Water Quality, Resource and Waste Management of TU-Wien. The goal of the visit was to acquire knowledge in the field of water quality management under joint supervision of Professors from TU- Wien, and University of Prishtina. The Master thesis was a result of the research visit.

Richard D. Sciacca has three decades of executive and senior management experience with Fortune 100-500 companies and healthcare organizations. He has developed several Trademarked market research voice of the customer (VOC) qualitative and quantitative methodologies. He excels at next generation product development and market segmentation for global companies. His marketing research has resulted in the successful development and launch of many of the world's most successful diagnostics instruments, assays, and product marketing strategies for companies. He founded and leads a marketing research consulting firm, INTMR LLC, as CEO for the past 14 years. He has recently co-founded and developed Vox

Populi Global®, Vox Populi USA®, Vox Populi Wisconsin® which are companies focused on identifying and translating the voice of the people for better governance and business.

Besmir Sejdiu is teaching assistant at University of Prishtina, Kosova, in the Department of Computer Engineering. He joined the Faculty of Electrical and Computer Engineering in October 2009. He received a Bachelor of Computer Engineering degree from the University of Prishtina in 2008; a Master of Computer Engineering from the same university in 2013, titled 'Data stream management system for water quality monitoring through wireless sensors' in 2013 under the supervision of Prof. Lule Ahmedi. MSc. Sejdiu has a long experience in a field programming and database design. He is also a teaching assistant at University of Mitrovica, Kosova. Some of courses that he teaches are: Database, Programming in Internet, Algorithms and Data Structures, Java Programming and Software Engineering. He has published several papers in different International Journal and Conferences.

Paolo Tagliolato graduated in Mathematics at the State University of Milan. At the same university he completed the Ph.D. degree in Computer Science in the area of multimedia information retrieval. His research focused on information retrieval, spatial analysis of cellular network data for urban applications (at the Politechnic of Milan), semantic spatial data infrastructures and environmental data management with open geospatial standards (at CNR), and development of Virtual Laboratories architecture as member of LifeWatch Italy ICT working group.

Syle Tahirsylaj renowned Professor of Meteorology in the Faculty of Agriculture of the University of Prishtina, has a number of relevant research papers in the field. During his career, Prof. Tahirsylaj was also a Director of Hydrometeorological Institute of Kosova.

Patrick Voland is a PhD student of geoinformation science at the University of Potsdam. He achieved a scholarship from the German Research Foundation (DFG) in its Research Training Group 1539 "Visibility and Visualisation - Hybrid Forms of Pictorial Knowledge". He studied Regional Science, and then continued as a master's student in Geoinformation and Visualization at the University of Potsdam. From 2001 to 2005, he worked for Atos Origin and from 2006 to 2014, he was a staff member of the University of Potsdam's Department of Geography. He is also affiliated with the company Motion Intelligence (Route360°), an advanced spatial analysis technology spin-off of the Hasso-Plattner-Institute Potsdam.

Index

A

Activity-Based Costing 252-257, 260-263, 266-268
agricultural 1-2, 6, 8, 25-26, 38-39, 47, 86-90, 93, 101, 136, 157, 198, 206, 208, 221, 224, 253, 255, 267
Air-Water Pollution Monitoring 215
ARTIFICIAL NEURAL NETWORK 158, 160, 164
automotive electronics 109, 113

B

Backpropagation 156, 158, 160-161, 165-168
big data 7, 50, 90, 105-106

C

car sensor 106-107, 111, 113, 116, 121-122, 125
Crop Circle 156, 158-159, 163, 165-168

D

data sources 131, 133, 135, 141, 147, 173-174, 185, 191, 253, 269
data storage 38, 40-43, 45-46, 49-50, 117, 123, 197
decision support 38, 40, 107, 254
decision support system 40
deployment guidelines 197

E

easy-to-use guides 196, 198, 233
environmental 8-9, 25, 39, 59-61, 105-108, 110-111, 122, 125, 147, 157, 196-198, 201-204, 206, 208, 213, 219, 221, 224, 233

F

farm management 38-39, 252-255, 257, 266-268
Farm Management Information Systems 38, 253, 266
food production 8, 25-26, 31, 34-35, 38

G

geovisualization 107, 126
global change 1-3, 8-9, 16-17, 20, 23-26, 31, 34-35
greenhouse gas 8, 34
GreenSeeker 156, 158-159, 163, 165-168

H

heterogeneous information 92, 101

I

information integration 86-87
information retrieval 87-91, 97-98, 101

L

long-term trends 58, 60, 78

M

Mind Genomics 1-2, 4-10, 12-13, 16
mind-sets 5, 16-17, 24-25, 34-35
mobile phone 108, 173-180, 182-188, 191-192, 208
mobile phone data 173, 175, 183, 191-192
Monitoring of Destruction Phenomena 217

N

N rate 157-158, 166

O

one-to-one relationship 146, 151
ontology mapping 87, 92, 94, 96, 101
Ontology merging 87, 92-93, 95, 101

P

performance 17, 38, 40, 42-43, 46-48, 50-51, 54, 100, 156, 158, 161, 164-168, 212, 253, 255, 260
phone data 108, 173, 175, 183, 191-192
phone network 108, 173-174, 178
Precision Agriculture 156-157, 208-209, 253-255, 257, 268

R

Resilient Propagation 156, 158, 161, 166-168
risks of misinterpretation 129-130, 136, 138, 140, 144, 150

S

SOLAP 129-134, 136, 140-141, 145, 147-148, 150-152
SOLAP datacubes 129-134, 140, 147, 150
spatial vagueness 129-134, 136, 152
surface water quality monitoring 62
Survey 183, 196, 204
sustainable food 8, 25-26, 31, 34-35

T

Triticum aestivum L. 163

U

urban spaces 173-175, 185, 191-192

W

wireless sensor 58-61, 69, 71, 76, 80, 196-199, 207-208, 213, 217, 221
Wireless Sensor Network 58-60, 71, 76, 197, 213

www.ingramcontent.com/pod-product-compliance
Lightning Source LLC
LaVergne TN
LVHW061546290926
849162LV00004B/929